SPARROWS POINT

MAKING STEEL—
THE RISE AND RUIN OF
AMERICAN INDUSTRIAL MIGHT

BY MARK REUTTER

SUMMIT BOOKS
New York London Toronto Sydney Tokyo

SUMMIT BOOKS
Simon & Schuster Building
Rockefeller Center
1230 Avenue of the Americas
New York, New York 10020

Published by SUMMIT BOOKS

SUMMIT BOOKS and colophon are trademarks
of Simon & Schuster Inc.

Designed by Irving Perkins Associates
Manufactured in the United States of America

10 9 8 7 6 5 4 3 2 1

Library of Congress Cataloging in Publication Data
Reutter, Mark.
Sparrows Point : making steel : the rise and ruin of American
industrial might / by Mark Reutter.
p. cm.
Bibliography: p.
Includes index.
1. Steel industry and trade—Maryland—Sparrows Point—History.
2. Steel industry and trade—United States—History. I. Title.
HD9518.S6R48 1988
338.7'669142'097—dc19
88-24873
CIP

ISBN 0-671-55335-6

To my parents and to P.J.

CONTENTS

INTRODUCTION

AGE OF STEEL

When the journalist John Gunther asked an expert "What is steel?" back in the 1940s, the answer he got was "America!" Four out of five manufactured items contained steel and 40 percent of all wage earners owed their livelihood directly or indirectly to the industry. Steel was everywhere, from the 70-story towers of the Golden Gate Bridge to the steel wool that housewives used to clean pots and pans. Steel was the building block of cities, the spine of transportation, the weapon of war. "The basic power determinant of any country is its steel production," Gunther wrote in *Inside U.S.A.*, "and what makes this a great nation above all is the fact that it can roll over 90 million tons of steel ingots a year, more than Great Britain, prewar Germany, Japan, France, and the Soviet Union combined."

Steel was the prototype and pacesetter of volume-based American manufacturing. The typical plant was huge, an aggregation of furnaces, coke ovens, rolling mills, and foundries, with a railway system of its own and not infrequently an ore and coal dock. The number of workers in such plants could run into the thousands. The size of steel products grew along with the nation. The 100-pound rail, 110-inch plate, the wide-flanged building beam, and the 130-ton navy gun were all wonders of their day. The consolidation of ownership kept pace with the development of the industry: by early in this century two corporations, U.S. Steel and Bethlehem Steel, controlled more than half of the nation's important plants, a majority of the workable supply of ore in the Western Hemisphere, much of the coking coal of Appalachia, and the means of transportation which connected the mines and mills.

This book is a history of the American steel industry, using as its focus the plant at Sparrows Point, Maryland. Located on a flat, marshy peninsula at the confluence of the Chesapeake Bay and Baltimore Harbor, this "waste-place" lay undeveloped until 1887 when an engineer named Frederick Wood walked along its perimeter and decided that it would make a superb tidewater port for the Pennsylvania Steel

Company. Over the next 60 years Sparrows Point became a monument to the nation's ability to make steel cheaply and profusely. By 1910 the steel rails rolled at the plant spanned the vastness of Mongolia, climbed the Andes in Chile, breached the pampas of Argentina, and descended into the tunnels of the London underground. Purchased by Bethlehem Steel in 1916, the mill expanded relentlessly until, in the 1950s, it became the single largest steel complex in the world. Out of the blaze and hiss of its 35 furnaces came the tail fins of Thunderbird convertibles and Chevy Bel Airs, the tin plate for Campbell's soup cans, the hulls of ocean tankers, the rods and wire for the Mississippi and Chesapeake Bay bridges, and a thousand and one other products that were sinew and container of our culture of bigness and abundance.

All this had political ramifications at home and abroad. Before and after the Spanish-American War the fate of ore-rich Cuba was linked to the order books of the steel mill 1,300 miles to the north. Eventually the quest for iron led American businessmen to more distant claims in Chile and the Orinoco Valley of Venezuela. The same tidewater access to the Atlantic that gave Sparrows Point a unique position in gathering raw materials gave it enormous strategic importance during the two World Wars. Presidential boards monitored the plant's output and the army and navy became its major customers. Making steel, Sparrows Point made history.

The mill held claim to another distinction: it forged a way of life for thousands of families. Pulled into Sparrows Point were a diverse group of people who lived in the company town, "the Pullman of the East," according to the U.S. Labor Bureau. When young black males began migrating north from Virginia, many came to Sparrows Point, and this migration continued with little interruption for three decades. In and out of the banging mills trudged the foreign-born, peasants from Russia and Hungary and skilled Bessemer blowers from Wales. The workforce grew from 3,000 in 1900 to 18,000 in 1929, to 21,000 during the Second World War and then to nearly 30,000 in 1957, when the plant's steelmaking capacity reached 8.2 million tons a year, or 15.5 tons per minute.

Like many corporate enclaves, the company and its operations were sheltered from public scrutiny. Even before production of steel began, the company was given the power to run its own police force by the governor of Maryland. Soon the mill and town were curtained off from the outside. This condition was not unique to Sparrows Point. Located on the fringe of metropolitan areas or in small cities, the mill capitals of America were imprisoned in their own blankets of soot and corporate control—they were the ash heaps that Gatsby and Daisy rushed by with eyes averted. Reports on the steel trade concentrated on corporate headquarters, not on the night-and-day blazings of Johnstown, Pennsylvania; Weirton, West Virginia; Gary, Indiana; or Youngstown, Ohio. "Ladies visit sweat-shops, but they never enter a steel mill," reporter Herbert Casson observed in 1907.

A strange and interesting paradox has emerged: Here in the United States where macro-scale manufacturing flourished, the actual sites of production have rarely been explored. Much has been written about the evolution and remarkable resiliency of corporate capitalism, ranging from Alfred D. Chandler's whiggish *The Visible Hand* to Harry Braverman's Marxist critique, *Labor and Monopoly Capital.* There have been excellent accounts of the strikes, union organization, and boardroom depredations that punctuated the rise of modern industry. Yet the mill capitals of America have remained largely invisible.

This book attempts to see a mill capital whole by re-creating the two worlds that overlapped at Sparrows Point: the high-powered executives who designed and managed the works, and the workers who lived out their lives in the mill and company town, developing a rich and complex culture of their own.

The first part of the book concentrates on the capital side of the equation. Drawing upon thousands of pages of private correspondence and business records, as well as court records, government inspection reports, and congressional testimony, it describes the businessmen and the business strategies that shaped the plant and impacted on domestic and international affairs. Frederick Wood built the original mill, a marvel of machine coordination, and Rufus, his brother, designed the town, where rank and social precedence were engineered into a grid of alphabetized streets. The book reveals the extraordinary private arms network of Charles Schwab, a mogul whose lifestyle gave definition to the phrase "conspicuous consumption," and his dour assistant, Eugene Grace, who battled presidents from Wilson to Truman and choked off innovation in favor of short-term profits.

A missing element in many accounts of industry has been the workers. Labor historian David Brody and others have emphasized the importance of cutting through generalizations about working-class life by examining specific communities undergoing economic change. In *Work, Culture and Society in Industrializing America,* Herbert

Gutman decried how little we know about the experiences of particular working people. The second half of this book speaks to these omissions, portraying the mill life created at Sparrows Point and describing, often in their own words, the perspectives of individual men and women. Charlie Parrish fought the color barrier in the blast furnace department a decade before the Montgomery bus boycott, Ben Womer lived out his life tending the behemoth open hearths, Mike Howard, Oscar Durandetto, and John Duerbeck helped organize the union, Marian Wilson inspected tin plate under the watchful gaze of Mrs. Alexander, and Bob Eney, raised in the shadow of the mill, desperately wanted to escape. Their experiences offer valuable insights into the evolution of working-class culture, the relationship between management and labor, and the local-level issues that led to unionization.

Today Sparrows Point has shrunken to nearly one-half of its 1957 capacity. The pipe, wire, and nail mills are padlocked and the employment rolls reduced to 7,900, the lowest level since the Depression. The steep decline mirrors that of the industry as a whole. Accounting for 65 percent of global steel production at the end of World War II, U.S. steel output has dropped to under 15 percent of world production. Cheaper and, in many cases, superior foreign steel has captured one-fifth of the American market, despite import protection, while aluminum, plastics, concrete, and paperboard have cut into the market itself. Despite a partial recovery in 1988, steel's troubles are deep seated and largely self made. At best, the popular solutions of the past few years—higher import restrictions, wage concessions, and relaxed environmental regulations—have prevented a disorderly retreat from becoming a full-fledged rout.

The present crisis has attracted considerable attention partly because plant closings have cut a path of unemployment across the Midwest and East, and partly because the industry's woes have spilled into the international arena, straining relations between Washington and its trading partners and adding to the trade deficit. Working beneath these reversals has been a stripping away of the security blanket that John Gunther's generation took for granted. Business gurus may argue all they wish about the vitality of computers and the growth of fast foods, but people still worry that the country's economic fundamentals are askew when such an important industry as steel is outclassed not only by Japan, but by former lightweights such as South

Korea and Spain—and when more people work for McDonald's than for U.S. Steel.

The symptoms of the disease have been evident for a long time. They first appeared when the industry was flush with profits and exercised nearly absolute dominance as market maker and labor colonizer. There were signposts along the way, warnings that no business could ignore without reaping the consequences eventually. Eventually is now.

PART I

BEGINNINGS 1887-1916

In various quiet nooks and corners I had the beginnings of all sorts of industries under way—nuclei of future vast factories, the iron and steel missionaries of my future civilization. In these were gathered together the brightest young minds I could find, and I kept agents out raking the country for more, all the time. I was training a crowd of ignorant folk into experts—experts in every sort of handiwork and scientific calling. These nurseries of mine went smoothly and privately along undisturbed in their obscure country retreats, for nobody was allowed to come into their precincts without a special permit.

—MARK TWAIN, A CONNECTICUT YANKEE IN KING ARTHUR'S COURT (1889)

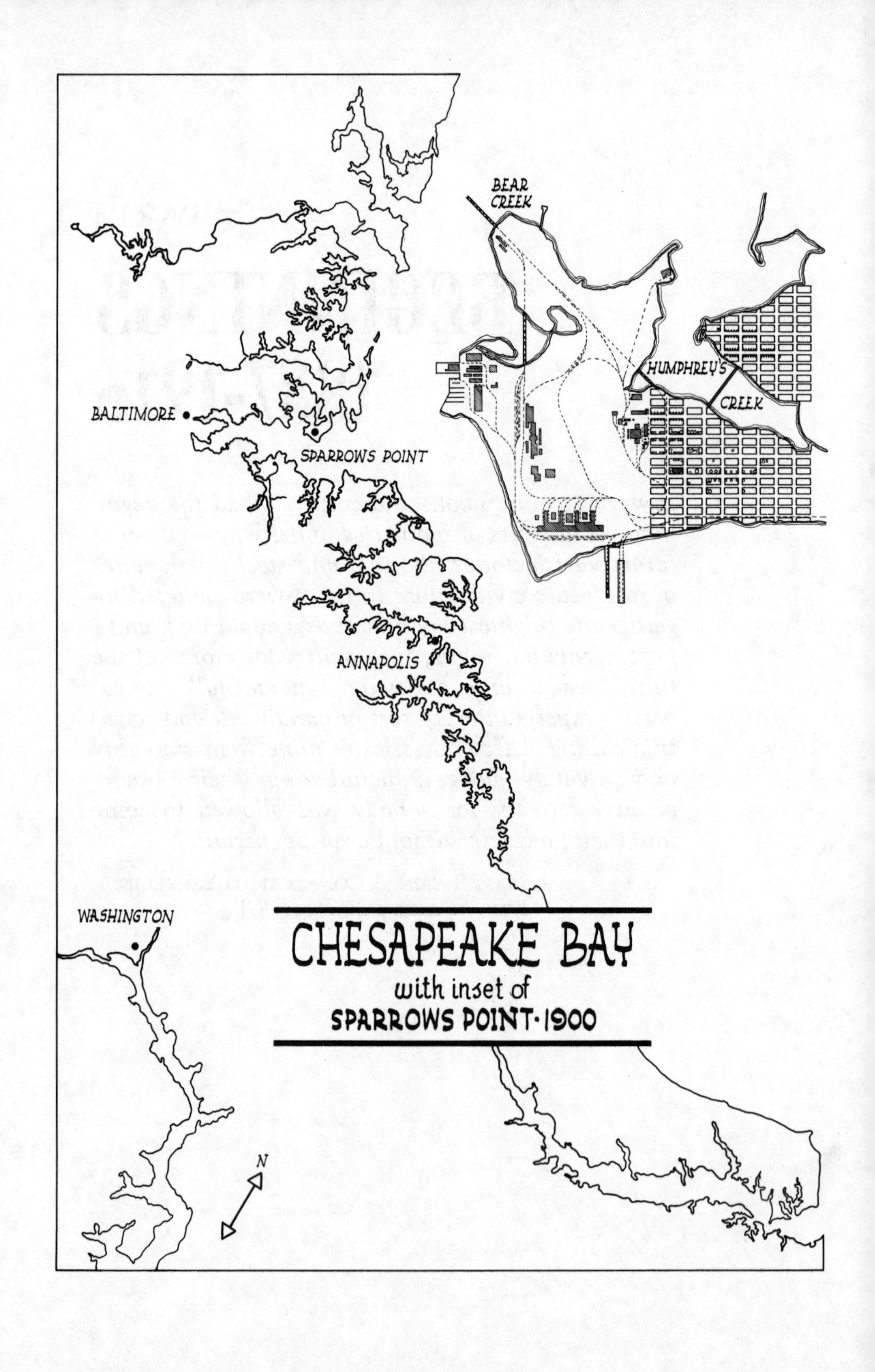
BEAR CREEK
HUMPHREY'S
CREEK
BALTIMORE
SPARROWS POINT
ANNAPOLIS
WASHINGTON
CHESAPEAKE BAY
with inset of
SPARROWS POINT · 1900
N

CHAPTER 1

GODDESS OF INDUSTRY

On Friday, May 30, 1890, two trains rushed to meet at a railroad junction east of Baltimore. One was carrying capitalists from Philadelphia. On board were Joseph Wharton, the nickel and iron king who had founded the Wharton School of Finance; H. H. Houston of the Pennsylvania Railroad; Charles Berwind of the sprawling Berwind-White coal consortium; D. B. Cummins, the Main Line banker and president of Girard National; Francis C. Yarnall, president of the Lehigh & Lackawanna Railroad; and 60 others. In Wilmington the train halted to pick up Benjamin Biggs, governor of Delaware.

From Washington came a contingent of Southerners led by Senators John S. Barbour of Virginia, J. C. S. Blackburn of Kentucky, and Arthur Pue Gorman of Maryland. In this group of travelers were such eminences as West Virginia coal baron Henry Gassaway Davis; Baltimore financier and philanthropist Enoch Pratt; Elihu Jackson, governor of Maryland; former governor Oden Bowie; railroad builder J. M. Hood; chief clerk of the State Department Sevellon A. Brown; chief of U.S. Army ordnance General S. V. Benet; and the archbishop of Baltimore, Cardinal James Gibbons.[1]

Swinging on a turnout at Bayview Junction, Baltimore County, the trains passed near the Canton docks made famous by the clipper trade with China before making a straight-ahead charge through Patapsco Neck, a finger of real estate that ringed the southeast edge of Baltimore Harbor. This was rural country, forested and sparsely populated. Not more than a couple of farmhouses could be seen through the beech, red gum, and oak trees. The trains rolled along, following the harbor, then the view cleared as the track crossed the headwaters of Bear Creek on a trestle three-quarters of a mile long. Ahead lay Sparrows Point, site of the day's celebration of the opening of the new steel mill.[2]

The dignitaries were greeted at the railroad depot by Frederick Wood and his brother Rufus, David Baker, James B. Ladd, Federick Webster, and Colonel Walter S. Franklin. The guests were led to a train of festively painted gondola cars to better observe the plant at close range.

Soon the party found itself at the tip of the peninsula, at the steel dock. Cranes were unloading ore into railroad cars from the S.S. *Dago* and *Wilylake*. It was explained that when the plant was fully operational about 20 boats of ore would be needed every week. The party proceeded to the stockhouse. Above them raw materials were sorted and transferred to narrow-gauge carts resembling mine lorries. The carts were carried to great elevators which hoisted the material to the top of the blast furnaces.[3]

The cables rasped heavily as the party—about 13.5 tons of human material outrigged in silk hats and Prince Albert frock coats—rose on the elevators until they reached the charging platform 85 feet above the ground. "The visitors were simply astonished," wrote a *Baltimore American* reporter who accompanied the group.[4] Arrayed before them were three other blast furnaces fully as tall as the one they had just ascended. Flanking these buildings were great sheds, opened at the sides and filled with machinery. Shifter locomotives were dragging flat cars in and out of the sheds. Suddenly, the men felt a violent rocking of the platform. The ground seemed to swell upward as a tremendous roar came from furnace B 200 feet away.

"Our workmen," an official announced, "have just shot off a blast." Air preheated to 1,400° F. was being blown into the base of the vessel, igniting the coke inside to start the process of reducing iron ore to pig iron. Bolts of fire flared up into the sky while, down on the furnace floor, employees could be seen darting about. With expressions of wonder the guests watched the procedure until the main blast was shut off 30 minutes later. "Unfortunately," said the reporter, "one of the employees was badly burned, but he was removed to a place of refuge and cared for."[5]

Passing from the blast-furnace department, the party went to the Bessemer house, a structure that looked like a medieval fortress. Gingerly the gentlemen mounted a platform to hear a description of the Bessemer converters under construction. When completed, the egg-shaped cauldrons would swing on their trunnions and produce 20 tons of molten steel in a blow. But today the converters lay silent and still, allowing the guests to walk unmolested beneath a ponderous geometry of bolts and rivets. Several hundred yards away, a rolling engine loomed up, its side flywheel cutting through the air like a whip.

At 2:00 P.M. the party was ushered into the pavilion hall. Clouds had rolled in from the bay and rain was starting to drizzle down on the peninsula, but the atmosphere was anything but gloomy. As the men made their entry, waiters emerged bearing trays of Chesapeake crab hors d'oeuvres, a winemaster poured claret into crystal decanters, and the luncheon honoring Sparrows Point burst forth in sumptuous style. After feasting on terrapin and whole spring chicken, plus relays of vegetables and potatoes in various sauces and stews, the 160 dignitaries pushed back their chairs, lit up cigars, and prepared for the rhetoric.

Luther S. Bent, president of Pennsylvania Steel Company, served as host and toastmaster. "I am happy to have as a guest a gentleman who has seen and watched this work from the beginning," he said in introducing Cardinal Gibbons.

The prelate expressed delight at the project. "It is said in the life of William Penn that he appropriated to himself, or to his own state, a small part of Maryland, and Maryland, with its usual amiability, allowed him to retain what he had taken," he told the men. "And now we behold a second invasion of our territory—the invader armed in the panoply of steel and with golden bullets. Be assured, gentlemen, that Maryland and Baltimore will be glad to be honored or afflicted with many such visits as this. I had from the first the fullest confidence in the success of this enterprise."[6]

Pointing to Bent, Cardinal Gibbons said he held the key to solving "the great problem of labor" in a practical way. "I have always spoken kind words for the laborers. To-day I am in the presence of capitalists and I feel sure that if anyone has the power to solve this great problem it is our friend who presides here to-day. He will gratify the capitalist by erecting these great furnaces and by producing large dividends; he will bring joy to the laborer by building comfortable homes, and paying them good wages. The best interests of each are the interests of both, and when the time for the happy consummation arrives, I shall be most happy to be the officiating clergyman to assure the union between them."

Blessed by the high priest, Sparrows Point collected accolades from other speakers. From Enoch Pratt: "If you should go over the whole of the United States, you could not find a place equal to Sparrows Point for such an enterprise." From Professor Ira Remsen, chairman of the chemistry department at Johns Hopkins University: "We have seen the most magnificent chemical experiments at this place that man ever beheld. The beginning of all that was in the laboratory." And from Mayor John Davidson of Baltimore: "Your prosperity is our prosperity; your interests are our interests."[7]

Felix Agnus, publisher of the *Baltimore American,* described the building of Sparrows Point as another example of capital harnessing the sleeping energies of the nation. "Out there on the harbor flows one of the prettiest of rivers, bearing on its bosom the great commerce of a great port. A few miles below this river joins the noblest of American bays, large enough to float the navies of the world. After you enter that bay you pass between the shores of lands glorious in their abundant fertility, and capable of supporting ten times their present population."

Applause rang out as the chop-whiskered publisher orated on. "Thousands of immigrants are arriving daily in New York with railroad tickets for the far West. This tide of travel must some day be diverted to this productive section, where good homes and all the convenience of a high civilization are in the reach of every worker." He recalled another exhilarating day in American history. Four years earlier the Statue of Liberty had been lit in New York harbor. Remember the fanfares of jubilation that swept across the country?

"The other day a big ship was coming up the Chesapeake. On board were some hardy Scotchmen, seeking their fortunes in the new world. Night came on, and as the steamship was making her way from the Severn to the Patapsco, the glare of a great light was seen.

"'Is that the Goddess of Liberty?' asked one of the passengers.

"'No,' replied the captain, 'that is the Goddess of Industry—Sparrows Point!'"[8]

Thanking Agnus for his inspiring words, Luther Bent offered a solemn pledge from the management of the Pennsylvania Steel Company. "You have seen only the initiatory steps of this enterprise. It will grow year by year through the force of steam, the hammer, brains and wealth." Then beckoning to his mill engineer, Frederick Wood, at the center table, he proposed a toast to America. "To advance is to be American," he declared, and 160 long-stemmed glasses saluted back in affirmation.[9]

*

Strange as it might first appear, there is no specific definition of steel. Rather, steel is used to cover a wide range of different products which have been refined from iron through intense temperature and slag reactions. To make steel out of iron, substantially all of the carbon is removed, producing a liquid mass of nearly pure iron, and then a controlled amount of carbon is reintroduced. Commercial forms of steel vary from soft or mild steel of about .1 percent carbon, to harder (stronger) steels of .7 percent for rails and wheels and 1 to 1.25 percent for spring and file steel; raw or "pig" iron, in contrast, contains 5 percent carbon or more.

The foundation material for both iron and steel is iron ore. Iron ore can be thought of as "iron rust." It is a combination of iron and oxygen, plus silica, phosphorus, and other nonmetallic impurities with a characteristic cocoa-red color. Except in fragments of meteorites found on earth, iron is not a free element, but tied up in the earth's crust with oxygen. For thousands of years iron was believed to be heaven-sent; in early Egyptian writings it was called *ba-en-pet*, interpreted as "hard stone from the sky."

Perhaps as early as 4,000 years ago the first iron was smelted from its ore by man. The discovery was almost certainly by accident: iron ore dumped in a crude furnace reacted with a "reducing agent," probably the charcoal of wood, removing the oxygen. Iron required no more heat to smelt than bronze (man's first metallurgical product), and iron ore was plentiful in ancient lands. Furnace iron was a pasty mixture of red hot metal and slag. The slag could be squeezed out by hammering.[10]

By heating the iron for a much longer time in a hotter fire one could produce steel. The Persians "steeled" iron by means of cementation. They placed bars of iron between layers of charcoal in pots, sealed the pots from contact with the air, and then cooked the pots in a hot flame. After ten days or so the bars had usually absorbed enough carbon from the charcoal to form steel. Slow, laborious, and not always successful, cementation was the chief means for making steel for hundreds of years. In the 18th century charcoal-fired crucible furnaces were developed in Sheffield, England, making it the world capital of steel. Still, the process was tedious and time-consuming; it took a week to make a batch of steel from iron. The technique severely limited the market. Steel remained a precious metal used mostly for fine cutlery, swords, and watch springs, while iron was the mainstay of the early industrial age.[11]

In 1854, Henry Bessemer's experiments with iron manufacture attracted the attention of Napoleon III. The French Emperor was interested in improving artillery guns so that they could withstand the heavy strains of better artillery projectiles. Without a stronger metal, the new projectiles would be worthless. It was agreed that Bessemer should experiment at the military testing grounds at Vincennes.

While he failed to produce the breakthrough sought by the emperor, Bessemer was inspired to undertake a thorough study of the question of metallurgy. Returning to England, he built an experimental converter. It was not long before he struck upon the idea of blowing cold air through a bath of molten iron. Conventional wisdom had it that cold air would simply cool down the iron, but Bessemer found the converse to be true: the mass ignited into a flame of stupefying intensity. Reaching 2,900 to 3,000° F., or twice the heat of "cherry-red," the flame burned out the carbon, leaving the iron in a nearly pure state.[12]

Although Bessemer did not understand all the chemical reactions that were involved, he realized he had reduced the time needed to manufacture steel from seven days to less than 30 minutes. James Nasmyth, the eminent scientist, immediately grasped the implications. "Gentlemen," Nasmyth said, picking up a small piece of steel used by Bessemer to demonstrate his findings at the British Association for the Advancement of Science, "this is a true British nugget; its commercial importance is beyond belief."[13]

At the end of the Civil War, a group of top railroad executives met in Philadelphia for the purpose of forming a company to produce commercial rails under Bessemer's patent. They included J. Edgar Thomson and Tom Scott, president and vice president, respectively, of the Pennsylvania Railroad; Samuel Felton, former president of the Philadelphia, Wilmington, and Baltimore Railroad; and Nathaniel Thayer of Baldwin Locomotive Works. Thomson and Scott bought most of the initial stock capitalization of $200,000.

Felton was named president. A transplanted Bostonian, he was a prominent figure in Philadelphia. During the war his railroad had been the only direct link between Philadelphia and the South. He was credited with getting Abraham Lincoln safely to Washington in 1861 for his inauguration by secretly diverting his train through Baltimore in the dead of the night. Felton purchased land on the Susquehanna River a few miles east of Harrisburg. Foundations for a factory town, later named Steelton, were laid in May 1866. Then disaster struck. The Bessemer converter intended for the plant was lost when the *Indus* was wrecked off the coast of Ireland and sunk. Over many months the machinery had to be reproduced from drawings sent from Britain. On May 26, 1867, the converter was started up—"blown in," in industry nomenclature—and three months later the first commercial Bessemer rail was rolled, or manufactured, for the Pennsylvania Railroad.[14]

During its first year of operation, the steel company produced 1,005 tons of steel. The rails it manufactured from the Bessemer process proved vastly superior to the iron rails upon which American railroads had depended. Iron rails fractured with chilling frequency. Sometimes, tearing loose from the track bed, they became dreaded "snakes" that twisted through the wooden car frames and pierced whatever stood in their way. On main-line track, iron rails required replacement as frequently as every six months. A steel rail could last ten years under the most punishing conditions, according to tests done by the American Society of Civil Engineers and the Pennsylvania Railroad.[15]

Steel rails represented an auspicious leap in technology, equal to if not greater in importance than the Westinghouse air brake. Prior to the Civil War, railroads were slow, light-weight carriers; after 1870 they began a 40-year spurt that transformed the lines into heavier, faster transporters that were of inestimable value to a nation undergoing industrialization. "Probably no other technological development," said two historians, writing in 1949, "has done so much to increase the capacity of the railroads and reduce their operating costs as this substitution of steel for iron rails."[16]

By 1880 Felton's company was rolling 113,000 tons of steel rail a year at Steelton, reaping profits that in one year netted its owners nearly $2 million, or 80 percent of capitalization. By this time rail-rolling had attracted some of the country's most venturesome capitalists. Most of the new steel firms were tied directly or indirectly to railroad interests. For example, 90 miles northeast of Steelton was the Bethlehem Iron Company. Prior to the Civil War it had manufactured iron rails for the Lehigh Valley Railroad. In 1873 the company moved directly into steel rail production. Located fifty-five miles deeper in the Poconos, Lackawanna Iron Company started manufacturing steel tracks in 1875 in association with the Delaware, Lackawanna & Western Railroad and the George Scranton family.[17]

Pennsylvania Steel dominated the Eastern district. Its sales were nearly a third greater than those of Bethlehem and 50 percent above those of Lackawanna. Andrew Carnegie possessed a similar grip over the Appalachian iron district following his purchase of Andy Kloman's Homestead mill, near Pittsburgh, in 1883. His closest rival in the region was the Cambria company at Johnstown, Pennsylvania. O. W. Potter's North Chicago Rolling Mill was supreme in the Western district, greater than the combined size of the other two steel concerns in Chicago, Joliet and Union. Overall, the "Big Three" of

Pennsylvania, Carnegie, and North Chicago controlled 45 percent of the national rail market, with eight smaller firms dividing the rest.[18]

*

The steelman's rule of thumb: 2 tons ore + 1.5 tons coking coal + 500 pounds limestone + 4 tons air = 1 ton of pig iron, the base element for steel. Only the air was free. All the other materials had to be transported to the mill, where three-quarters of their combined weight either went up in smoke or were combined to form slag. This was why raw materials were high on the steelman's list of preoccupations. Perversely, though, by his way of thinking, iron ore and coking coal were never found in sufficient quantity near one another.

In Pittsburgh the steel masters had all the coal they needed close at hand, but they had to go to the Upper Peninsula of Michigan to get good grades of ore at Marquette. This ore was boated some 700 miles from Lake Superior via the Sault Ste. Marie Canal and thence 140 miles by rail from Lake Erie to Pittsburgh. Even when this Michigan ore was blended with less expensive ores from southern Ohio, the Pittsburgh ironmongers paid a penalty of $2 to $3 per ton of ore for their trouble.

In the East, workable sources of ore were spotted around Pennsylvania and New Jersey. The largest of the deposits was at Lebanon, Pennsylvania, 28 rail miles east of the Steelton plant via the Cornwall & Lebanon. The low transportation costs of Cornwall ore counterbalanced the high cost of coking coal. Located on the wrong side of the Appalachians, Steelton had to pay the costs of a 250-mile haul from the Connellsville coking belt. Despite volume price breaks from the Pennsylvania Railroad, the trek was an expensive proposition.[19]

From a decade of experience in the cost complexities of material handling, Samuel Felton's son-in-law, Luther Bent, who had taken over the reins of management as general manager of Pennsylvania Steel, was convinced that the best route lay in capitalizing on the dawning efficiencies of water transport. While railroads had opened up the interior and made shipments of ore and coal feasible, Bent was sold on the idea that costs were moving against inland manufacturers. To depend on railroads and railroad freight rates for all raw material shipments seemed self-defeating. Bent believed that a producer who could get his ore by water could gain an important competitive edge.[20]

The truism that fortune favors the prepared mind was borne out a year later when a business associate, steamboat owner Alfred Earnshaw, mentioned to Bent that his agents were hearing rumors of important ore deposits around Santiago, Cuba. Bent assigned his top engineer, Frederick Wood, to investigate. While much will be said about this remarkable man, it may be remarked, by way of introduction, that he was no Samson of the forge. Wood stood 5 feet, 6 inches tall and weighed less than 140 pounds. On his passport to Cuba, he was described as having "gray" eyes, "indented" chin, "dark" hair and "fair" complexion.[21] He was born in Lowell, Massachusetts, on March 16, 1857, the younger son of William Wood, a weaving-room foreman at the Boott Cotton Mill, and Elizabeth F. Kidder. After attending public schools in Lowell, he entered the newly opened Massachusetts Institute of Technology. Excelling in mining engineering, he was selected as the assistant to the renowned metallurgist, Robert Hallowell Richards. Within months of graduation Frederick joined Pennsylvania Steel, where he won a reputation as a boy wonder and was rewarded with a steady round of promotions.[22]

When Wood walked off the boat at Santiago in April 1882 he was greeted by Earnshaw's agents and several other businessmen. Local guides were hired by the agents and a supply train of donkeys assembled. Thus equipped, the young engineer proceeded into the foothills of the Sierra Maestras. In the first half of the century, the lands had been under cultivation, with extensive sugar, coffee, and cacao estates in the hands of French planters. On some hilltops the Spanish had mined copper and as late as 1850 Santiago was the copper-producing capital of the New World. But greedily drilling shafts as much as 300 feet below sea level, the Spaniards had destroyed their treasure. Seawater had gushed into the shafts, forcing abandonment of the deep mines. During the first War of Revolution (1868–78) the plantations also had been wrecked, and the land around Santiago had relapsed into wilderness.[23]

His first stop was the property of José Ruiz de León, 20 miles east of the city. Some ore had been mined there, but Wood wasn't impressed with its quality. So he retraced his steps, examining the banks of streams that dropped off the inland peaks. Climbing these slopes was an exercise in traveling north. Every 350 feet ascended in the tropics was the equivalent of moving 100 miles north of the equator. The land was 700 to 1,300 feet above sea level, which made the district one of the healthiest in Cuba. After several days he made his way up the Río Juragua. He recorded his initial impressions in his log book:

> I ascended this hill diagonally from one of the branches of the Rio Juragua in a South Westerly direction and traversed about 600 yds. before arriving at the brow of the hill which faces the sea. All the ground passed over was strewn with blocks of loose ore varying in size from a few pounds to many tons. At the brow of the hill a compact mass of ore is exposed in situ for a width of 165 ft. and to a depth of more than 60 ft., with every indication that the real width of this deposit is very much larger.[24]

His hunch was right. Over the next few days he found 1,000 acres of sidehills honeycombed with veins of ore, 15 million tons in all. While the earth is full of ore, most of it is commercially worthless owing to its low iron content or impurities. The ore Wood discovered, however, was just about right. It was a hard hematite rock of 62 percent iron. Other metallic and nonmetallic impurities were few, especially phosphorus. Phosphorus was the bane of steelmaking. It made the metal brittle and was extremely hard to get rid of. (There was an old saying that phosphorus was to metallurgy what the devil was to religion.)

Returning to Santiago, Wood telegraphed his findings to Steelton and booked passage on the next boat to the U.S. In six months he was back on the island, leading a team of engineers to the Juragua site. Plans were made for surface mining and for a railroad line to carry the ore from the deposit to Santiago Harbor 17 miles away. Since the land was unclaimed jungle, mineral rights could be secured free of charge by simply filing an application at the land office in Santiago.[25]

Concurrent with these preparations, Luther Bent and Alfred Earnshaw were busy conducting high-level negotiations with Spain. Although the businessmen wanted the ore, they didn't want to pay Spain's export taxes on it. Once again, fortune smiled on their endeavors. At the very moment they approached the Spanish government, the Cuban economy was on the skids due to a worldwide glut of sugar. With sugar prices plummeting and civilian unrest on the rise, the prospect of an investment by an important U.S. company was tempting to Spanish authorities. The Philadelphia business community had been involved in the Cuban economy since the 1850s, and it did not prove difficult to persuade Luis M. De Pando, governor of the province, of the wisdom of supporting tax and tariff breaks to attract Yankee capital. The governor's endorsement of the venture was apparently crucial. The royal decree that was issued by the Crown of Spain on April 17, 1883, was an eye-raiser. It provided that for a period of 20 years iron mining companies in Cuba (of which there were none at present) "should be free from all tax on the surface area of all claims of iron or combustibles" and that "ores of all classes should be free from all export taxes." Other provisions were phrased in a way that could only benefit the Americans, to wit: "that mining and metallurgical companies should be free from all other impost; that for a period of five years the mining companies should be exempt from the payment of duties on all machinery or materials required for working and transporting the ore."[26]

In sum, Spain traded all rights of royalties and export fees until 1903 in return for a mine, a rail line, and about 700 jobs. Twelve hundred miles to the north, in Washington, D.C., Bent's agents were active. It wasn't hard for them to gain the backing of U.S. Senator Jim Cameron of Pennsylvania, who was closely affiliated with steel and railroad interests in the state. Under his influence the new tariff schedules that were passed by Congress in 1883 reduced the tariff on imported iron ore from 20 percent "ad valorem" (20 percent of value stated) to a flat 75 cents per ton of ore imported. Since the Cuban ore was valued at the time at $6 a ton at East Coast ports, it was apparent how valuable the tariff alteration was to Bent and his associates. Frederick Wood's internal papers indicate that about $200,000 in import duties was saved during the first six years of mining.[27]

To share the risk of the project Pennsylvania Steel enlisted Bethlehem Iron Company. They formed Juragua Iron Company, Ltd., a 20-year limited partnership established under the laws of Pennsylvania. The two steelmakers put up a total of $1,395,000 for development of the mines and railroad. Shipments were handled by Earnshaw's steamers from Santiago to Philadelphia. The partnership's board of directors included Earnshaw, Bent, and Wood of Pennsylvania Steel, and Richard Linderman and Joseph Wharton of the Bethlehem company. On July 17, 1884, the narrow-gauge line from the mine to Santiago Harbor was finished and hailed in a ceremony by Governor De Pando, whose name was affixed to the bridge over the Aguadores River. Total shipments that year amounted to 25,295 tons. In the next year they jumped to 80,716 tons and in 1886 to 112,074 tons.[28]

To exploit the ore with utmost economy, the company sought a tidewater plant where the rock could be converted directly to pig iron rather than shipped at a penalty to Steelton. Again Frederick Wood was commissioned by Bent to investigate possible sites. His prospecting showed Baltimore and, specifically, Sparrows Point, to be the best location. One hundred and fifty nautical miles from the Virginia Capes, the peninsula was closer to Cuba by 100 miles than any spot around Philadelphia. It was 65 miles closer to the bituminous fields of western Pennsylvania. Reckoned by the steelman's rule of the thumb, Sparrows Point was a hub where the raw ingredients of ore and coke could be assembled at low cost.[29]

The geography of the land was typical of the shoreline of the Chesapeake Bay. The peninsula, along with the whole of the bay and the Eastern Shore, once had lain beneath the ocean, and the sifting action of the sea had ground rock and rolling hillsides into heavy, compact layers of clay. Contrary to popular legend, the peninsula was not named after the bird, but after Thomas Sparrow, an English colonist who was granted the parcel in 1652 by Lord Baltimore. Although Sparrow had high hopes to build a settlement, nothing came to fruition. Only a single farm family had settled there in 1814 when British warships sailed up the Chesapeake and, moored off Sparrows Point, pummeled Fort McHenry with cannon fire, inspiring Francis Scott Key to write "The Star-Spangled Banner."

The harbor cut Sparrows Point into roughly square dimensions of 1½ miles in length and breadth. The uniform flatness of the terrain made it excellent in Wood's mind to craft a mill. With sufficient fill the land could be made to conform to virtually any shape. The surrounding waters were deep, 22 to 25 feet in the main channel. Some dredging, to be sure, would be needed along the shore, but this was of minor consequence. So it came to pass that under Wood's recommendation the board of Pennsylvania Steel purchased the peninsula in 1887 from five local landholders for a grand sum of $57,900.[30]

*

The surveyors landed on Sparrows Point on March 1, 1887. Within days incipient signs of industry could be seen across the horizontal expanse. Superintendent David Baker was there, slopping around the icy pools of Fitzell's Bog with a team of assistants. His assignment—to take detailed test borings of every square yard of surface along the southern shore. He and his assistants stuck long augers into the peninsula and then removed the sticks to examine the soil inside. Off to the northeast, bobbing through the dormant jumbles of marsh grass, was Frank Fahnestock, chief surveyor. He stopped at regular intervals to train his telescopic lens on Bear Creek's twisting coves. In a land long under the sway of squalling seagulls, a dozen modern-day Leatherstockings were plumbing the depths and marking off the contours of Sparrows Point.

The men explored every nook and corner of the peninsula, then reported back to their chief. Wood worked alone in an unused cabin at Holly Grove, a defunct summer resort. From survey readings and test borings he began making sketches in his notebooks. Some encompassed the whole peninsula, while others were limited to a small quadrant. With *X*s and squares he plotted out the major components of the plant. He *X*ed out the sheds and furnaces. Between the *X*s and squares he drew the pipelines that would distribute the mill's water and fuel.

Wood rearranged the parts and tinkered with the interweaving woof of pipe and track, often scratching out one result and starting again on the next page. His aims were unabashedly ambitious: he wanted to design a plant that would improve upon the standard mill layout pioneered at Steelton by famed civil engineer Alexander L. Holley. He also took some time to rough out the town. His idea for the village was a network of streets which crossed the flatlands in grid formation. Along the streets he drew a company store, a schoolhouse, and the village railroad station, together with rows of houses.[31]

In this manner Wood celebrated his 30th birthday on March 16, 1887. After several more days of daytime and nocturnal scribbling, he returned to Steelton with a working plan to present to Bent and the board of directors. The task of building the plant had fallen to him, and it was a monumental undertaking. He called on his brother for assistance. Almost nine years older, Rufus Wood was not trained in steelmaking. His professional life had been spent in Boston, where he was an accountant for a hardware company. Rufus quit his job to become on-site supervisor and paymaster at Sparrows Point, handling orders that his brother sent down from Steelton.[32]

The first summer was terrible. Hot weather blanketed the peninsula, hatching swarms of mosquitoes that invaded the temporary labor camps at Holly Grove. The crews struggled to drain the bogs, using sump pumps powered by windmills. Coal torches were lit at night in an effort to ward off the winged pests. But what surprised Rufus most was the prevalence of disease in the Chesapeake lowlands. Ague was a fever commonly associated with the early stages of malaria. So was "the chills." Both waylaid the workforce and, while rarely fatal to the sufferer, slowed down progress. Rufus dispensed

Kidder's Cathartic Pills to the sick and lamented the mounting casualties in letters to his brother:

> [JUNE 24, 1887] Mr. Doty I am sorry to say suffers almost continually with the ague and several other men are down with the chills at times, but I think it is in some cases the result of their imprudence.
>
> [JULY 14] This morning J. G. Fahnestock had not reached the brickyard before he was seized with a severe chill and had to be carried home and put to bed.... The heat for the past three days has been almost beyond endurance, averaging 95 in the shade, and I have had to-day a pretty sharp warning that I must keep out of the sun when possible. Violent pains in the head, accompanied with nausea and an indescribable sensation of sinking, have nearly unfitted me for any work to-day, but I have managed to creep around slowly.
>
> [JULY 18; from the wharf following a boat trip from Baltimore] We have just escaped with our lives from the most fearful squall that I ever experienced, lasting twenty five minutes, during most of which we expected to go to the bottom, and were prepared for the capsizing with life preservers about us.
>
> [JULY 21] One man is dangerously ill with convulsions, and many now suffer from dysentery. Green scum forms on the water surface after it is left standing for a few hours.[33]

Through it all Rufus kept order. On June 22, 1887, he wrote to Frederick, "I have written to the County Commissioners requesting the appointment of John Campbell as commissioner officer of the law, one of whom is seriously needed in this mixed community of ours, now about 100 persons. Especially to prevent the sale of liquor, which I find Cunningham [the night watchman] has been freely dispersing to anyone who applies."

The first building completed was the brickyard. It began producing 30,000 bricks a day. A permanent wharf, 900 feet long and 100 feet wide, was finished next, and railroad tracks were laid upon the pier. The first locomotive was landed on July 18. Soon carloads of supplies arrived, and track was extended to all parts of the property. In August the Sparrows Point Company Store was started and Frederick L. Webster placed in charge. In October ground was broken for the blast furnaces. Granite pilings from Port Deposit and sandstone from Virginia were transported in sailing vessels and landed at the pier.

Soon so much material arrived that it was found necessary to build a transfer slip alongside the pier. Heavy materials were brought in railroad cars towed on barges to the slip and carted to any point desired by the shifter locomotive. The massive steel framework and boiler plates were manufactured at Steelton and brought to the Point by rail—from 1888 to 1890 the average tonnage of material was 15 carloads per day.[34]

In the second summer sickness was greatly reduced ("No more chills and fever at Sparrows Point!" proclaimed a newspaper report) as a result of the successful drainage of the marshes and careful attention to sanitation. Over 100 houses were finished and the streets were graded. Construction of the Sparrows Point Railroad also moved forward. Yards were laid at Colgate's Creek and the entire road was ballasted with rock and oyster shells. Traffic interchange agreements were worked out with the Pennsylvania and B&O lines. Soon a dozen passenger trains connected the plant with Baltimore's Union Station each day.

From the brothers came a close accounting of monies spent: $1,087,918.23 for the first two blast furnaces, $819,332.50 for the Bessemer unit, $91,971.30 for the docks, $1,263,054.99 for the rail and blooming mills, $80,994.77 for streets and water. In all, nearly $7 million was paid for the mill and $700,000 more for the establishment of the company town. This compared to an investment of $10,477 for the average factory built in Maryland at the time.[35]

Concurrent with construction, important understandings were reached between the company and the political bosses of Maryland. Before the state legislature convened in 1888, Luther Bent was at the doorstep of Governor Elihu Jackson, a member of the Democratic machine in Annapolis popularly known as "the Ring." Like nearly all Maryland statesman, Jackson was deeply involved in railroad politics, spending a good portion of his time as governor overseeing the fortunes of the Baltimore & Eastern Shore Railroad, of which he was treasurer.[36]

Pennsylvania Steel did not want to play railroad politics; it wanted to be master of its own political destinies. Exactly what Bent said to Jackson is not known, but the results of his efforts could be seen in the gentlemen's agreements reached between the state and the steel company. They gave Bent's organization total political hegemony over the company town, vesting in the manufacturer the right to control everything from the hiring of teachers at the local schools to the enforcement of town justice and collection of residential garbage.

The apparatus of policing and law enforcement was handed to the company through the Maryland Special Police Act. Passed by the legislature in 1880, the law permitted five classes of business—railroads, ironmaking mills, coal companies, canal companies, and steamboat companies—to appoint "special police" with the approval of the governor. Under the plan hammered out between Jackson and Bent, Baltimore County Sheriff George Tracey agreed to vest all local patrol and arrest functions on Sparrows Point to a police department to be established by the company. The company in turn paid the salaries of the local police and assumed all related costs. In effect, Governor Jackson sanctioned a separate and distinct police unit for the company's land, one in which neither he nor any other public official would have supervisory control. The arrangement was barren of safeguards against arbitrary action by the company police save for a provision calling on the police not to engage in election fraud.[37]

The governor's favors did not end there. He agreed to oppose any attempt by the legislature to incorporate Sparrows Point as a town, a measure that might require the election of a town board of commissioners. Obtaining Jackson's backing was important, but Bent also realized that there were other mouths to feed. So he hired J. Alexander Preston. He was one half of the Baltimore law firm of Preston & Bowie. The other half was Oden Bowie, a "Ring" founder whose term as Maryland governor had been notable for its shenanigans. Retired from politics, Bowie was president of the Baltimore & Potomac, the Pennsylvania Railroad subsidiary that had gotten its right of way thanks to a Bowie-engineered franchise. The Pennsy was among an assortment of racetracks, railroads, and Baltimore banks on the client list of Preston & Bowie. Pennsylvania Steel joined the lineup.[38]

During the 1888 legislative session, Preston reported back to Bent about his efforts on behalf of management. He said he had held discussions with Isaac Freeman Rasin, the party boss of Baltimore, traveled to Baltimore County offices in Towson, and consulted with Senator Arthur Pue Gorman on Capitol Hill. His mission was to keep the Ring apprised of the plant's progress and to keep management out of the way of "fat-fryers"—bills cooked up by legislators to force businessmen to pay bribes for their defeat. The festivities burned bright around State House Circle, causing Preston to jest to Bent early on, "The Clerk of the Court of Appeals, has asked me, to drink and dine with him, and knowing Annapolis whiskey, I drop this line, to post you, for fear it may kill me on sight."[39]

Successfully navigating the company's agenda through the shoals of Annapolis, winning approval for a bridge at Bear Creek and a local temperance law sought by the businessmen, the lobbyist was paid well. Bent offered him $75 a day and another $50 a day for expenses. Precisely what kind of expenses would cost $50 a day when a good meal could be had for $3 or $4 was apparent in their correspondence. After he returned from Towson once, Preston wrote cheerfully to Bent, "I never keep account of my expenses, and never charge them. I think $200 [for cash expenses] would be fair, but will be satisfied with your view of the matter." Whatever it took for him to perform his services was paid in cash. Here there was no close accounting.[40]

Rufus Wood fretted about the lobbyist. Unaware of the details of his arrangement with Bent, he wrote to his brother on March 20, 1888, "I was accosted by Preston on the street this P.M. and he said that he was going to Annapolis to-night with Rasin and thinks the matter [the temperance bill] will be all right. Mr. Taylor said one disadvantage of having Preston to represent us in this matter is that, being a great lobbyist with plenty of boodle in his pockets for most projects that he advocates, the Ring will not believe him when he tells them there is no money in this case and will not work for it." The Ring, though, did work for the company and one finds evidence of favors returned. Walter R. Townsend, reading clerk of the general assembly, was given the task of preparing the company's property tax bills. Joshua Frederick Cockey Talbott, Maryland insurance commissioner and soon to be U.S. representative, later headed a partnership that purchased slag from the company that was resold to the Baltimore County roads department at a markup.[41]

A special relationship had been set in place, one that would serve the interests of the steelmaker for decades to come. Whatever the company wanted by way of special permits and other favors, the company got. Emblematic of the political ties that linked Sparrows Point with Annapolis was Bent's appointment of Preston & Bowie as chief corporate counsel in 1889. Over the next 25 years, Preston, his son Alexander Preston, and Oden Bowie's son and successor, Carter Lee Bowie, would oversee political questions for the company. Having great lobbyists on State House Circle was reassuring to the Goddess of Industry.

CHAPTER 2

CREATING THE WORKS

Not long after the blast furnaces started lighting up the night skies of the harbor, Luther Bent turned over the Sparrows Point complex to his protégé Frederick Wood. On June 27, 1891, Bent signed the incorporation papers creating the Maryland Steel Company of Baltimore County, a wholly owned subsidiary of Pennsylvania Steel, and appointed Wood president of the enterprise. Wood, in turn, promoted his brother Rufus from cashier to general agent, the number-two slot of the company. The two Woods sat on the board of directors together with Bent, Walter S. Franklin, and Edgar C. Felton of Steelton. With this shuffle Bent stepped back from day-to-day management. As president and senior statesman of the steel and mining conglomerate, he said he was content to savor the rewards of "our Maryland adventure" from his office at the Girard Building, Chestnut and Broad Streets, in Philadelphia.[1]

A neat division of labor was drawn through these appointments. Frederick handled all aspects of rail production while Rufus presided over the company town. Pursuing each sphere of corporate responsibility with customary forethought, the brothers forged a system of factory production and community life of far-reaching influence. The word *system* is critical to the understanding of the brothers, for they saw themselves not as swashbuckling capitalists nor as Wall Street tycoons, but as supremely organized industrialists. Their objective was to develop a centralized management that would fuse business and social life in a way that was rational, orderly, and free of any dissenting voice. Not surprisingly, their byword was *efficiency.*

Yet they sought something more: in their private letters (Rufus in particular wrote profusely), they discussed the type of worker to be created. He would be "a sober, industrious" individual, according to Rufus, a man who would labor mightily in the mill and live honorably in town. The rough-and-tumble world of steelmaking would not interfere with his finer instincts. Gentlemen in town and toilers on the job were what the Wood brothers envisioned, and, in keeping with Victorian times, they spun a fine web of moral rectitude over the mill life they fashioned.[2]

In broad outline Frederick and Rufus were typical of the men who were recruited by large corporations to run their businesses. Precursive technocrats, their aim was to achieve success for the corporation rather than for personal gain. They undertook the complex process of turning a beachhead of machines into a far-flung, integrated unit of production, and they did so in deliberate, discrete steps that made it possible for power brokers to stitch together larger and larger production units after 1900. The brothers built the economic and social foundations on which the later achievements of Sparrows Point rested, and, if their roles were chiefly unknown outside the world of steel, they were by the same token not part of an impersonal marchstep of industrial progress. Given great latitude by Luther Bent and the board of directors, they were the "big bosses," and the choices they made (or didn't make) in mill technology, hours and pay, and prevention of accidents were of utmost consequence to their workforce of 2,500.[3]

The most dramatic imprint was made by Frederick Wood. Wood was one of a comparatively small number of engineers who shaped the landscape of mill and furnace. Integrated rail-rolling was as great a source of future shock then as computers are today. It could make a product quicker than previously thought possible; it was mechanically eye-catching; and almost immediately it began to alter contemporary concepts of factories. As E. C. Potter noted in a speech published in *Iron Age* in 1898, "The rail is to-day the cheapest finished product in the whole domain of iron and steel manufacture, and is at the same time the most difficult to make. It requires an expenditure of at least $3,000,000 before a single rail can be economically turned out. One of the most complicated and delicate operations known to metallurgy is carried on at the tremendous rate of over a ton of product per minute, day after day, within limits that will not permit of a variation of more than five one-hundredths of 1 percent either way from the standard."[4]

Using the most advanced tools of his time, Wood applied two principles that would become hallmarks of this century's manufacturing. They involved size and speed. In the case of mill apparatus, Wood was an extreme proponent of scale, not infrequently pushing furnace burden and blast to new limits. And regarding speed, to slash production costs, he innovated "throughput" methods which stamped the plant as unique in the business.

*

The great achievement of the 19th century, according to many students of technology, was the invention of the method of invention. Underlying the advances of the "industrial revolution" were discoveries of a verifiable approach to analyzing materials and for calculating new forms of energy. This required a change in human thought.

In ancient mythology, fire and blacksmithing were presupposed to be under the sway of gods or tied to humans with supernatural powers. Through the European Renaissance spirits were thought to lurk in the flames of forges. In a 17th-century best-seller, aptly titled *Natural Magick,* Giambattista della Porta told of appeasing sensitive spirits to make iron. If the metal was treated with "sympathy," he wrote, it would become soft and pliable. "It delights in fat things, and the pores are opened by it, and it grows soft: but on the contrary, astringent things, and cold, that shut up the pores, by a contrary quality, make it extreme hard." Like his contemporaries, the alchemists, who sought to transmute baser metals into gold and find an elixir of life, della Porta tended to anthropomorphize—to ascribe human properties to nature around him, including to inanimate things.[5]

A key accomplishment of modern inventors was to remove human characteristics from the realm of materials and machinery, a process capped by the remarkable advances in metallurgy. In David Landes's arresting image, "Prometheus was unbound." By tinkering with blasts of air drawn through a bath of molten iron, Henry Bessemer had discovered a way to make steel. By experimenting with grooved rollers and steam engines, other inventors had perfected the hand-rolling mill, a quantum leap over the trip-hammer method of working metal. Then, too, the development of coking coal to smelt iron ore in the blast furnace brought another stage of steelmaking to an advanced state. Unexpectedly, it was a handful of American engineers who put all the stages together.

It was unexpected because before the Civil War America notoriously lagged behind Europe in the basic and applied science of metalmaking. For decades the British were the leaders in nearly every aspect of metallurgy, with Robert F. Mushet, Henry Cort, and Charles William Siemens joining Bessemer in seminal achievements. Germany, Sweden, and France also boasted sophisticated metalworking industries. While imitating, importing, and otherwise freely borrowing from the Europeans, the U.S. companies embarked on a campaign to make steel in mass quantity. Actively encouraged by their railroad sponsors, the integrated railmakers became enthralled with the possibilities of joining together, or "integrating," the heretofore stop-and-go process of steel manufacture.

Removing delays in the assemblage of raw materials, smoothing out kinks in the rolling mills, "hard-driving" the furnaces to their rated capacity and beyond—such was the philosophy of steel-mill engineers, and they did this mechanically, empirically, sequentially, stringing together a geography of sizes and shapes that had no analogue in nature. Noted a bedazzled William Dean Howells in *Atlantic* magazine, "Yes, it is in these things of iron and steel that the national genius most freely speaks."[6]

Frederick Wood first applied his engineering cunning to the layout of Sparrows Point. The *X*s and squares in his early drawings were translated into a highly-compressed network of processes. Rails emerged within eyeshot of the blast furnaces where the ore was smelted. From end to end less than a mile separated the ore docks from the yards where the finished product was shipped. Overall the facilities occupied less than 400 acres of the property. In a technical analysis of Sparrows Point *Cassier's* magazine said with admiration, "The works as a whole exemplify concentration and centralization in a pronounced degree." *Iron Age* chimed in with accolades for a design that incorporated the essence of the "American practice."[7]

What did not fail to impress trade journalists was the grandiose scale of the constituent parts. This was especially true with the Bessemer converters. The two converters installed by Wood were by far the largest ever constructed. Each had the capacity to produce 20 tons of steel in a single blow. This placed them at twice the capacity of the latest units of Andrew Carnegie, no slouch himself in the pursuit of bigness. So pleased was Frederick with his machines that they were posed for official photographs before they were bolted into place in the Bessemer house. In one picture a young workman was shown leaning against the vessel, while a grizzled veteran in floppy hat was perched above on a ponderous ledge of cast iron.

For reception of hot pig the converter was placed in a horizontal position. Then it was turned up and a blast of air was introduced through small holes at the bottom of the vessel. The air ignited the contents, burning out the metallic impurities at 13 times the temperature of boiling water. During the pyrotechnics a first vesselman with colored glasses watched the color of the flame as it roared out of the mouth. In a few minutes the flame changed from violet to orange and finally to white. (If the flame was too intense, the vesselman added cold scrap to the melt; if the heat was too low, there was a technique known as side-blowing where the converter was tilted to one side and air was blown partly over the contents.)

The white flame signaled that the impurities had escaped into the air. Any more combustion would waste the iron itself. So the vesselman and his helpers threw several levers and the converter creaked forward, letting the molten metal flow back into the central chamber. A ladle was then brought up by a traveling crane, ready to receive the contents. Beside the main ladle was a smaller ladle. It held the molten spiegeleisen that regulated the final level of carbon. The desired amount of spiegeleisen was poured into the big ladle together with the purified iron.[8]

Frederick Wood was able to install converters of such magnitude because he had solved a production problem which had stumped steel-mill engineers for 35 years. The problem was how to transport the liquid steel to the soaking pits where it was reheated for rolling. The complexity of the problem was compounded by the fact that, if cooled down too rapidly, the steel would form a scab on the sides of the handling equipment, causing severe damage. Prior to the opening of Sparrows Point the universal practice was to pour the steel into a pit where a series of empty iron vessels known as "ingot molds" were arranged. Then the molds were pushed to the soaking pits by gangs of sweating men.

Wood cut short these hazardous and labor-intensive maneuvers by inventing a system known as "casting on cars." First the molten metal was poured into the ladle. Next it was carried across the Bessemer floor by an improved crane unit to a narrow-gauge railroad designed by Wood. Through a hole in the bottom of the ladle, the steel was discharged into the molds standing upright on flat cars. When the molds were filled, the train was drawn out of the building by a chain-cable device. A switching locomotive picked up the load and hauled it to the soaking pits.

Eliminating the bottleneck at the Bessemer furnace, Wood's invention placed new demands on the soaking-pit equipment. This led to another burst of creativity. The old method of "stripping" steel at the pits was downright primitive. Gangs of men took wood beams and battered them against the molds until the steel was freed. Wood had a better idea. Experimenting with various lifting devices, he and a collaborator, Pittsburgh inventor Henry Aiken, came up with an improved stripping crane or "ingot extractor." It was a hydraulic crane that lifted the molds off the steel vertically. Use of side lugs that locked into the crane's hooks was a clever touch that expedited matters. All told, the extractor saved an hour of production time.[9]

Iron Age reported that Wood's car-casting and stripping system "has completely metamorphosed the old Bessemer shops, increasing output and reducing cost of manufacture." Nearly all existing steel plants adopted the process, and mills built since Sparrows Point incorporated the design. The magazine went on to say that the invention "has been characterized by eminent authority as the most important improvement in Bessemer steel practice since the invention of the Holley bottom [in 1874]."[10]

The same sense of product flow directed his design of the rail-rolling mills. After being heated to a uniform rolling temperature, the ingot was carried on a cable system to the first set of rolls. These were the blooming rolls. The rolls were made of very hard steel and were driven by a massive steam engine. By manipulating a set of levers, the roller crew reduced the ingot as it went back and forth through the rolls. With each pass the ingot became longer and thinner until it was stretched into a bloom 20 feet long and 8 inches square.

Emerging from the rolls, the steel did not stop for reheating as in the past, but kept right on moving on a conveyor adapted by Wood from a procedure developed at Duquesne. At the end of the shed the bloom entered a shearing unit where a giant rotary blade dropped down and cut off the ragged ends. Next the bloom went on to the receiving tables of the rail mill. The rail-mill rolls were powered by Wood's pride, a 2,500-horsepower Porter-Allen engine. The engine had 48-inch steam cylinders and a flywheel 22 feet in diameter.

Back and forth, up and down, lengthening to 180 feet between the rollers, the rail assumed its characteristic flanged T-shape, traveling at speeds of up to 25 miles an hour. At the final set of rollers it was dropped by a lever-operated manipulator onto an endless-chain conveyor where it passed through another shear cutter and headed for the cooling beds before straightening and drilling to accommodate the bolts for the end joints. What had been crude ore only a few hours earlier was soon to be finished as six lengths of rail.[11]

*

Scholars have variously described the tools of 19th-century engineering as the "manufacture of power" (Elting Morison), the "intensification of heat" (Peter Temin), and the "economies of speed" (Alfred Chandler). All were used to good effect at the plant. Here production was not confined to a single building but took place inside a constellation of specialized structures. A product was moved to the next substation of manufacture as swiftly as possible. Cotton and textile mills had relied on steam or water power to augment production; at Sparrows Point vastly enlarged power was complemented by dramatically increased speed. "It is a question if anywhere in the world steel in commercial form is produced in a shorter period than at Sparrows Point," remarked *Cassier's*.[12]

The governing metaphor of the whole enterprise was that of a speeding train. Within the precise scheduling of raw materials and coordination between mills one could detect the structural continuities between the "American practice" of making steel and the method of American railroads to signal, switch, and expedite freight and passenger movements. The lingo of railroading was present throughout the process. Furnaces were said to be running with their "throttles wide open" when the blast equipment was fully engaged, and the blooming and rail machines were nicknamed "roll trains." Wood had transposed the economies of fixed-rail transportation into the factory, and the factory, at least in those industries increasingly identified with mass production, would never again be the same.

Interestingly, what made this sophisticated design practical was steel. In erecting the mills at Sparrows Point, Wood and his colleagues adapted a special type of building pioneered by the railroads: the train shed. Of massive balloon framing with trusses, girder spans, and arched ribbing, it represented a radical departure from the cramped wood and masonry factories of yore. The British inaugurated the form, and U.S. rail systems carried it forward. In addition to spanning a dozen or more tracks and platforms, the post–Civil War train terminals housed waiting rooms, ticket rooms, restaurants, foot passageways and stairs, servicing equipment, offices, and sometimes even hotel rooms. They were among the most complex structure of their time.[13]

Frederick Wood was aware not only of the railroad stations in Philadelphia, Boston, and Harrisburg—he had gone through King's Cross,

Manchester Victoria, and other British terminals while on a tour of the country's steel manufacturing sites in February and March 1889. The rolling mills at Sparrows Point were steel-framed sheds of train shed dimensions. Attached to the upper rooflines were clerestory windows and smokestacks for venting heat and dust. Such wide-beamed enclosures liberated a steel engineer like Wood from the stacked-floor confines of the traditional factory building and permitted him to stage a multitude of operations—cranes swinging overhead, ingot trains chugging below—in his buildings.

So fundamentally different was this sort of unified system that a new word was needed to describe it. To knowledgeable writers terms like factory and mill seemed wholly inadequate for places like Sparrows Point. The word *factory* implied a single building or a discrete unit of manufacturing, while the term *mill* originated from grinding grain and other agricultural activities. What also seemed outdated was *shop,* used by Maryland Labor Commissioner Thomas Weeks when he surveyed Baltimore factories in 1886. Slowly a word crept into the popular vocabulary which captured this new industrial form, and one found it increasingly used in the Baltimore press. Sparrows Point was not a factory, but a *works,* a word which encompassed the huge shapes and interlocking conveyances by pinpointing their overriding function.[14]

And what was the role of the workman at the works? Publicly, Wood was very circumspect about this matter. As he described it to General Agnus's *American,* his job as company president was "to break in the armies of men required for each department." How he planned to do this he never openly mentioned because he feared his agenda might provoke controversy if it were publicized. In Baltimore a ten-hour weekday and briefer Saturday work schedules had been the established norm since 1836. In that year laborers had stopped work and paraded through the streets with drum and fife, proclaiming to all concerned that ten hours should constitute a day's labor thereafter. "The conflict was decided in a week in favor of the workingmen, and for 50 years, men have as a rule worked but ten hours a day in Baltimore," wrote Johns Hopkins professor Richard Ely.[15]

Dispensing with such conventions, Wood took the continuous workings of his machinery as the model for working hours on the peninsula. Employees were divided into two shifts so that the plant could operate around the clock. Men toiled 11 hours a day for a week of day shift (on Thursdays and Saturdays ten hours), and 13 or 14 hours a night for a week of night shift. When the shifts changed on Sunday the night shift worked 24 hours straight through. On the following Sunday the night crew had the day off as the day crew toiled from 7:00 A.M. Sunday to 7:00 A.M. Monday.

The 24-hour shift was as unique to American steel manufacture as the integrated production system: no other steelmaking nation pursued such hours of labor with such vigor. Defenders of this practice said the mills had to be operated continuously to avoid the high costs of banking furnace fires. However, in Britain, Germany, and France, workmen were not subjected to such tests of endurance. General hours of labor were nine or ten hours. Few workers picked up their furnace shovels on Sundays. Nor were foreign producers so stingy in the granting of days off. At Sparrows Point, there were no vacations and two holidays a year. The holidays were Christmas Day and July 4, and on those days of enforced leisure the workmen were not paid.[16]

In addition to long hours, Wood established a numerical procedure to monitor a worker's output. Aided by the invention of the time clock, company timekeepers compiled every day the hours and tonnage of the crews. "It is [the timekeeper's] duty to see that every man in his department is at work in the morning," noted a contemporary account. "At noon he confers with the under-bosses, and before 6 o'clock he gets every man's time for the chief mogul. A man found standing about idle, or one caught smoking while at work, is likely to be docked for the day."[17]

Not content to stay in his office, the "chief mogul" surveyed the mills in rapid-fire tours. In his pocket notebooks Wood scribbled down orders, observations, and triumphs of tonnage recorded to one-thousandth of a ton:

> —Directed J. D. Campbell to reduce carpenters to 10. Bartols gang to 8 men.
>
> —Week ending May 13 used 24.5 cars of 18 tons each per day.
>
> —Will cross bracing, end of mill, interfer[e] with pipe crane of soak pit crane?
>
> —Keep acct. of furnace scrap used.
>
> —Wm. Gray left boiler shop without notice.

—Let Baker have access to time books to see how men work.
—High Water Record: Oct. 18, 93—best 12 hr. turn in billet & rail mill—640.375 tons of rails.
—Why are not daily reports sent regularly from the Store?
—Two houses for Hungarians.
—10,000 ton option. Ship promptly.
—Write Hemphill about converters.[18]

An engineer of the onrushing train, Wood had little patience with those who might tire from his pace. He believed that the laborer should be industrious and aware of his station. "None is more exacting, according to the men, than President Wood," a reporter for the *Baltimore Herald* wrote in 1893. "He does not hesitate to call anyone to account for what he supposes to be a malfeasance, no matter how slight it is. He often meets with a rebuff from some new laborer, who is not aware of his identity." The reporter gave this instance:

> The company is now engaged in grading a piece of railroad running through the property, and uses iron ore as ballast. The men have to wheel the ore up a steep embankment, and, as it is very heavy, the barrows are never filled. Mr. Wood was inspecting the work and reprimanded one of the men, who was staggering up the incline with a heavy load, for not having his barrow filled. The laborer, a new hand, threw his wheelbarrow, load and all, over into the ditch. Then squaring off at the energetic little president he wanted to know what in the h—— he had to do with it. Before Mr. Wood could answer the laborer turned on his heel, and going to Mr. Campbell, the boss, asked for his time.
>
> "What are you going to quit for?" asked Campbell.
>
> The man explained that some little fellow rather smartly attired had come and interfered with him, and he did not propose to put up with it. Campbell at once surmised who the little fellow was, and said, "I'll have to give you your time whether I so desire or not."[19]

In 1895 Frederick Wood compiled an internal report for Bent which showed how much steel costs had been cut back. Comparing the years 1895 and 1875, Wood listed the major costs of producing a ton of pig iron in the blast furnace. As usual his figures were precise to the penny:

COST OF MAKING PIG IRON (PER TON)

	Steelton in 1875	*Sparrows Pt. in 1895*
Iron ore	$12.79	$6.26
Coke and coal	$7.04	$3.50
Limestone	$1.33	$0.21
Machine repairs	$0.86	$0.15
Labor	$3.69	$1.18
Total Cost	$25.71	$11.30

Reductions in the costs of materials, men, and machine repairs were increased as the pig iron was manufactured into rails at Sparrows Point. Through the efficiencies of the 20-ton Bessemer converters and the casting-on-cars system, the cost of making steel rails from pig iron had been shaved to $11.82 a ton from $34.71 in 1875. All told, the company could produce a ton of finished rails for $23.12, as against $60.42 20 years earlier.[20]

Although the cost of materials had decreased appreciably, the leading factor in the drop in costs was labor. It had gone down by 65 percent. At first glance this fact appeared incongruous, given the perception of machines as the releaser of man's energies and creator of wealth. Yet, in the act of substituting horsepower for human power, Wood and his fellow engineers had altered the relationship between the steelworker and his work. Whole categories of jobs were eliminated under Wood's factory system (over 100 positions in the automated Bessemer house), and the skills required among those who remained were for the most part downgraded.

In 1875, the year Wood used as the base point in his report, about 40 percent of all mill jobs at Steelton were classified as "skilled." Pay of $4 a day or more was not uncommon. Legend even had it that the master rollers, upon whose split-second timing the early mills depended, came to work in horse-drawn carriages, so lucrative were their occupations. While the story was apocryphal, it expressed a common perception that skilled workmen were overpaid and thereby impeded progress. At Sparrows Point the skilled ranks had been thinned to 25 percent, and the entire definition of "skilled" was undergoing a revision.[21]

In place of the highly skilled, highly paid craftsmen of 1875 were machine tenders and day laborers. The position of master roller had been eliminated, and the position of roller made a salaried foreman's post. With the roller (foreman) directing, operations were carried on by four so-called "roll hands" supplemented by operators who did specific duties. Among them were two oilers, five tablemen, two rail crops, four machinists, two machinist helpers, one hot-saw operator, one greaser, and a floating group of eight to ten laborers. Overall manpower needs were limited to 38 to 40 employees a shift, or about half the number in 1875.

Where skilled workmen could not be displaced, labor savings were achieved through increased output. In 1875 a Bessemer crew at Steelton produced 80 tons of steel a day. In 1895 a crew at Sparrows Point could make 300 tons a day. In most skilled jobs, hourly wages had not changed between 1875 and 1895 despite enormous gains in productivity. Wages for semiskilled workers at Sparrows Point actually fell behind those in nonmechanized foundries in Baltimore. According to a survey by the Maryland Bureau of Industrial Statistics, machinists and boilermakers in city foundries received 22 to 34 percent more an hour than at Sparrows Point. Had it not been for the long working hours, the semiskilled men would have earned considerably less than their counterparts in Baltimore. By working 12-hour days, Point employees were able to achieve rough parity with city workers on an eight- or nine-hour schedule.[22]

The wages of unskilled laborers also eroded with the onslaught of mechanization, dropping from $1.25 a day in 1875 in Steelton to $1.10 in 1895 at Sparrows Point. As a result, by working every day of the year except the two unpaid holidays, the unskilled laborer could expect to earn a grand sum of $399.30 annually. This was a considerable shortfall from the $500 to $600 a year believed by a U.S. Bureau of Labor report to be minimal to maintain an "American standard of living" for a family of five.[23]

The introduction of improved processes had not alleviated the working classes from the anxiety of material want. Commenting on this development, Thomas Weeks, the Maryland labor commissioner, complained, "Where large capital commands the most perfect machinery, there is but little scope for the average man, and he is reduced to a life of toil, keeping pace with the machinery of the shop, going when it goes, and stopping when it stops." In a pointed aside to the Maryland legislature, Weeks said, "While glowing reports of increased products, sales and shipments have been generally accepted as evidence of the individual prosperity of our citizens, yet such facts often hide under a gilded exterior very many hardships, perils and disasters suffered by the industrial masses, and which it is often within the scope of legislation to alleviate and remove."[24]

*

Such an arrangement could only work in a union-free atmosphere. Behind the "high-tech" accomplishments of the railmakers was the rout of unionism at their mills. In 1862 a fraternity of Pittsburgh ironmongers had formed a national organization called the National Forge. In 1866 the union had 600 members; in 1873 it had grown to 3,331. The Amalgamated Association of Iron, Steel and Tin Workers was organized when delegates from the National Forge, United Nailers, and two other unions declared themselves a unified body in 1876. The union had a policy of limiting its rolls to craft employees and denying membership to unskilled and semiskilled steelworkers. Outright discrimination was practiced against immigrant and black workmen by many local lodges. Nevertheless, there was a growth in membership among skilled employees, especially after Carnegie Steel instituted an industry-wide wage cut in 1883. In 1888 the Amalgamated claimed 15,000 dues-paying members. It was considered to be the strongest union in Samuel Gompers's newly formed American Federation of Labor.[25]

But its strength was illusionary. Its membership was concentrated in the iron-puddling and rolling mills that were coming under competitive challenge by the cut-rate integrated producers. At first the rail manufacturers had tolerated the union. They had little choice, since union members dominated the rolling crews. Interestingly, the tonnage-based wages negotiated by the Amalgamated fit in well with the early economic structure of the railmakers. When orders dried up, the steelworkers took a drop in pay. Their tonnage rates helped the industry ride out several bad panics in the 1870s. Yet, mindful of the future, the companies viewed the union as an impediment to mass production. Introduction of machinery was one way to undermine the power of the Amalgamated. Another way was to locate a new plant at a remote site where the company's power could dominate local government—a site like Sparrows Point.

At the mill the uneasy truce between management and members of the Amalgamated employed in skilled jobs began to unravel in less than a month after the beginning of operations. Responding to a small strike on June 21, 1890, by a group of employees seeking higher pay, Frederick Wood issued a memo to his timekeepers requiring that all workmen in the future sign a "Conditions of Employment" slip. While not an outright banning of union membership, the slip listed "agitation" and "insubordination" as sufficient grounds for discharge.

Over the next year the Amalgamated lodge at Steelton grew restless. In July 1891 it presented Bent and Edgar Felton with a petition calling for a 20-percent increase in wages. Bent and Felton refused to discuss the new scale or recognize the union as a bargaining agent for employees.

About 2,000 men then went out on strike. The Pittsburgh headquarters of the Amalgamated refused to back the walkout, saying it was a local matter that had not been authorized by the national body. This left the strikers out on a limb. Bent threatened to cut them off, warning darkly that he would rather shut up the mill and transfer all operations to Sparrows Point than submit to an organization that tried to impose unreasonable conditions on him. His threat was successful: Unionism was equated with subversion by the Harrisburg press and local merchants were panicked by Bent's announcement. Within two weeks the strike was over and the Amalgamated union was no longer a factor in the mill.[26]

Frederick Wood seized the opportunity to declare Sparrows Point an "open shop," a euphemism for nonunion employment. Baltimore's labor newspaper, *The Critic,* denounced the company as a union buster that should be driven out of Maryland—ABANDON HOPE ALL YE WHO ENTER HERE, said a headline about the plant, recalling Dante's words at the gates of Hell.[27] But the paper's rhetoric was in vain. The Amalgamated was losing its muscle in steel mills around the country. At Sparrows Point very few of the 2,500 employees were willing to risk dismissal over the question of unionism. The Amalgamated took its final stand against the railrollers a year later. After fighting off Pinkertons, the sheriff, and the Pennsylvania state militia to keep out strikebreakers, the Amalgamated succumbed to superior force in the bloody battle of Homestead.*

*Shortly before his contract with the Amalgamated was to lapse in 1892, Andrew Carnegie transferred managing authority of Homestead to his associate, Henry Clay Frick. Frick ordered a wooden fence 12 feet high to be built around the mill with lookout posts for guards. Thus fortified, he shut down the plant and announced that it would not reopen until union members

When Frederick Wood wrote his report to Bent in 1895, unionism had been routed out of every railmaking mill except as an underground movement. Unionism would never again gain a foothold in this industry, vowed Axel Sahlin. In an essay prepared for the British Iron Trade Association, Sahlin, general superintendent of the company, elaborated on management's complaint against unionism. His statement touched on many of the arguments that were commonly used by industry:

> Labour Unions practising the policy of impeding progress and curtailing output, are not tolerated in the American blast furnace works. As far as the writer knows, such an organisation is to be found in only one rather unimportant plant located in the centre of a great city, and the experience with the union at this place has been such as to make owners and managers resolutely set their face against all attempts to introduce similar societies at their works. Agitation amongst fellow-workers or expressed leanings toward unionism are, therefore, considered good reasons for discharge as soon as the opportunity offers. It is not the men themselves that the manufacturers fear, but the unscrupulous agitators who, with a despotism never equalled by employers, domineer over the men, and who, without the knowledge, judgment or ability to conduct a large business, presume to dictate how such an affair shall be managed. With unions such as are known in some other lands, I venture with emphatic statement that blast furnace work as successfully carried out to-day in America would be impossible.[28]

The cost of labor, then, had been reduced by 65 percent through a combination of corporation policies. The unblinking concentration of

---

accepted a wage reduction. On the morning of July 6, a group of Pinkerton guards were spotted by a union lookout coming down the Monongahela River. They were confronted by the strikers and a gunbattle broke out. Nine strikers and seven Pinkertons were killed. Over the next two months Homestead became the site of open warfare before the strikers were crushed by the combined weight of Frick and the Pennsylvania National Guard. In November 1892 the mill was reopened as a strictly nonunion shop. "We had to teach our employees a lesson, and we have taught them one that they will never forget," Frick wrote to Carnegie with satisfaction.

the engineer, the penny-pinching of the timekeeper, the labor savings of new machinery, the substitution of unskilled for skilled labor, and the forcible removal of the union—all of these factors had caused profound ramifications for the steelworker. As a unit cost he had been cut, trimmed, rolled down, and manipulated as fully as the steel he handled. Having first conquered the volatile chemistry of metal, the steel masters had gone on to subdue the most valuable ingredient of the process. And the same technological system that yielded long hours and low wages placed another burden on the employee, the threat of physical harm.

*

Before the industrial revolution human catastrophes were largely the result of natural events. Floods, storms, tornadoes—forces outside of man's control—attacked human lives and destroyed property, and this led philosophers to discuss the role played by chance and coincidence. Thus, notes historian Wolfgang Schivelbusch, "In Diderot's *Encyclopédie,* 'Accident' is dealt with as a grammatical and philosophical concept, more or less synonymous with coincidence."[29]

The age of invention altered this equation, causing what Schivelbusch calls the rise of the "technological accident." In this type of accident the process was reversed: instead of destruction coming from some external source such as a flood or tornado, devastation was wrought internally. When a machine malfunctioned, it destroyed itself with its own burst of power. In America this phenomenon came to public attention dramatically with the railroad accident. Here a system that multiplied human power a thousandfold suddenly became an agent of human devastation, a death trap for passengers and crew alike when a rail split or a boiler exploded. The greater the power of the man-made equipment (higher speed, greater boiler pressure, etc.), the greater the degree of potential calamity in the case of error or malfunction. "One might say," said Schivelbusch, writing of the perils of 19th-century train travel, "that the more civilized the schedule and the more efficient the technology, the more catastrophic its destruction when it collapses."[30]

And so, too, with the modern American steel works. New and novel dangers posed by the flammable chemistry and speed of the operations produced a steady toll of technological violence punctuated by spectacular wrecks. Shortly before midnight on Sunday, December 6, 1891, one such disaster struck Sparrows Point:

BLOWN INTO ETERNITY.

TERRIBLE ACCIDENT AT SPARROWS POINT IRON WORKS

"With the violent explosion," wrote a reporter of the eruption of blast furnace B at 11:15 P.M., "hot bricks were thrown out in all directions, and a flame of fire came forth and enveloped the men, setting them and their clothing on fire. The noise of the explosion was like the deep boom of a heavy cannon, and the shock was so great that it caused dwellings a great distance away to vibrate. This, coupled with the blowing of the tug whistle, caused all the residents to rush from their houses in alarm."[31]

The residents found that Arthur M. Austin had been hurled 30 feet by the blast. Death was instantaneous. A native of England, Austin had come to Sparrows Point three months earlier to learn the process of steelmaking. He was 21. George Braidwood, age 35, died several hours after the blast. "The poor fellow inhaled the flames and his life went out in torture," said the newspaper. John Lynch, age 30 and a machinist helper, was found with a fractured skull. He lingered on until 6:00 in the morning, and was then prepared for burial by George Dunn, the company undertaker. Andrew Pugh, age unknown, lived until the next evening. He had serious head contusions and had burns over half his body. Pugh left a widow and an infant child.[32]

Not only in death but in survival molten iron caused devastation to human flesh. Again from the press accounts:

> Thomas Miller, of Linwood, N.C., one of the colored men at the hospital, tells a horrible tale of his sufferings after the explosion occurred. He was thrown into a deep ladle used for carrying molten metal. The ladle was almost red-hot, and he slid into it head foremost. He had to climb up the sides, and in his frantic efforts to get out burned most of his clothing off and roasted the flesh on his hands. After he climbed out he fell over on a pile of hot cinders and brick, and severely burned his neck and face. His arms and knees are also badly burned.
>
> Augustus Bell, another of the colored men, is suffering much from burns about the hands and face. He comes from Virginia.
>
> W. T. Tubman, the other colored man, was painfully burned

> about the face and head, and both his hands are a mass of blisters. His burns are very serious.
>
> Washington Sapp, the white man at the hospital, is a freight brakeman, and was struck on the back of the head by a flying hot brick, which threw him heavily across an iron bar. He fell with his face on the bar, breaking his nose and receiving a contusion on the forehead. He was also burned about the body by hot ashes.
>
> None of the men are fatally injured, though it will be a long time before any of them fully recover.[33]

The aftermath of the explosion was significant for what it said about the company's response to accidents. Under Maryland law all suspicious deaths were heard by an inquest jury formed of local citizens. On the day following the accident, a jury composed of Sparrows Point citizens was organized. It included John Smith, superintendent of the street department, and Patrick Regan, a blast-furnace foreman. Wood told the jury the accident was "unaccountable." Apparently, he said, water had leaked from a pipe at the bosh and, coming in contact with gas inside the furnace, had caused the explosion. At some earlier point, he noted, it had come to management's attention that some furnace pipes around the bosh had become badly worn and parts of the firebrick walls inside the furnace were damaged. Because of the dangers of this condition, Wood had ordered workmen to begin "blowing out" the furnace prior to repairs. This procedure was nearly complete when the blast occurred. Furnace superintendent David Baker told the jury that he had ordered the furnace bell lowered when he heard a deep suction of air inside the vessel. He said he managed to step behind a pillar for protection. The men on the furnace floor, however, were out in the open when the explosion occurred seconds later.[34]

Because company procedures had been followed and no human error was committed, Wood asserted that the cause of the explosion was a mystery. In terms of the immediate cause of the blast, Wood was probably right. Steelmakers did not have equipment then to monitor chemical reactions inside a furnace. But in terms of the larger issue of plant safety, Wood's testimony raised a number of questions. Why, for example, wasn't the furnace shut down when the damaged pipes and walls were first discovered? Did the delay in repairs contribute to the water leak? Should the company have restricted workmen from the exposed floor area when the furnace bell was lowered?

The inquest jury did not address these questions. Instead it issued the following statement after brief deliberation:

> We, the undersigned jury of inquest, do find our verdict in the case of the death of Arthur M. Austin, George Braidwood and John Lynch [Andrew Pugh, the fourth victim, was still alive] to be that they came to their death by an explosion of gas at furnace 'B,' Sparrows Point, the explosion, from the evidence heard, being unaccountable and unavoidable, and for which the Maryland Steel Company is not accountable.[35]

The words *not accountable* were of great importance to the company. By being cleared of responsibility, the company was protected from damage suits arising from the explosion. At the time Maryland liability laws required proof of "gross negligence" by an employer for a victim of an accident or his surviving kin to collect damages. That the company was protected from damage suits was demonstrated when the widow of George Braidwood filed suit in Baltimore Court of Common Pleas. She asked for $10,000, claiming negligence. In 1892 the suit was removed to Baltimore Superior Court for trial. But the trial was never held and the case languished on the court docket. In May 1899 the case was still on the docket. From this point the suit cannot be traced in court records, but it appears almost certain that the widow did not collect damages since in the rare cases where a verdict was rendered against the company Wood kept copious notes of the matter. He had no notes on the Braidwood suit.[36]

The handling of the blast and its aftermath was typical of a pattern that would last through 1914, when the passage of Maryland's workman's compensation law (one of the nation's first) set up compulsory employer insurance for work-related injuries. Repeatedly, the inquest jury ruled that a death was unaccountable or due to the victim's carelessness. Repeatedly, the company offered men with broken limbs small cash settlements or payment of hospitalization costs in return for signed waivers clearing the company from potential liability.[37]

Why didn't Wood employ his engineering talents to the task of making his mill safer? The answer appears to be best answered by a statistical table filed among his papers. Every year Dr. James S. Woodward, chief of the Sparrows Point hospital, tallied up accident costs. The following report for 1902 was ordinary:

DISTRIBUTION OF ACCIDENT EXPENSE ACCOUNT	
MEDICAL EXPENSE:	
Hospital expense	$1,908.19
Medical supplies	792.72
Surgeons' salaries	2,480.00
Total	$5,180.91
Cost per accident	$ 2.22
ALL OTHER EXPENSES	
Board paid	269.71
Donations	571.42
Releases	269.80
Funeral expenses	638.50
Legal fees [Preston & Bowie]	2,060.86
Artificial limbs, eyes, etc.	150.00
Total	$3,960.29
Cost per accident	$ 1.69
TOTAL YEARLY COST	$9,141.20
TOTAL COST PER ACCIDENT	$ 3.91
TOTAL NUMBER OF ACCIDENTS	2,336[38]

Compared to $7.5 million in sales for rail and billet shipments that year, the $9,141 tab for accidents was inconsequential. To be sure, accidents cost the company money in terms of repairs and lost production time, but the human costs were so scant (0.12 percent of revenues) as to provide little financial stimulus for Wood to direct his attention to shielding moving parts of machines or directing his supervisors to discipline careless or reckless workers. Weighed against the absence of external incentives for improved safety—there were no laws setting standards for plant safety in Maryland—Wood's disinterest in eliminating mill hazards made eminent business sense. Securely rooted in the economic classicism of Adam Smith and the social philosophy of Herbert Spencer was the principle that a "master" had no legal responsibility to his "servant" other than the payment of market-level wages. Common-law doctrine decreed that because a workman chose his occupation, he assumed the risks of bodily injuries at work. In a world of the survival of the fittest, the burden of factory accidents fell squarely on the victim.[39]

William D. Heiges was a Sparrows Point iron molder who had been permanently injured by a piece of iron that flew from a drop machine. In court testimony James McAfee, the drop-machine boss, was asked whether precautions had been taken to protect Heiges and other molders from the pieces of iron. "No, sir, not at that time there wasn't...We just yelled 'Heads up.'" A jury in Baltimore awarded Heiges $2,500 for damages caused by employer negligence. Through Preston & Bowie the company appealed, and in 1896 the Maryland Court of Appeals rendered a new verdict. It said that even if the workplace was hazardous and "with reasonable care [could be] made more safe," the employer was protected from damage suits because a worker who accepted the job accepted the job's "ordinary dangers" too. "If a servant has knowledge of the circumstances under which the employer carries on his business, and chooses to accept the employment, or continue in it, he assumes such risks incident to the discharge of his duties, as are open or obvious." To decide otherwise, said the court, would deprive the workman of his right to manage his own affairs.[40]

Viewed from the perspective of the reforms made a generation later, the issue of safety was less a question of technology outstripping society's knowledge of machine hazards than of a failure by businessmen and public leaders to acknowledge the human costs of industrial progress, a willed obtuseness which was couched in solemn expressions of liberty and freedom. Under the laws of the time the workman was granted a new freedom—the freedom to assume the dangers of machines over which he had little or no control.

So along with the ingenious machinery and the low unit costs there was a third aspect of the new steel works little known by a public that enjoyed the convenience and creature comforts of steel. As journalist Herbert Casson noted after a tour of steel country, "Much sympathy has been expended, and rightly, upon those who are compelled to work in sweat-shops. But a sweat-shop is a haven of safety and rest compared to a steel plant. There is little public opinion with regard to the perils of a steel mill, for the reason that few outside of the trade know anything personally of the conditions that exist."

Casson went on, "As I have gone from one steel city to another, I have felt more often like a war correspondent than like the writer of a story of peace and prosperity. The steel business is not all dividends any more than war is all flags and music. There is a stern side—a side which ought to be made brighter by the steel kings."[41]

CHAPTER 3

DESIGNING THE TOWN

My dear Fred,

Have you thought any further about father's and mother's birthdays, which are not far off? . . . I know they would like to hear from you.

Reminding Frederick of the date of his parents' birthdays, buying chrysanthemums for a business associate, and fretting about his health ("I hope you took no cold from your exposure yesterday," he wrote on a wintry day)—Rufus Wood performed the domestic chores among family and friends. With similar vigor and care over details, he managed the town affairs as general agent.

One half mile from the blast furnaces stood the town. For an average of 12 hours a day steelworkers labored in the mills and for the other 12 hours a majority of them resided in the village. Superintendents and general foremen also lived there, to be close to the mill in case of an emergency. The hamlet represented an aspect of capitalism that was becoming increasingly associated with giant manufacturing concerns: paternalism. In the Midwest, paternalism had crystallized in Pullman, Illinois, a workmen's community south of Chicago erected by sleeping-car magnate George Pullman. In South Carolina the Pelzer Manufacturing Company had a company town named Pelzer, and in New England company-owned boardinghouses were as common a sight in textile towns as church steeples.[1]

Altogether about 100 company towns had been established in the 1880–90 period. To this roster came Sparrows Point, a town which the U.S. Bureau of Labor called "the most noteworthy example of a complete steel community, planned, constructed, and controlled by a steel company."[2] In spite of a population of over 3,000, which made it the second-largest community in Baltimore County, town residents had nothing to say about local governance. Normally some form of electoral democracy existed in a company town; normally there was an elected mayor or at least town commissioners. But owing to the gentlemen's agreement worked out between Maryland's Governor Jackson and Pennsylvania Steel, all ordinary governing powers were held by the company, which gave extraordinary power to the small man with a silver goatee who shared the executive office atop the company store with his brother.

From the start Rufus was a fanatical believer in education. What he called "modern" education: learning not confined to rote memorization and *McGuffey's Readers,* but based on the practical "doing-it" methods advocated by psychologist G. Stanley Hall; learning through drawings, calisthenics, and marching to the notes of the piano.

What he found in Maryland was a far cry from educational industriousness. Rufus's sigh was almost audible in a letter he wrote to Frederick castigating "the backwardness of Southern peoples," a backwardness that did verge on the scandalous. The Maryland legislature still had not passed a compulsory education law, despite records which indicated that the number of children attending public schools did not even constitute a majority of those of school age. (No compulsory education law would be passed until 1916, when children aged 7 through 13 were legally required to attend school.) Dr. Joseph M. Rice called Baltimore city schools grossly mismanaged and pedagogically marginal, "a product of the ward politicians," he said in 1892. However, they were demonstrably superior to those in Baltimore County. In one year the county school board distinguished itself by allocating a total of $175 for the education of the "colored student population" in the county.[3]

Such shortsighted policies would not do for Rufus. The steel mill needed workmen who could understand written instructions and were proficient in basic mechanical skills. Under his guidance the company built two schoolhouses in town, one for white pupils and the other for black pupils. In return for the company's outlay Rufus was named trustee of the schools by the county school board, which gave him a free hand in implementing a number of innovative programs. The first public kindergarten south of the Mason-Dixon line was opened by Rufus in 1892. "Go to the white and colored kindergartens," instructed a company brochure which the general agent helped to write, "and little tots in the fives and sixes are looking at blackboard pictures such as birds, dogs, cats, and other animals. Then they try to draw the pictures in outline at their desks. In that way the figures on the board make an impression on their minds that they remember far better than the name in the book as it is spelled out in print. In this way they learn to read the alphabet far more quickly."[4]

Between the first and fifth grades, schooling followed the standard fare of the "three R's," with algebra, geography, history, and bouts of *Westlake's Common School Literature.* After fifth grade the boys and girls entered into separate one-year programs courtesy of Rufus. The boys were instructed in the use of lathes, presses, and other tools in a manual training program, while the girls spent their afternoons in the "domestic sciences" wing of the school building.

"On her little one-burner alcohol stove," the company brochure noted, "a girl cooks a breakfast of oatmeal, omelette or bacon and eggs, perhaps hot biscuits or toast—every part of the menu palatable and done properly. A wife who is a good cook is a 'joy forever' to her husband, though she may not be a 'thing of beauty.'" Apart from cooking good meals the girls were instructed in the art of bedmaking ("spreading the sheets and blankets without a wrinkle and tucking in the coverlet so as to show its ornamental needlework"). In a model laundry room the girls practiced washing and ironing. There were lessons in room arrangement and decor. Prizes were given out for the best clothing designs. "Only a look at some of the drawings of these ten-to-fifteen year old daughters of the toilers is needed to show their refined taste and artistic ideas in dress. Many of the girls wear costumes of their own design entirely made and trimmed by their own hands."[5]

A "human modeling studio" was how Rufus described his domestic sciences program, and the words seemed to sum up his mission at Sparrows Point. He sought to improve the children of "the toilers," to mold a future generation whose frugal female heads would be perfect mates to the young men who "are so thoroughly fitted in the use of wood worker's tools that when graduated, they are placed on the roll of employees."[6]

Education was one route by which the general agent hoped to build an improved working class. Religion was another. Leasing the grounds for church buildings at the nominal sum of $1 a year, he encouraged the organization of church congregations which eventually numbered seven (Methodist, Evangelical Lutheran, Presbyterian, Catholic, and Episcopalian for white residents, and Baptist and Methodist for black residents). Raised a Unitarian like his brother, Rufus took pride in the religious activities of the community, though he sometimes took it upon himself to enhance the Scriptures with a more secular text. In 1893, for example, he invited the Rev. Dr. George Edward Reed, president of Dickinson College, to come to town to preach a sermon from the Episcopalian pulpit. Rev. Reed's sermon was titled, "Every Man Is a Capitalist." Supplementing these sermons were lectures given by the general agent himself. "Last night," he wrote to Frederick, "I gave a somewhat informal talk to the people on the scenes of London, with lantern illustrations, which were very brilliant and successful."[7]

Yet of all the moral principles Rufus held dear, none was more important than the need to rid the working class of its dependence on alcohol. While his brother and Luther Bent viewed the ban on liquor in the town as a business proposition (an "economic gain," said Bent), temperance was a holy crusade for Rufus. Venturing forth on payday in black suit and necktie, he soberly tracked down "liquor trafficking" in the vicinity to a tavern on North Point Road named Dorsey's. "I am glad to see that today's *Sun* confirms the rumor that Dorsey has gone to jail in default for the $500 fine imposed on him," he wrote to his brother after one expedition. "There seems to have been more disorder and disturbances here in the past two weeks than for a long time before, and it appears that a good many cases of bad women, especially across the creek, have just turned up, and we are just making investigations to see if these evils shall be checked."[8]

On a number of occasions he ordered his driver to take him to Highland Heights, where 50 saloons had sprung up near the Sparrows Point Railroad. On Saturday nights steelworkers freed from the mills rushed to the district, forsaking the wholesomeness of Rufus's town with a vengeance. Many returned on the last train loaded to the brim. Rufus beseeched Baltimore County officials to shut the saloons down, noting indignantly, "The owners of these resorts entrap the employees of this company, entice them into their premises, sell them liquor without limit, rob them of their earnings, and deprave them in body and mind."

Rufus rejected the idea propounded by *The Critic* that harsh working conditions and the 24-hour Sunday shift might contribute to the overindulgence. Rufus wished to change the workman, not the works. So while managing to contain his horror at the company's violation of the Fourth Commandment, he looked at his education system as a way to curb the problem in the next generation. "The greatest hope for the mitigation of this curse to the human race," he said of alcohol and the working class, "lies in the promotion of education in our public schools, teaching our children to beware the ruin of health, reason and morality by this besetting sin."[9]

*

Unlike the classic steel hamlets of Homestead or Johnstown, the town was not pressed to the mill gates nor did it climb helter-skelter up steep, eroded hills. There was space on Sparrows Point and there was a palpable design that laid stress on planning over haphazard building. Rufus after all had been raised in that monument to Yankee planning, Lowell, Massachusetts. Lowell was the place where the concept of a planned industrial community began in America. It was established in 1823 at the juncture of the Concord and Merrimack Rivers, 25 miles north of Boston, by elite Boston tradesmen including the Cabots, Lowells, Jacksons, and Duttons. It soon became a place of pilgrimage. Charles Dickens recorded his favorable impressions of Lowell in *American Notes;* Ralph Waldo Emerson preached Transcendentalism there from the stage of the town Lyceum. (Dickens was not so taken by Pittsburgh. "Hell with the lid lifted," he called it.)[10]

By the time Rufus was born on November 8, 1848, Lowell was in the hands of a favored class that reported not to local residents but to the Cabots and Lowells of Boston. He grew up with his brother a stone's throw away from the Boott Cotton Mill where his father was a weaving-room foreman. Given this background, it made eminent sense that he would fall back on Lowell for inspiration. In implicit and explicit ways he tried to transfer the "Yankee El Dorado" of his boyhood to the shores of the Chesapeake.

Sparrows Point was erected on the same building plan as Lowell. Very important gradations paralleled the grid of rectangular blocks that started at the harborfront. First came the houses of the company officers and mill superintendents. Then came the dwellings of the foremen and skilled white workmen. Next were the double houses and cabins of the low-paid black laborers. The demarcation of class (and of race, since Rufus did not direct his reformist zeal to the separatist traditions of "Southern peoples") was writ in paving stone and gravel in the street plan.

The plan was quite astonishing in its coded simplicity. All houses were built along the east-west streets, which were lettered. They started at A Street on the shorefront and ended at K Street. The letter of the street a resident lived on denoted his position in the mill, which in turn affected to a remarkable degree whom he would associate with in town. Different "streeters" occupied different social worlds.

The superintendents lived on B and C streets. Joining them were the town notables: Joseph Blair, the principal of schools and justice of the peace, and the company physician, Dr. James Woodward. "Bosses' Row" ended at D Street. This was the first big divide in town and it was rendered physically tangible by the schoolhouse at D and 4th and St. Luke's Church at D and 6th. North of D Street was the area for the white workmen and foremen. The neighborhood included more than 150 houses and was bounded by an arm of the harbor known as Humphrey's Creek. In addition there was a small annex near the foundry listed as West E and West F streets. At 5th and F Streets an 838-foot-long wooden trestle crossed Humphrey's Creek. The footbridge ran over to the black settlement, a group of 64 houses clustered on its own promontory at H, I, J, and K streets. (Following the logic of the layout, there was a G Street, but it was under Humphrey's Creek.)[11]

Within these designated districts there was no mistaking who presided. The bosses' houses were large and well appointed, but the largest and best appointed was the general agent's. Rufus headed the town by virtue of occupying the "big house" at 702 B Street.* This handsome three-story colonial was designed by Baltimore architect Benjamin Owens. There was a porch in the front and a screened veranda that overlooked a private garden on the east lawn. The front door opened to a staircase that circled up in two ovals to the top floor. There were 18 rooms in all, including a paneled library (Rufus, a history buff, tells of sitting down with a big book on the Crusades), two parlors, dining room, seven bedrooms, a kitchen, a pantry, and a summer kitchen.[12]

For Rufus the rules of perspective brought into focus the natural order of town living. If he pushed aside the curtains of his rear windows, he could observe the town before him. Only the company store and church steeples were higher than his third-floor perch. Ahead of him the dwellings marched in neat files for two-thirds of a mile to the far north side where the small houses of the black families cut a swatch between the creek and woodlands. And turning to the south, Rufus could gain inspiration from his unobstructed view of the outer harbor: over 12 square miles of water flowing blue and clear past Fort Howard and merging in brisk chops with the bay, a bonny sight which reminded him of the sea approach of his beloved Boston.[13]

*The property on A Street was intended originally for Frederick Wood. However, with a family that included two boys and four girls in time, Frederick opted to buy a 12-acre property in North Baltimore. He named the property "Carolinden" after his wife, and commuted to the Point by train.

The general agent lived in the house with his family: his wife, (the former Ruth Meredith of Harrisburg, Pennsylvania), two sons, his mother, and two servants. Rufus had been a bachelor in his prior incarnation as a Boston accountant, and he did not marry until he had established himself at the Point. The marriage took place in 1890 when he was 42. His bride was 21. Their first son was delivered by Dr. Woodward in June 1892, and a second son was born five years later. His mother, Elizabeth Wood, moved into the house following the death of her husband in Lowell. She lived there until her death in 1903. The family employed an Irish-born servant named Sarah Lynch and a black cook named Emma Hicks.[14]

Alongside Rufus's property were the residences of other ranking executives. "Some of the latter are as striking in architectural design as they are ostentatiously furnished," remarked the *Baltimore Herald* reporter. "Those not occupied by the bosses are used as boarding houses."[15] According to the daughters of the future superintendent of orders, who grew up first on C and then B Street, Bosses' Row was a land of pleasant living far removed from the straining machinery of the mill.

"We had eight bedrooms (one was for the maid) and that did give us a lot of space," recalls Elizabeth McShane, who was interviewed when she was 76.

"And downstairs," adds her sister, Katherine Roberts, "we had two living rooms, both quite big, a big dining room, a kitchen, a bathroom—"

"You called the company for everything," Elizabeth says. "They would paper and paint the house, completely, and do over the floors regularly."

"And your parents entertained by their position."

"Oh, yes. I don't think anyone had a better childhood than we had growing up. We played with a dozen boys and girls from up and down the street, on our side of town. It was a very concise place."[16]

Based on Rufus's notes and ledger, little expense was spared for comfortable accommodations for management. His residence on B Street cost $5,035.61, a goodly sum for the time, and for at least the first year he paid no rent on it. Other houses on B and C streets cost an average of $3,233. They were rented out for $22 to $25 a month, including steam heating and modern electric lights.

The names of the managerial class—Crane, Reese, Parker, Martin, Hummel, Bassett, Thomas—bespoke the same lineage as the brothers

Wood. With the exception of a Scottish-born superintendent, all 12 department heads were American-born. Five of them were born in Pennsylvania, one from Massachusetts, one from New Hampshire, three from Maryland, one from Michigan, and none from south of the Potomac River.[17]

In the white area north of D Street the average cost of a house was $967. The best houses were the sturdy red brick rowhouses along E Street. Measuring 13 feet, 2 inches wide by 28 feet long, they had six rooms, a small front porch, kitchen annex, and back yards separated by high wooden fences. With half the number of rooms of B and C Street dwellings, they made for crowded living. The common practice was to let out at least one room to boarders to help with the rent. By way of amenities there were few, remembers an early resident: "We had coal-oil lamps; potbellied stoves in the living room and the cook-stove in the kitchen for heat. The hydrant [for drinking water] was at our back gate." Along several blocks were freestanding cottages, their half-timbered gables and scalloped shingles adding a Victorian, almost seaside, touch to the straight-ahead files of rowhouses. Rents throughout were reasonable, starting at $5.50 for the little five-room cottages and going up to $12 a month for the more imposing dwellings. There were also several boardinghouses and a hotel for transients at 4th and E.[18]

In terms of ethnicity, the district was the abode of native-born Easterners, nicknamed "buckwheats" because of their rural or small-town origin, sparingly mixed with Irish and English immigrants. Seventy-six percent of the adult males said they were born in this country, according to the data taken for the 1900 census. Once again this was Yankee territory: 92 percent of its American-born residents were born in states north of the Potomac River, mostly in Pennsylvania. Education was spotty. Some had nine and ten years of formal schooling, others as little as two.[19]

Among skilled men especially, the steel company seems to have transferred wholesale the native-born Pennsylvania Germans who had formed the backbone of Steelton's workforce. Pennsylvania Germans, or "Dutchmen," had a reputation for hard work, conservatism, and habits of obedience to authority. "Amongst these men there is hardly ever the question of a strike, dictation to their masters [or] curtailing output," remarked former works superintendent Sahlin. Fairly typical of the population was that found on the 400 block of E Street. Living there were the families of the assistant store manager, a master mechanic, six foremen, an engineer for the railroad, a salesman for the company store, and a widow who took in boarders. The surnames—Hufman, Yestadt, Strasbaugh, Gerhart—were unmistakably Pennsylvania German, mixed with a few of Irish descent like Mackley and McCrosky.[20]

Across Humphrey's Creek the houses for black families were a grade below the white-side rowhouses, erected for $753 apiece. They were a study in unadorned functionalism: whitewashed, pine-planked double houses, each unit measuring 12 feet, 3 inches wide by 26 feet long. Two and a half stories high, they had 3 rooms on the ground floor, two bedrooms on the second, and a third bedroom tucked beneath the eaves. There was no running water and outhouses were located near the rear alleys. On average each of the 64 so-called "family" houses held 11.3 people, or nearly four persons per bedroom, according to the 1900 census. A typical example of the conditions was found at 612 I Street. Nine adults slept in the space of three small rooms—three married couples (one with a baby) and three unmarried men. To have that many adults squeezed into the same house was possible only because the men worked on different shifts.[21]

"It was crude living," remembers Florence Parks. "You got your water out on the street and you couldn't put any wallpaper on the walls, the plaster being so rough and unfinished." Her parents had moved to the Point from rural North Carolina at the turn of the century. Her dad was employed as the assistant baker at the company store and her mother was a maid " 'cross the crick."

"My mother worked for the superintendents. Her name was Nannie (the white people called her Nan) and they loved her cooking. She'd put the food on the white people's table—veal cutlets and other nice meats, oyster fries, cream of asparagus, artichokes, cream of cauliflower—and they'd just say, 'Oh, Nan, it was just delicious." My mother, you see, was an expert cook, and any food they had left over, any food they didn't serve, my mother got and that's what we ate. Now it was my job as a girl to clean the silver. I'd go across that bridge in the afternoon and get working 'cause everything had to be just so. Polishin' and polishin' and keepin' it sparkling—it was a job. Something should have been paid extra for doing that, but my mother got not a thing extra for cleanin' any silver."[22]

The black workmen were part of the great migration from the South that began in the 1890s. According to the census report, 85 percent of the adult black males in the town were born in the South, mostly in

Virginia, 13 percent in Maryland, and 2 percent north of the Mason-Dixon Line. "The rule was, if you were a country boy from the farm, they wanted you, especially if you got relatives here. They didn't want the city Negro 'cause they thought he'd be uppity," Florence says. Ninety percent of the men were employed as laborers at $1.10 a day. There was a small contingent of semiskilled black men in the riggers' gang and in the transporation department. A few of the blacks worked as waiters, janitors, or company store employees in town. Seventy-four percent of the males were 30 years or under. Twenty-four percent reported they could not read or write.[23]

While lettered streets formally ended at K Street, that did not end the town. There was a fourth side to the town, "Shantytown," home of unmarried immigrant workers and single black males. Sited alongside the steel furnaces, it boasted no lettered streets. The shanties were originally built as a temporary labor camp during the mill's construction, but they lingered on. There were 337 shanties in all. The U.S. Bureau of Labor condemned the buildings as filthy and cramped. Slapped up in rows, they were divided into rooms of 10 by 14 feet, according to the bureau. The number of occupants varied according to the productive activity of the mill:

> When the plant is operating with a full force, four men are allotted to a [room], in which they eat, sleep, and do their cooking and washing. The furnishings are primitive in the extreme. For sleeping purposes there are two 2-high bunks, the space between the lower and upper being slightly larger than required for a man of ordinary size to crawl in. The men provide not only mattresses and bedding, but also the stove and the fuel necessary for cooking and heating. If they do not desire to cook for themselves, meals can be had at reasonable cost in the several mess houses located among the several groups of shanties. Some of these mess houses are operated by men employed in the company's office. The water for cooking and washing is obtained by the occupants of these shanties from outside hydrants.... A number of the men were interviewed relative to the weatherproof conditions of the shanties, and these men stated that repairs were not kept up and that sometimes the shanties were wet and cold.[24]

To understand why Shantytown existed, one needed only to consult the names recorded in the census. Living in the shanties were John

Billy and Stav Olix; also Nicky Moltec and John Rabbitt; Joseph Galla and John Withowich; Kasmish Schluiski and Joseph Balsjacania. In all, two-thirds of the white residents were listed as "al" (alien) by the census taker.

The immigrants not housed in these cubicles were placed in other out-of-sight locales. Several four-room cabins were built by the company at the north end of the peninsula for sawmill workers in 1887. In 1900 the census taker found 22-year-old John Hudok living with his new bride, Lizzie, age 19, in one of the tar-paper cabins. John's cousin, George Hudok, also lived in the cabin, as well as six other workmen. All of them reported they were from Hungary. All of them were furnace laborers.[25]

Shantytown confirmed what has been hinted at in the private papers of the Wood brothers—a deep disdain for the immigrants who were beginning to figure prominently in the steel works of Pittsburgh and Chicago. As works superintendent Axel Sahlin noted tartly, "Of the crude immigrants who arrive, and are willing to take it [steel work] up, only a moderate percentage have the necessary strength and intelligence to fit them for the work."[26]

Management could indulge in their prejudice against the Eastern European because they had found a source of cheap labor—the Southern black man. No other steel mill outside of Birmingham relied so heavily on blacks for general labor. The first black workers had arrived at Sparrows Point at the dawn of its development, in July 1887, their train fare from Pennsylvania paid for by the company. The formula used by superintendent Sahlin went roughly as follows: the Hungarians weren't bad as furnace hands ("being without friends or home they feel that on their own efforts depends their existence"), but for really hot and dirty work none could equal black country boys. They were especially valuable in the hot Maryland summers which "make it difficult for white men to perform such heavy manual labour," Sahlin said. "I have personally known and supervised hundreds of coloured iron workers capable of competing with white labour in any country." So in time more than half of the shanties were reserved for unmarried black employees, who were charged $1 a head per month in rent.[27]

*

At Lowell, Massachusetts, "a strict system of moral police" had been introduced into the company town. Close supervision of employees was believed extremely important, Henry A. Miles, a Lowell clergyman, remarked, because factory productivity depended on the morality of the operatives in their home lives. "Without this, the mills of Lowell would be worthless," he stated. "Profits would be absorbed by cases of irregularity, carelessness, and neglect; while the existence of any great moral exposure in Lowell would cut off the supply of help from the virtuous homesteads of the country." Accordingly, foremen like Rufus's father maintained lists of male and female operatives and kept track of their habits on and off the job. Persons who did not adhere to the local temperance rule or otherwise comported themselves in a manner considered immoral were entered into a central registry in Lowell. They lost their company housing and were blacklisted.[28]

Rufus devised a similar, if more elaborate, system to reward the good and punish the bad at Sparrows Point. The cornerstone of his system was a triptych of form letters. The first involved an application for housing, the second stated the terms of the lease, and the third administered evictions. To gain admittance to the town, an employee had to ask his supervisor to fill out the requisite form:

> **Clerk of Time Office:**
>
> The bearer, __________, has a family and wishes to rent a house from the Company.
>
> I can recommend him as a good workman.

Upon receipt of the letter, Rufus's pay clerks checked the workman's employment history by using records of the mill timekeepers. Special attention was paid to regularity of work habits. "It will require some careful selection to see that the most desirable ones get the houses," Rufus noted. For many families Rufus himself could evaluate their desirability. Sometimes, though, a background check was needed on a person's character. In this case Rufus summoned his own "moral police," the four-person police department. The chief figure in the police department was John "Boots" Campbell. Because he rented a farmhouse near the black neighborhood, he was the de facto symbol of law and order. Once a year he went door to door, taking a poll of residents, recording the names, sex, age, and conjugal condition of each household for review by the main office.[29]

The house lease made a family's tenancy conditional on the employment of the male head. Occupancy of a house, it was specified in the lease, was only "for and during the time he continues in the employment" of the company. If a worker severed his contract of employment, no notice to vacate the house was required to be given by the company. So in case of a strike, an employee was required to vacate the premises at once.[30]

Similarly, the commission of a major infraction at work (dereliction of duty, intoxication, insubordination, unexcused absences, and agitation) resulted in the loss of a company house. Evictions, while said to be small in number, were executed with terse swiftness via company form letters:

> **R. K. Wood, cashier**
>
> You will please instruct time keepers that __________ is discharged. Notify to leave house.
>
> F. W. Wood
> Pres.

Posted rules further looped the knot of control. There were rules by the general agent regarding the licensing of dogs in town; notices on the erection of backyard sheds; letters giving a resident permission to cultivate a vegetable garden or use company property for a speed-cycling track. Extraordinarily energetic, Rufus left his mark everywhere, and everywhere were pleas from workmen seeking his guidance or good favor. On May 24, 1893, an ornately scrolled (if misspelled) letter arrived on his desk, signed by three petitioners:

> **G.U.O.O.F. No. 3339**
>
> The object of our Order is known the world over. To our past record we point with prid, and our future is full of hope. Knowing as we do, whatever benefits the Company of Sparrows Point either directly or indirectly, benefits each and every individual of said town, and the purer and better each invidual makes himself, the more the Company's interest is enhanced, and also knowing your philanthropic spirit, we ap-

peal to you for a room in the YMCA Building. Hoping you will give us an early and favorable reply

We are
Yours Respt.
Anthony Thomas
Edward Dikes
Page Chapman

While wholesome living was undoubtedly a major goal of Rufus's endeavors, there also was a bottom line, and in this regard his training as an accountant never abandoned him. While doing good, he never forgot to do well: the steady ring of profit resounded from every commercial activity he undertook.

After taking into account the company's donations for the unkeep of the schools, Smokers' Hall, and several other services, Sparrows Point was still a consistent moneymaker. For example, at the end of 1897 Rufus toted up rental income and costs over the previous four years for his brother's perusal:

	House Rents	*Total Costs**	*Net Income*
1894	35,099.73	18,900.78	16,198.95
1895	34,168.02	23,226.00	10,942.02
1896	33,715.45	25,986.03	7,729.42
1897	38,659.04	21,480.65	17,178.39

*dwellings, street, sewer, water, general

These numbers show that during the previous four years Maryland Steel collected $141,642.24 in rents, spent $89,593.46 in costs, and retained $52,048.78 in net income. Such superior results were largely due to the company's payroll-deduction system in which rent was deducted before wages were issued. In this way the problems of late rents, partial payments, and uncollectables were eliminated because the tenant in effect prepaid his coming months' rent with his prior months' paycheck. The company also saved money by having its own staff and carpenters do maintenance work instead of relying on independent contractors.[31]

But Rufus was not greedy. As overseer of his flock, he made it a point of honor to keep house rents uniformly reasonable. And they were: based on available data, Sparrows Point housing was 20 to 25 percent lower than equivalent housing in Baltimore. Proportionally, rent for the lowest class of housing was neither higher nor lower than for the best housing in town. Thus, the company's interests and the "people's interests" converged.[32]

The company was able to attract a reliable workforce to the peninsula and the residents got cheap rent, which in turn made their pay envelopes look a lot better. Such an outcome produced a sense of mutual self-interst in maintaining the status quo. Moreover, within the strict segregation of classes, there was always the promise of individual mobility. A "good workman" could step up an alphabetized street or two within his social group. In a little town of 3,500, such benchmarks were readily visible. They could be measured and envied and sought out by others. The result: "There is great demand for the best houses," Rufus marveled, "8 to 10 applications for every vacancy."[33]

A similar system prevailed at the sprawling company store on C Street. Backed by the credit of the steel company, the store could purchase provisions in large lots at discounts unavailable to smaller merchants. What's more, the store had next to no losses from bad accounts since all goods were purchased on scrip. Upon entering the company's employ, a worker received a store book. When he desired to make a purchase the book was taken to the credit desk and stamped, "Credit so much." No more than the amount entered in the book could be bought. Before payday the book account was deducted from the worker's wagers. (Paydays were on the Saturdays following the 15th or 30th of the month.)[34]

On both ends of the transaction, then, the store had a distinct advantage over merchants working in a free marketplace, which guaranteed profits since no independent retail store was allowed on company property. Although the Wood brothers maintained that the store was run on a break-even basis ("for the convenience of the community and not as a source of material profit to the Steel Company," said Frederick), their business papers indicate a somewhat different situation. The store consistently netted 10 to 12 percent a year, and some of these earnings were used as a bonus plan for company executives. The arrangement worked like this. When the store company was formed in August 1887, 500 shares of stock were issued. The parent corporation was given 359 shares, while the remaining 141 shares were divided among five officers—Luther Bent (50 shares), Eben F. Barker (50), Frederick Wood (20), Edgar Felton (20), and Rufus Wood (1).

The men paid nothing for the stock, but got back a cash dividend of $50 to $100 per share per year. Between 1887 and 1901, when the store declared $375,000 in dividends, the executives shared $105,750, and the remainder was paid to the Maryland Steel Company. Bent and Barker picked up as much as $5,000 a year, while Frederick Wood and Edgar Felton each got $2,000. Although Rufus's dividend from one share of stock wasn't much, he drew a $1,000-a-year salary as director of the company store. Unobtrusively, then, some of the hard-earned wages of the residents found their way into the pockets of the high command.[35]

*

The town, to be sure, represented more than the sum of its earnings. For the Wood brothers its significance lay in the economic management of their manufacturing business. The community closed yet another circle of operations. As more families entered into dependence on the company, the interests of the company were consolidated. Had the residents been allowed to go their own way in town, owning their own property and free to shop where they wanted, it would have inevitably created a division of interests and perhaps made them consider their home life and their work in the plant as separate spheres. Company power was strengthened in direct proportion to the slackening of the workers' financial independence. Reckoned in this light, town operations held a lot in common with the works: they helped reduce and stabilize the human element in the cost of production.

Such a structure was at once radical and conservative: radical in its comprehensive organization of people and conservative in its social objective of freezing each class into place. If the village's scrubbed good looks harked back to Lowell and a blue-blood heritage, it also foretold the future. In its configuration of physical space and arrangement of social groups, one could see in outline the school of urban planning and its schemes of slum removal and moral uplift through monument-sized building. ("Make no little plans; they have no magic to stir men's blood.... Make big plans; aim high in hope and work," enjoined Daniel H. Burnham, the founder of U.S. urban planning.)[36]

Company towns expressed in mortar and bricks a strongly embedded sentiment among the respectable and powerful. This was "the iron law," according to Budgett Meakin, that the best way to handle problems arising from mass industrialization was in the benevolent yet firm control of the employed by the employer. By owning all the houses and unifying local government through the offices of "the better class," added the influential *Baltimore Sun*, Maryland Steel had surmounted the social pathologies that were gnawing at Baltimore and alarming the middle class. The attitude of the paper, and that of its Sunday editor, Henry Louis (H. L.) Mencken, was put forward in a five-part series effusively composed and prominently published in 1906. The opening articles dealt with the importance of steel, how iron ore, mined 1,300 miles away, was scooped from the holds of mighty ships and transferred into the blast furnace. After conveying the steps of steelmaking and duly reciting all the facts and figures of rail production, the newspaper focused on the company town:

LIFE AT SPARROWS POINT
IS SIMPLE AND PLEASANT

The Company Does All The Worrying—Saloons And Politics Barred.

One beneficial by-product of the company town, the *Sun* emphasized, was the cessation of union-inspired strife between capital and labor. Another was the propagation of orderly civic life which appealed to the Teutonic sensibilities of editor Mencken. Indeed, the polluting crassness of electoral democracy, with its stupid citizenry and electioneering "frenzies" in the lower-class wards of Baltimore, aroused the greatest rhetorical ire of the writer, who was unnamed and who might well have been Mencken himself. The paper beckoned:

> Pass from the tranquil administration of the town and enter the turbulent political atmosphere of Baltimore, with its warring factions and strife at the primaries. About 2,000 of Baltimore's 500,000 population are day residents of the Point, and yet it remains itself. The race problem it has practically solved by putting the blacks on a far side of broad waters of a creek apart from the whites. Negroes form one-quarter of the population and yet the relations between black and white are quite pleasant—perhaps because they do not meet in conflict at the polls.[37]

The article was in praise of the successful emplacement of the ideas expressed in the orations of Cardinal Gibbons and newspaper mogul Felix Agnus: the sense that, freed from the shortsightedness of politics and the pie-in-the-sky of labor agitators trying to "preach the brotherhood of man and the unblessedness of capital," the engine of industry could bring mankind to a higher plane of development. The article was little concerned with the voices of the workmen themselves (not a single workman or family was quoted); it was mythmaking the paper was engaged in. Here was a fetching story of the smoky tower of steel as kind father and patron saint of its people. If the story could not stand up to scrutiny, that did not deter the paper nor editor Mencken from indulging in unfounded sentimentalities. The nitty-gritty of profits and the larger question of unhealthy dependency encouraged by scrip at the company store was not an issue of the day. What the paper found relevant and highly entertaining was the "jingle" penned by the superintendent which summed up the benign motives of the enterprise:

> *Behold a house of many wings,*
> *Wherein is done all sorts of things,*
> *As selling goods and building ships,*
> *And making clothes and mending rips,*
> *Repairing shoes and baking bread,*
> *And trimming coverings for the head—*
> *A regular Pandora's chest,*
> *Where people come who want the best.*[38]

Within this setting hundreds of families settled on the peninsula, hundreds went to work across the marshes at the mill, and hundreds raised their children and performed all the perfunctories of daily life. And within this setting rectitudinous Rufus labored on, applauded by respectable opinion and driven by his own bouts of self-righteousness.

CHAPTER 4
FOREIGN AFFAIRS

All the while Sparrows Point hummed efficiently, turning out rails for the domestic market as well as for overseas sales in Argentina, South Africa, China, and Australia, the clangor of steam shovels and the picks of laborers resounded from the mountains of Cuba. What Frederick Wood had started in 1882 with the discovery of ore near Santiago had mushroomed into an important mining complex with heavy investments by the Pennsylvania Steel Company. These investments were part of a bigger context, that of the subtle yet powerful influence of private business interests on U.S. foreign policy during and after the Spanish-American War of 1898.[1]

From first to last, the war with Spain was full of surprises. Begun as a heartfelt protest by Americans against Spanish misrule of Cuba, the war ended as an act of national expansion. Spain was expelled from the New World and American rule was extended to the Caribbean and Far East. President William McKinley had not wanted to go to war, but war was proclaimed nonetheless in 1898—"in the name of humanity, in the name of civilization," he declared to Congress—under the public emotions whipped up by the press of William Randolph Hearst and Joseph Pulitzer following the mysterious sinking of the battleship *Maine* in Havana Harbor.[2]

A flotilla of army regulars and volunteers, including the fashionable Rough Riders under the command of Leonard Wood and Theodore Roosevelt, was launched from Tampa Bay to defeat the Spanish in Cuba and aid the native "freedom fighters." As the flotilla approached the island, American schoolkids chanted the battle cry popularized by Hearst in rallies staged in New York, Baltimore, and Boston:

> *Cuba, Cuba, bow, wow, wow!*
> *Libre, Libre, chow, chow, chow!*
> *Vengeance, vengeance, down with Spain!*
> *Yankee, Yankee, remember the* Maine![3]

Fully as surprising as the display of martial fervor by Americans was the actual course of the land war in Cuba. Determined to zero in on Spanish strength at Santiago, the U.S. military command embarked on their campaign at the precise spot where Pennsylvania Steel and other American capitalists had situated their iron-ore domains. When the Second Army Division landed on the island on June 22, 1898, they landed at the headquarters of Pennsylvania Steel's mining enterprise at Siboney. The Fifth Army of General Shafter simultaneously invaded Daiquiri, port of an adjoining mine works.

After joining forces, the troops made their way toward Santiago through the same jungle paths used by Frederick Wood 16 years before. They captured the Juragua mines at Firmeza and defeated enemy forces at Aguadores Bridge before the main body trudged north to San Juan Hill, where Rough Rider Roosevelt captured the limelight by charging enemy positions on horseback, a pistol salvaged from the *Maine* clasped dramatically in his hands. A couple of days later the Spanish garrison in Santiago surrendered and the land war was over.[4]

In freeing the Cuban people from the yoke of colonial oppression, the U.S. Army also liberated some valuable Yankee investments. The Daiquiri mine was not owned by sleepy Spaniards (as assumed by the traveling U.S. press corps and repeated in subsequent histories of the war), but by the famous Baptist *conquistador* of U.S. oil, John D. Rockefeller. In 1889 Rockefeller and members of his Standard Oil party—Colgate Hoyt and C. W. Harkness and the Ely brothers of Cleveland—had purchased mining rights to 2,500 acres and had built the port town of Daiquiri as part of their continent-wide quest to corner iron ore through techniques perfected in the oil trade. In the previous year the Daiquiri-based Spanish-American Iron Company had mined 206,029 tons of ore, as against 246,530 tons at Siboney-Juragua.[5]

From the sketchy information available it appears that the landing of troops at Rockefeller's Daiquiri and Pennsylvania Steel's Siboney was not so much a matter of planned conspiracy as a question of the convergence of self-interests. (Frederick Wood's papers, for instance, do not disclose any correspondence between his company and the army.) It seemed a case of the companies building such modern facilities in a backward country that their military importance was paramount. General Shafter needed the docks and fresh water supplies as much as the companies needed his army to resume mine operations, which began shortly after the surrender of Santiago.[6]

The Treaty of Paris of 1898 handed to America the largest increase in territory since the purchase of Alaska in 1867. Cuba, Puerto Rico, and the Philippines were placed under its guardianship. The territory comprised 163,000 square miles, equal to the area of California and Connecticut combined. In the first year of the American occupation of Cuba, which lasted until 1902, the country was under the authority of Major General John R. Brooke. In late 1899, President McKinley named Major General Leonard Wood as military governor as part of a general shakeup where Russell Alger was jettisoned as secretary of war and Elihu Root assumed the post.[7]

General Wood and Secretary Root were expansionists imbued with a belief in racial superiority and a sense of invincibility brought on by victory over Spain ("a splendid little war," harrumphed fellow expansionist Secretary of State John Hay). Internally General Wood strove for the fuller development of Cuba and better sanitation to control "yellow eyes" (yellow fever). Externally he lobbied successfully for tariff reciprocity that lowered the rates on Cuban imports, making the island's goods more valuable to American business. He called his task "the building up of a Republic, by Anglo-Saxons, in a Latin country," writing to Elihu Root, "We shall sometime own or at least must always control the destinies of Cuba."[8]

In the matter of economic development, though, a congressional roadblock stood in the way. It was the Foraker Amendment. To affirm that the United States had invaded Cuba on the basis of democratic principles rather than economic self interests, Congress had attached the amendment to the Army Appropriation Bill of March 1899. It stated, "No property, franchises, or concessions of any kind whatever, shall be granted by the United States or by any other military or other authority whatever in the Island of Cuba during the occupation thereof by the United States."[9]

General Brooke had made the error of being too literal-minded. In his governorship during 1899, he had infuriated American businessmen by interpreting the amendment at face value and refusing to grant concessions to petitioners who knocked at his door. General Wood (no relation to Frederick Wood) approached the matter with more bureaucratic creativity. To help facilitate railroad construction on the island, he and the War Department seized upon a device called the "revokable permit." The military office in Havana issued revokable permits for rights of way requested by private interests across public lands and roads. What created the biggest stir was the Cuban Railway, organized by the American-born railroad builder Sir William Horne. Eventually going the length of the island from Santa Cruz to Santiago, the 360-mile line was constructed without resort to the forbidden "concessions." It opened up virgin stands of timber and attracted the venture capital of such noted Americans as E. H. Harriman, J. J. Hill, and Henry M. Flagler, the Rockefeller man who developed Miami and Key West through the Florida East Coast Railway.[10]

Yet even before the gambit of revokable permits, General Wood had appeased U.S. iron interests. In an administrative ruling on February 8, 1900 (Civil Order No. 53), General Wood ruled that mining claims by foreigners were not covered by the Foraker ban. Wood made this determination by splitting semantic hairs, notably between the meaning of mining "claims" and mining "concessions." More tellingly, the order stated that any governmental restriction on mining would cause "serious prejudice to many individuals" and "would be positively detrimental to the interests of the island in the highest degree."[11]

General Wood went on to give an indefinite continuance of the existing mining claims at Juragua and Daiquiri (which were set to expire in 1903). In addition, he exempted all future mining claims from Cuban property taxes or mining royalties in an addendum that was not published with the original ruling. The practical effect of this ruling was to place future Cuban governments in the awkward position of incurring the official displeasure of Washington if attempts were made to impose taxes or other fees on American mining interests.

Two foreign policies intersected during this critical point of U.S.–Cuban relations. Even before a constitution was formed or elections held, Cuban politics had been compromised. Writing 30 years later, former U.S. Ambassador Harry F. Guggenheim pinpointed the 1898–1902 period as the time when "American" first became a distrusted word. "Our lack of a well defined foreign policy in Latin America, the inconsistency and, at times, ineptness of our diplomacy there at that period, our apparent hypocrisy in carrying out imperialistic activities in spite of benign official dicta to the contrary—all were firmly fixed in the minds of the Cubans."[12]

Within a week of the issuance of Order No. 53, Frederick Wood arrived in Santiago to survey new mining sites firsthand. Unfortunately, the notes of his trip were not part of his surviving papers, but in his letter files there was correspondence indicating his intent to pick up as many claims as possible under the new ruling. From Havana, for example, the company's agent, N. B. C. Nitze, signed over documents to Wood on properties of potential value, writing: "I hereby assign to F. W. Wood of Baltimore, Md., the accompanying option on the Buenavista and Carbonera iron properties, granted to me on this date by Antonio Gobel of Havana, Cuba."[13]

In the same period Evans Dick, president of the underwriting house of Dick Brothers & Company of Philadelphia, corresponded with Wood in Santiago. Apparently responding to possibilities that Wood had outlined, Dick wrote on February 14, 1900, "You must appreciate that for you and me to make any money on this we must have a firm option which will run along a sufficient length of time to let me get a syndicate, have the property examined and to float a new company. It should take at least ninety days to do all this."[14]

A year later the proposed syndicate came into being as part of a general recapitalization of the steelmaker. The Pennsylvania Steel Company (of New Jersey) was formed with an authorized capitalization of $50 million. Over 250,000 shares of common and preferred stock, at $100 par value, were floated by Dick Brothers and Drexel & Company on Wall Street, bringing in a total of $27.2 million in funds. The steelmaker's oldest sponsor, the Pennsylvania Railroad, backed the syndicate by purchasing $15 million in stock. The Philadelphia & Reading picked up a large stake and, by invitation only, additional shares were made available to a group of affiliated individuals, including Baltimore manufacturer R. C. Hoffman, railroader Frank Firth, and company lawyer J. Alexander Preston.[15]

Frederick Wood was named senior vice president and Edgar Felton president of the holding company. (Wood continued as president of Maryland Steel, the subsidiary owning Sparrows Point.) Other key figures included Drexel partner Edward T. Stotesbury, Effingham Morris of Girard Trust Bank, and coal and railroad kingpin George F. Baer. All were placed on the board of directors.[16] These changes would prove crucial to the fate of the Cuban fields. With mergers shaking up the steel business in the Midwest, Pennsylvania Steel sought to get a bigger slice of the valuable Cuban ore. Now the company had both the funds and the backing of high-level businessmen to accomplish the task.

*

On April Fool's Day 1901 John D. Rockefeller sold off his ore grounds at Mesabi to the United States Steel Corporation. U.S. Steel—popularly known as the "Steel Trust"—was a collection of Midwest properties consolidated into a holding company by J. Pierpont Morgan. Centered in Pittsburgh with tentacles flung west to Chicago and north to Lake Superior, the Steel Trust incorporated all of the properties of Andrew Carnegie, all of the Chicago mills of O.W. Potter and John W. Gates, all of the Rockefeller mines and Lake steamboat lines, and a phenomenal number of other Midwest mills—120 in all—that manufactured iron pipe, beams, canners' tin plate, hoops, wire, and bridge parts. Hitherto these mills had developed independent of the Bessemer steelmaking group. This was "integration" on a grandiose scale, the stuff of superlatives in an age where some of the great trusts of the 20th century were being born—AT&T in communications,

United Fruit in perishables, American Tobacco in cigarettes, General Electric in machinery, and National Biscuit in mass-produced crackers and sweets. Biggest of the big was U.S. Steel.[17]

The Steel Trust constituted the seizure of the labor and machine efficiencies of the "works" for the financial benefit of a few insiders. Even the conservative Supreme Court would later agree that the Trust had been formed for the "illegal purpose" of acquiring a monopoly. (Possession being nine points of the law, illegal intent was ruled to be insufficient grounds for dissolution of the Trust.)[18] Where there had been six rail rollers active in the Midwest district in the 1880s, there now was one corporate owner, capitalized at $1,402,846,817, or 2½ times the gold and silver currency then in circulation in the country.[19]

Morgan did not propose to pay this stupendous price out of his own pocket. That would have been foolish. He let the public pay. His investment syndicate issued a blizzard of stocks and mortgage bonds and, with everyone excited about steel, they were snapped up in every corner of the country. When everything was sorted out it was found that only half the issued securities actually represented tangible property. The rest—$726 million—was excess capitalization, or "water," according to an exhaustive post mortem performed by the U.S. commissioner of corporations. A major beneficiary of this inflationary offering was the House of Morgan. The banking syndicate pocketed $62.5 million in cash, underwriting fees and stock shares in the transaction while not building one new blast furnace or offering a single new product line to the public.[20]

Morgan was not alone in reaping a fortune. In selling out to U.S. Steel, John Rockefeller made in the vicinity of $80 million (the exact amount was never disclosed), the "little Scotch pirate" Andrew Carnegie got a fabulous $225 million for his steel mills (or $46 million more than he privately admitted they were worth), and a new breed of barons emerged on the American scene—the "Pittsburgh millionaire." Instant millionaires thanks to Morgan's creative financing, 30 former Carnegie officers began a spree of consumption that enraged reformers and unionists. In an age when the average steel wage still stood at $2 a day, one Pittsburgh millionaire hired a special train, stocked it with champagne for his many guests, and rolled nonstop east across the country from San Francisco. "In seeking for gold, we have forgotten some of the vital principles of civilization," Jacob Schonfarber, a Baltimore unionist, commented bitterly before the U.S. Industrial Commission that year.[21]

On the same April Fool's Day Pennsylvania Steel (of New Jersey) acquired the assets of Rockefeller's Spanish-American Iron Company at Daiquiri, Cuba. There was no public hoopla, for the chief strategists didn't want it.[22] Frederick Wood and Edgar Felton were distrustful of the glitter of Morgan's Wall Street and the brash young men of Pittsburgh. But in purchasing Daiquiri they were motivated by the same goals of consolidation and removal of competition. Sharing half ownership of the Juragua deposit with Bethlehem Steel, and with legal claim over the undeveloped Buenavista, Carbonera, and other properties, Pennsylvania Steel controlled the "gates" of Cuban ore—the only source of foreign iron of consequence for U.S. producers—as unmistakably as the Steel Trust dominated the supply and pricing of domestic ores of the Great Lakes region. (Later Bethlehem acquired the Juragua deposits in full from Pennsylvania Steel.)

Paralleling the north-south division of hemispheric supplies, a "friendly understanding" was effected between Morgan's company and Pennsylvania Steel. It was denoted in Wood's papers as:

U.S. Steel ....67.22 percent
Pa.–Md. Steel.......................... 10.40 percent[23]

What was being described here was a collusive market pact under the broad leadership of U.S. Steel. Rail sales were to be divided up according to the above yearly tonnage shares. With 77 percent of the business handled by the two firms, the remaining 22 percent was to be split among five other companies—the three Eastern mills of Lackawanna, Cambria, and Bethlehem (members of the original rail-rolling fraternity); Colorado Fuel & Iron, a newcomer established in Pueblo, Colorado, by the ubiquitous John D. Rockefeller; and Tennessee Coal, Iron and Railroad, a consolidation of Birmingham iron companies in Alabama.[24]

The fact that market pooling was forbidden under the Sherman Anti-Trust Act was of as little worry to the participants as the Foraker ban had been to the McKinley administration. The Rail Pool was an open secret in business circles, its existence alluded to in *Iron Age* as frequently as steel executives stoutly denied any knowledge of the arrangement. Frederick Wood, for one, simply lied about the arrangement. In June 1901 he informed the *Baltimore Sun* that his company was "outside" the pool when the truth was that it was a charter member.[25] Business *über alles,* not ethics, was the impulse that inspired the enlarging empires of iron and steel.

With rail sales divided up, Pennsylvania Steel was free to concentrate on the extraction of ore at the best possible price. Frederick Wood kept close tabs on the engineers at Daiquiri. His papers between 1902 and 1905 are filled with notations of the campaign to wrest the mineral wealth from the Sierra Maestras. The company's opening of Lola Hill was a case in point.

Lola Hill was a cone-shaped promontory that rose above Daiquiri Valley five miles from the coast. Because of the steepness of the terrain, efforts to tap the ore veins by Rockefeller's miners had been unsuccessful. Development of steam-shoveled mining, however, would help to tame the recalcitrant mountain. Two steam shovels were shipped from the U.S. in 1901 and painstakingly dragged up the hill on mine tracks. Hand shovels were thrown aside as the big machines moved three tons of rock in their buckets. Five other steam shovels were placed on the hillside, along with dynamite and equipment houses. An article in *Iron Age* noted that at any one time two dozen mine crews worked along notches that ran up the hill like blades of a giant corkscrew. In all, 2,360 tons of ore and waste were handled a day.[26]

Conveyors drew the ore down to a railroad yard (the waste was pitched into the hollows) for the short ride to Daiquiri and the long journey to Sparrows Point. To a steelman it was not the distance that counted; it was the cost of mining and freight. Harry Huse Campbell, the chief metallurgist, pointed out that the company could dig ore in Cuba, transport it over one thousand miles to America, and make pig iron for less than one-half cent per pound. Campbell identified the key parts of ore handling as:

—steam shovels at the rock face
—railroads to the shipping port
—a crushing plant and gravity system for delivering ore to the shipping pier
—a high-pier dock with gravity pockets
—a fleet of special ore carriers[27]

Like the steel works, the mining and transport operation was an interlocked pipeline featuring a large and regular flow of materials. "Links in a chain of specialized equipment" was how Campbell described it.

By bringing the same economies of integration into the ore fields, cost performance was enhanced greatly. This was underscored by the rising tide of cash dividends declared by the Cuban subsidiary to Pennsylvania Steel between 1901 and 1906:

1901	none
1902	$120,000
1903	$240,000
1904	$336,000
1905	$384,000
1906	$840,000[28]

If anything, the figures understated the importance of the mine-to-mill shipping bridge, for after deducting the capital costs of opening up Lola Hill, ore was delivered to Sparrows Point at 10 percent below the open market price. This saved the manufacturing arm of the company as much as $400,000 a year. What's more, as shipments from Cuba advanced, the unit costs of railmaking decreased. In this period the break-even point was 200,000 tons of rolled rails a year. For every ton produced above the 200,000 figure, profits inched upward. For example, in 1906 when Sparrows Point produced 366,370 tons of rails, net profits averaged a healthy $3.28 per ton.[29] This tonnage, by the way, was enough to lay a double track from New York City to Sacramento.

There were more linkages as one surveyed the industrial scene. Eastern railroads benefited. The more ore shipped from Cuba, the more trainloads of coal rumbled down from Appalachia on the Pennsy and B&O Railroads, increasing revenues. And the more finished rails from the mill, the more traffic was available for railroads in a commodity they both shipped and consumed.

In 1907 the steel company and its backers took another big step. Alexander Cassatt, president of the Pennsylvania Railroad, approved plans for the opening of the Mayari deposit on the north coast of Cuba. On July 1, 1907, the railroad purchased $3.82 million in 20-year bonds to finance the venture. Again the timing was favorable. Earlier the Spanish-American subsidiary was awarded mining rights for the entire Mayari plateau, 27,850 acres, by Estrada Palma, first president of Cuba following the end of U.S. occupation in 1902. As part of the deal, Palma suspended all property taxes and royalties on the mine property—the same provisions established under Order No. 53 of former Governor Wood. Furthermore, a reciprocity tariff treaty entered into by the U.S. and Cuba had dropped import duties on Cuban ore to a record low—32 cents a ton.[30]

The record of President Palma was not one of greed or cupidity; rather it was a case of too much trust placed in his patron and too little understanding of the island's darker forces. Installed by Americans, Palma believed in Americans. He was therefore willing to grant liberal terms to U.S. interests. But when conservative politicians staged an armed revolt in 1906, Palma found his *gran amigo* very fickle. Worried that U.S. interests were compromised by the disorder, President Theodore Roosevelt rushed U.S. troops to the island, invoking the rights of intervention under the Platt Amendment.*

The troops splashed down on September 30, 1906. Eventually a force of 6,400 was stationed on the island. The U.S. Army of Cuban Pacification became the de facto policing force on the iron ranges. Assigned to protect U.S. property and police the local population, they administered justice and patrolled the mine towns on a regular basis. They did road work, conducted population surveys, and studied sites for future army installations.[31]

In this political context, the return of U.S. administration and use of U.S. troops to guard American-owned property, the Mayari mines were developed by Pennsylvania Steel between 1907 and 1909. A railway line was strung 13 miles through the lowlands to the base of the plateau, and the plateau was scaled by means of an inclined railroad. The inclined railroad was a breathtaking sight. It consisted of two standard-gauge tracks rising up on a roller-coaster gradient from elevation 130 feet to 1,620 feet. Between the tracks ran a cable weighing 123,000 pounds.[32]

Metal-block barneys attached to the cable eased a loaded car down the grade. The trip took about 12 minutes. By using the laws of gravity to maximum advantage, the engineers arranged for the loaded cars

*Passed as a rider to the 1902 Army Appropriation Bill, the Platt Amendment stipulated in Article III, "That the government of Cuba consents that the United States may exercise the right to intervene for the preservation of Cuban independence, the maintenance of a government adequate for the protection of life, property, and individual liberty, and for discharging the obligations with respect to Cuba imposed by the treaty of Paris on the United States, now to be assumed and undertaken by the government of Cuba." The amendment was one of the first instances where forcible intervention was defined as a legal right.

to lift empty cars up the cliff on the second track. Through counterbalancing the need for hoisting equipment was greatly reduced. From the base of the incline, the railroad carried the ore to Nipe Bay. While cutting through wild glades and dense undergrowth, the jungle railroad maintained high standards. At no point was the ruling grade over 0.5 percent or curves greater than 6 degrees. The bridges were designed by the Pennsylvania Railroad and constructed wholly of steel.[33]

At Nipe Bay the jungle gave way to a battery of kiln stacks. The ore was carried into the kilns, where its moisture was reduced. It was formed into pellets, transferred to the wharf via electric buggy cars, and loaded onto the boats by means of ore bridges with trolley grabs extending out 90 feet over the water. The wharfs were designed to withstand the worst storms off the Atlantic and other environmental assaults.[34]

The tapping of Mayari was emblematic of American industry's achievements overseas in the first decade of the century. From the opening of banana plantations in Honduras to the establishment of rubber colonies in Indochina and mine smelters in Mexico, business was on the move. "Americanization of the world" was how the British journalist W. T. Stead described the phenomenon in his best seller of the same name. "Earth-hunger and the scramble for markets" was the more jaundiced view of John Hobson, the radical economist. For Frederick Wood, the fact that the Pennsylvania Railroad chose tracks made of Cuban ore for its important Pennsylvania Station expansion project in New York was proof enough of how well his mines had been fitted into the demands of American steelmaking.[35]

*

One of the constants since Columbus's time was the exploitation of cheap manpower in Cuba. Slaves clanked their fetters in the shafts of the early copper mines around Santiago. They were among the 900,000 Africans brought over to the island in shackles. A system of bound labor was also established. Coolies from China were the first to be imported. They were indentured for eight years to the purchaser, who was required by the contract to pay 20 or 30 cents a day in wages and supply a stipulated food ration, clothes, lodging, and medical attention. So many abuses existed as to make bound labor de facto slavery, both for the Chinese workers, who constituted the large majority, and for workers recruited from the Canary Islands and rural Spain.

Before this system ended, in 1873, an estimated 150,000 workers had been added to the Cuban workforce, chiefly for the benefit of colonial agriculture.[36]

Such primitive social conditions offered the officers of Pennsylvania Steel the opportunity to inculcate American values in the Cuban working class. But the company towns and worker training programs of Sparrows Point were never part of the company's agenda in Cuba. From the commencement of mining, the native population was sidestepped in favor of the agents of their oppression—the Spanish soldier. Whenever it was possible, Spanish soldiers were hired as company miners. This remarkable situation was abetted by corruption. The soldiers were paid 70 cents a day in wages, while their regimental officers appropriated their pay as soldiers. "As a man could earn considerably more mining than soldiering, and the miners had better rations, all parties were satisfied with this arrangement," claimed one chronicler of the practice.[37] The Spanish-American War, of course, severed this flow of able-bodied workers. The Yanks were back to square one, the Cuban peasant, which was not to their liking.

Sparrows Point labor had been constructed on the basis of race, of white buckwheats over Virginia field hands, and for a while Frederick Wood considered instituting a similar system in Cuba, with the ironic twist that the American Negro would constitute "the better class" of workmen. In 1899 he weighed a proposal to ship blacks from Sparrows Point to Cuba. He wanted them to stimulate the local "laggards" and lower tonnage costs. Then another solution presented itself. Why not import Spaniards? They spoke the island's language and would not demand the "American wages" of black steelworkers. So, having been shipped out of Cuba as soldiers, Spaniards were shipped back as laborers.[38]

Labor agents were hired in Spain to scour the rural provinces for men to be hired under a variant of the "bound labor" system. There were some changes in detail, to be sure: pay and rations were better, and instead of serving for eight years the miner signed on for a maximum of three years. However the dynamic was the same: the purchase of alien labor with little option but to work under the conditions imposed by the employer. Although the Spanish recruiter got his commission up front for every miner that was signed, the miner was required to work for 150 days before he received the agreed-upon $30 compensation for his voyage over. If discharged, he faced a blacklist at other American enterprises which exchanged discharge lists with the mining company.[39]

Not surprisingly, it was the poor who accepted these terms. Shipped by steerage to Havana, most of the recruits were "ordinary" peasants —"youths and single men who sell their scanty property of tools and utensils at home to pay their passage to Cuba," according to Victor S. Clark, an American academic who visited the iron mines. He spoke of their docility in a report for the U.S. Department of Labor:

> Some American employers consider them the best unskilled laborers of Europe. They are physically robust and not addicted to many of the vices of laborers of the same class in the United States. They are more docile than the latter, and fully as intelligent for many kinds of service. Unlike the Cuban, they are frugal, seldom gamble, and often allow their savings to accumulate in the hands of their employers. They are not quarrelsome, and do not usually carry concealed weapons. The newly arrived laborers have no small holdings like the Cuban peasant to relieve them from the necessity of steady labor.[40]

According to Clark's account, there was a definite pecking order among the Spaniards. Galicians were the most numerous and were employed as common laborers. Next in numbers were the Asturians of the Cantabrian Mountains. Found to be good office workers, they were engaged as clerks. The Catalonians were the Pennsylvania Dutch of the Latins. Industrious by habit (there was a Spanish proverb that Catalonians could "win bread from stones"), they were employed as steam-shovel operators, machinists, and railroad engineers, and were considered very valuable. Similar to Sparrows Point, promotions were made from within the group. The best of the Galicians were made into gang leaders; a resourceful Asturian could climb to assistant clerk or even timekeeper. Single men were boarded at barracks, while the families of skilled workers were given space in cottages. Food was served in *fondas* (mess houses).[41]

The officers of the company were all American-born Caucasians save for Pedro Aguilera, a Cuban. Every year they were given a four-week "sabbatical" in America. This was in keeping with current beliefs that overexposure to tropical light could cause severe biological distress to Caucasians. The idea that Anglo-Saxons could quest and conquer in the tropics but never physically labor found expression in books such as *The Effects of Tropical Light on White Men* (1905) by Major Charles E. Woodruff. An Army surgeon, Woodruff argued that if exposed to excessive beams of light, a "bold, energetic, restless and domineering" white man could devolve into a personality that was "resigned and religious and imitative." In time, Dr. Woodruff averred, the large human brains of blond, blue-eyed Aryans could even shrink to the inferior size of Negro and Mediterranean peoples. Flourishing concepts of biological determinism (the *Effects* book sold many copies in Britain as well as the United States) were fundamental to the thinking of the steel company. To keep the skin color and the brain power of its officers intact, it reimbursed the costs of the men's sabbaticals to the north while leaving "dark-skinned" Spaniards to dig the ore.[42] The notion of suntanning as a status symbol and of the Caribbean as a carefree retreat for cold-climate habitués was yet to be invented.

Base pay for the miners was $1 a day for a 10-hour day, a common wage for nonagricultural labor on the island—and considerably below the minimum wage of 13.5 cents an hour at Sparrows Point. Although the $1-a-day wage was 30 cents more than the company had been paying to Spanish soldier-miners before the war, the operation still managed to get out ore for 5 cents less a ton. This finding helped inspire a work speed-up program called *la tarea* (the task). It based a day's wage on output. Mine crews were required to fill a given number of cars of ore or waste a day. If output went above this limit, the crews received a small bonus, typically 15 or 20 cents a day.[43]

Even among the Spaniards, mining was not a popular occupation, according to *Iron Age* editor Charles Kirchhoff, who visited the mines in 1907 as a guest of Pennsylvania Steel. The work was physically demanding despite the mechanical contrivances, and several of the work rules grated on the Spaniards' love of conviviality. "The 'American' is not, on the whole, more popular in Cuba than he is in Mexico or any other Spanish-American country," he wrote.[44] Most of the men elected to return home at the end of their contracts. As a consequence, recruitment in Spain had to be nearly continuous in order to secure the manpower needed to haul off the mountains of ore bound for North America.

CHAPTER 5

SECRET SUBMARINES

In the years 1909 through 1913, Frederick Wood concentrated on rounding out the facilities at Sparrows Point. An open-hearth shop was installed in 1910 which increased annual capacity by 40 percent. The blast furnaces were completely overhauled and equipped with improved blowing machines. Although business varied by season, daily employment hovered around the 4,000 mark. Three thousand worked in the steel mill and another one thousand at the commercial shipyard.

Earnings were stable and at times plentiful. "The company is in a strong financial situation and has continuously reported a heavy surplus over its fixed charges," said *Moody's.* The biggest news around town had been the death of Rufus Wood. On May 16, 1909, he had collapsed of a heart attack and died immediately in his house. He was 60. Appointed in his place was the 30-year-old son of Luther Bent, Quincy, who administered town affairs in accordance with Rufus's precepts.[1]

Outside of town the company's place in the world of commerce was secure. Maryland's largest industrial employer continued to win important contracts from the railroads. The market for rails reflected the *lateral* growth of railroads after the turn of the century. Following waves of frenzied building, railroads had gotten around to the task of tying their network into strong arteries. Main lines were double- and triple-tracked to expedite traffic and new freight yards were built to handle tonnage which had quadrupled in the quarter century between 1885 and 1910.[2]

As part of the rebuilding campaign, thousands of miles of lightweight track were replaced by heavier rails. In the 1890s the 75-pound Pennsylvania "Big Head" rail (weighing 75 pounds per yard) was considered the acme of weight and design. But accelerating demands for speed, heavier car loadings, and bigger locomotives prompted major changes. The 100-pound rail designed by the Pennsylvania Railroad had become the standard for main-line track. The heavier track had a gratifying effect on sales for Sparrows Point—it meant that more steel was used per mile of track laid.* With the advent of the 100-pound rail, a full ton of steel was needed for every 30 feet of track, or 176 tons for every mile, not counting bolts, spikes and the other metal fastenings.[3]

The Rail Pool established under U.S. Steel leadership functioned as a stabilizing force in the trade. Within its traditional geographical zone of the East, Sparrows Point supplied hundreds of miles of track for major building projects, including the B&O Railroad's main-line relocation project near Paw Paw, West Virginia, the Pennsylvania Railroad's Hell's Gate route in New York, and projects of the Western Maryland Railway, Reading Company, Atlantic Coast Line, and Interurban Rapid Transit (IRT) subways in New York.[4]

Sparrows Point was further nurtured by an export cartel known as the International Rail Association. It had been under operation since the turn of the century by German, French, Belgian, Austrian, and British steelmakers, who were all highly dependent on the export market. Individual members included Germany's Stahlwerks Verband, the Belgian Steel Works Union, and Britain's John Brown & Company. The American participants of the cartel were U.S. Steel, Pennsylvania-Maryland, and Lackawanna. According to courtroom testimony by William E. Corey in the steel dissolution case, the participants agreed on two overall principles: that they should refrain from selling rails in each other's domestic markets, and that business in so-called "neutral countries" (that is, countries without their own rail mills) should be divided at a price agreed upon by the member companies.[5]

The latter principle was difficult to enforce, even though the pool had established a clearinghouse in London to keep track of orders and sales prices lest a conspirator might cheat on his confreres. The Germans in particular were considered highly aggressive—their profits were generated on the narrow margin between contracted raw materials and exported finished products—and they would go outside the cartel if it meant they could reach a separate pact. Posturing by the

*In 1902, only 645,000 of the 2.9 million tons of rails rolled in U.S. mills were 85 pounds or heavier. By 1913, 2.3 of 3.5 million tons were 85 pounds or heavier.

Germans on an early Mexican National order greatly annoyed Frederick Wood. "I have thought that $26.00 at tidewater would be about the proper price to name," he complained in a letter to Moses Taylor, vice president of Lackawanna Steel. But without cooperation from "the other side," he said, "it would probably be necessary to bid equal to $24.00 or $25.00 delivered at Mexican ports to secure the business.... What the outcome is likely to be, we are unable to predict, as there have been several breaks since the matter was first taken up."[6]

In spite of such frustrations, the pool greatly assisted Wood in finding markets for excess output when domestic orders were low. Generally speaking, the American mills got first crack at rail orders originating from South America, Central America, the Caribbean, and the Far Pacific basin, while the Continental and British railmakers had precedence in Scandinavia, Western Russia, the Mediterranean basin, colonial Africa, and Asia. Under this global scheme Sparrows Point sold rails to various Mexican lines, Argentine Great Western, Costa Rica Railroad, Cuba Railway, Victorian Railways of Australia, Chinese Eastern Railway (the Chinese segment of the Trans-Siberian route), and several South African railways.[7]

On average, 24 percent of Sparrows Point sales went into the export market. This meant 80,000 tons on a good year. That brought in about $2 million in extra annual revenues. On very good orders, for example when Chinese Eastern Railway purchased 107,550 tons of 64½-pound rails, the plant did very well. The company even landed sales it had not anticipated, thanks to the power of the pool and the good offices of U.S. Steel's export division:

> Mr. E. C. Felton
> President, Pennsylvania Steel Company
> Philadelphia, Pa.
>
> *PERSONAL*
>
> Dear Sir:—
>
> We have just learned that the Cambria Steel Company have reported a sale of 1100 tons of 60 lb. rails to Hugh Kelly & Company for shipment to Cuba at $26.75 Fob. New York.
>
> In view of our understanding with Mr. Stackhouse that the Cambria Company would not need any export business, we think that arrangements should be made to transfer this order,

and should therefore be obliged if you would kindly take up the matter with Mr. Stackhouse and advise us the result.

Yours very truly,
J. A. Farrell
President, U.S. Steel Products Export Co.[8]

*

One hundred forty miles north of Sparrows Point, in smoky Lehigh Valley, the other major steelmaker of the East had broken out of the industry mode. To win more orders for its steel furnaces, the Bethlehem Steel Company had become a major supplier of military goods for the U.S. and foreign governments. Bethlehem's origins as an arms merchant started modestly enough. In 1886 the company had built a small mill for the manufacture of armor plate at the suggestion of W. C. Whitney, then U.S. secretary of the navy.

Armor plate was a new type of steel that contained small increments of nickel (3.5 percent) and chromium (1.5 percent). The resulting plates were heat-treated so as to present a hard-fibered surface that was resistant to penetration and at the same time did not shatter under shock. In 1893 the Krupp Company unveiled its patented armor plate. Tests before ordnance experts in Europe showed a 20- to 30-percent superiority over previous hulls. Joseph Wharton and Andrew Carnegie had secured the exclusive American rights for this process. Although they paid a stiff license fee to Krupp, their investment paid off. Against the aging Spanish fleet, U.S. warboats had proved impregnable—the victories at Manila and Santiago testified to the technological superiority of steel-armored vessels.[9]

Bethlehem used the revenues from navy sales to move into artillery gun production. Here 1900 was a critical date. That year "La Compagnie de Bethlehem" demonstrated at the Paris International Exhibition a new type of heated tungsten-chromium steel for machining. The extra-strong steel, co-invented by Frederick W. Taylor, a Bethlehem engineer and soon-to-be guru of "scientific management," enabled heavy cuts to be made for big guns at a high rate of speed. Soon there was scarcely a gun plant in the world not equipped with "high-speed" tool steel of Bethlehem origin. "It is no exaggeration to state that no single invention has made such a revolution in metal-working methods [for armaments] as the introduction of Taylor-White high speed steel," pronounced an artillery publication.[10]

Into this scene entered Charles M. Schwab. Losing an acrimonious turf battle at U.S. Steel against chairman Elbert Gary in 1903, Schwab resigned as president of the company and assumed financial and managerial control of Bethlehem Steel. On his first official tour of the plant in August 1904, he buttonholed a representative of the *American Iron and Steel Institute Bulletin* and said, "I shall make the Bethlehem plant the greatest armor plate and gun factory in the world." Expanding on his intentions two months later, he told the *New York News Bureau:*

> I intend to make Bethlehem the prize steel works of its class, not only in the United States, but in the entire world. In some respects the Bethlehem Steel Co. already holds first place. Its armor plate and ordnance shops are unsurpassed, its forging plant is nowhere excelled and its machine shop is equal to anything of its kind. Additions will be made to the plant rather than changes in the present process or method of its manufacture.[11]

To keep his vow Schwab had to go outside America and sell his arms to the rest of the world. While in America Congress played Scrooge to the War Department, over the border and across the seas there were military rivalries and wars, budding dictatorships in South America, and a "big navy" horse race between Britain and Germany. In short, there was no lack of demand if one showed enterprise. In Europe and Asia every important nation had one or two gun manufacturers. On the Continent there were Skoda of Austria-Hungary, Krupp and Deutsche Waffen of Germany, Cockerill of Belgium, and Schneider-Creusot of France, and in England there were Vickers and Armstrong. The foundries of Creusot had supplied arms to French monarchs since Louis XIV, while Alfred Krupp, founder of the Krupp dynasty, had first earned his stripes as "Cannon King" in 1842, 12 years before Henry Bessemer undertook his experiments on artillery iron for Napoleon III.[12]

In business outlook and policy the arms makers of Europe were no different than their colleagues in rail-rolling: they sought comfort in numbers. While they worked closely and apparently patriotically with the military departments of their respective countries, they had joined together in pooling arrangements. They had patent pools to restrict access to the market. They divided up the market of "neutral" nations lacking gun plants, and they effected "nonaggression" pacts regarding the sanctity of their own sales spheres. The first pool, the so-called "Harvey syndicate," licensed patent rights on different processes in the manufacture of armor plate. Many disputes later, the Harvey syndicate was replaced by the "Armor Plate Pool." It provided a comprehensive structure for regular meetings and allocation of orders. All pool meetings were conducted in private and the participants were sworn to a code of silence which included denying that the cartel existed.[13]

For Charlie Schwab and his organization the breakthrough year was 1906. Meeting in Europe with the Armor Plate Pool, Archibald Johnston, chief munitions salesman, demanded parity with the Europeans. He called for an over-sixfold increase in allotment.

"We want two million dollars," Johnston said evenly. "We are not here to haggle and engage in any undignified bargaining." Basil Zaharoff, the Vickers spokesman and leader of the Europeans, thought it over. Then he rose in reply. "Well, gentlemen, Mr. Johnston has made his proposal, and I believe there is no good in arguing against it."

When Johnston cabled back the news to South Bethlehem, Schwab was flabbergasted at first. "Message all garbled, incredible amount mentioned. Avoid code, send message straight." But when he realized that the message was correct, he was "delighted."[14]

Over the next few years Schwab focused his considerable energies on arms sales. He traveled to London, the arms bazaar of the world, and made several swings through the capitals of Europe. He sold guns to a score of needy governments, most of them in the U.S. sphere of Central and South America. A later statement showed that he consummated arms sales with Chile, Guatemala, Greece, Imperial Russia, England, and France. His reputation gained such stature that he had private audiences with Kaiser Wilhelm in Berlin.[15]

In 1910 Schwab closed a deal for armor, armament, ammunition, spare parts, and accessories for the navy of the Argentine Republic. The $10 million order was reported as the largest foreign military contract awarded to an American firm to date.[16] In the same year the company was named contractor for the arming of two gunboats ordered by the Cuban navy. According to a letter filed in Frederick Wood's business papers (his company also was interested in the gunboats), the new president of Cuba, José Miguel Gómez, was to be paid approximately 10 percent of the $950,000 contract price. The correspondent was quite explicit on this point:

> In order to carry out the business, it is necessary to dispose of a certain percentage for expenses (*derrames*).... The English house of Stetten has offered 15% and a German house 12%, but they will not get the business as the President desires to favor an American house. I think that the business can be secured with less 10% for the said expenses.[17]

Next to the word *derrames* was written, "leakage, lavishness, i.e. rake-off." In spite of his correspondent's urgings ("I will do all that is possible that whatever percent Mr. Wood offers may be sufficient for these expenses," said G. Petriccione), Wood refused to bid. A month later Bethlehem won the contract in conjunction with Cramp's Shipyard of Philadelphia, beating Vickers and another German yard. On October 11, 1911, the gunboats were launched at Philadelphia in a ceremony attended by President Gómez's son-in-law, who was secretary of the Cuban navy.[18]

The Cuban contract was a good illustration of the difference between the two companies. One was cautious and conservative; the other jumped at every contract even if it required *derrames* on the order of $95,000 based on the contract price and suggested bribe by G. Petriccione. Schwab cultivated the image of Bethlehem as a feisty young upstart; he went out of his way to collar reporters with a story of his struggle to get a bank loan for a new mill. His rejection of conventional steel markets and his thinly veiled contempt for Elbert Gary and his "Sunday-school morality" earned him a renegade reputation within the industry, a sense that his bouts of salesmanship violated established norms of behavior. "For years steel men have regarded Bethlehem, not as a steel company, but as an ordnance concern—the American Krupp's," said *Moody's* reporter Arundel Cotter.[19]

In 1913 Schwab moved further into the arms business with the purchase of the Fore River Shipbuilding Company of Quincy, Massachusetts. A proven contractor of navy warships, Fore River controlled the patent rights to the latest offensive weapon on the seas—the torpedo submarine. The yard placed Bethlehem in the position of being the only U.S. shipbuilder equipped to build and fully arm submarines. To Schwab the shipyard made a perfect fit with his business in South Bethlehem, furnishing an outlet for armor plate, structural shapes, ordnance, and forgings. Although the shipbuilding company had $4.8

million in stock outstanding, Schwab was able to whittle the purchase price down to $750,000 because the owners were in deep financial difficulties.[20] Just how much of a bargain Schwab had gotten would become apparent after war broke out in Europe in 1914.

*

World War I was an object lesson in the unexpected interplay between military orthodoxy and rapid advances in military warfare. In the years before the outbreak of hostilities, the generals of Europe had devised schemes to deal surgical "knockout" blows against each other. The von Schlieffen Plan of Germany was founded on an almost mathematical obsession with troop movements across Belgium and France. The French, meanwhile, had their own game plan. This was "Plan XVII"—a drive to the Rhine, a cutting off and isolation of the German right wing, and a triumphal march to Berlin.[21]

In the first weeks of war the belligerents followed their separate scripts, the Schlieffen "cycling" offensive pedaling troops Paris-bound and the Allies clinging to their own timetable. The Germans were on the verge of carrying the day when the unanticipated reared up to thwart the scheduled. After two weeks of retreat, French and British troops took a stand at the Marne River outside of Paris. They fought ferociously. Three days of battle concluded in a German retreat.

"The Miracle of the Marne" of September 1914 would have extraordinary consequences. Deprived of quick victory, the Germans began to dig trenches to stabilize their positions. The French and British, unable to overrun these positions, also dug in. Soon two parallel trenches ran across Belgium and France from Switzerland to the English Channel. In seeking to outflank one other, the belligerents had become ensnared in a defensive war. "Dead battles, like dead generals, hold the military mind in their dead grip," Barbara Tuchman noted grimly. The supreme commands of Europe found they had prepared themselves for the last war.[22]

It was at the onset of the phenomenon of trench warfare that Charles Schwab was summoned to London by the British War Office. A secret cable from the War Office was received at Schwab's residence in New York City on October 20, 1914. Late that day he booked passage on the S.S. *Olympic* under an assumed name, Alexander McDonald, accompanied by his munitions salesman and vice president, Archibald Johnston. In spite of the camouflage, a ship reporter for *The New York Times* spotted him boarding the vessel. Schwab told the reporter that he was ill and taking the sea cure. Two days later, on October 22, an item to that effect appeared in the *Times:*

> Because he was ill and wanted to be left alone, Charles M. Schwab, President of the Bethlehem Steel Corporation, sailed yesterday for Glasgow on the White Star liner Olympic under the name of Alexander McDonald. He said he was going over for the sea voyage and returning on the next ship.[23]

Once in London, Schwab and Johnston entered into negotiations with Lord Horatio Kitchener, secretary of state for war. With the battlefront already outstripping the capacity of Britain's own gun plants to supply arms, Kitchener proposed that Bethlehem enter His Majesty's service as a munitions supplier. The secretary of war was fully prepared to enter into a long-term contract with Schwab so long as the steel mogul pledged not to sell arms to Britain's enemies, a reference to the company's shipment of several big guns to the kaiser before the outbreak of hostilities.[24]

Schwab gave his word of honor and Kitchener gave his order, a smorgasbord of guns, shells, and parts that totaled a remarkable $40 million. Kitchener's order of 200 high-powered guns, 600 gun caissons, 800 carriages and 1 million rounds of shells meant that the mill would be running for months at full gallop.[25]

The ordnance contract, though, did not end Schwab's mission to London. On November 2 and 3 he conferred at the Admiralty with Sir John Fisher, first sea lord, and Winston Churchill, first lord of the Admiralty. Owing to Sir John's dislike of idle chitchat, the question he and Churchill had been mulling over was popped immediately to Schwab. The shortest period a submarine had ever been built in was 14 months. Could he make delivery in six months?

Better than that, Schwab answered gamely. In a matter of minutes the threesome had reached an agreement in principle for Bethlehem's construction of 20 torpedo submarines at a stiff $500,000 apiece, or twice the cost if the vessels were built in peacetime. Given use of the Admiralty's special code (a privilege he would be granted throughout the war), Schwab cabled South Bethlehem to confirm the feasibility of the arrangement. Company vice president Henry S. Snyder was given the responsibility to arrange a meeting with officials of the Electric Boat Company of New York about subcontracting the engines and other motor parts. To simplify engineering it was agreed that the British submarines would be modeled on the H-class design which the Fore River shipyard had produced successfully for the U.S. Navy. The idea was that the vessels would be delivered to the Admiralty at Boston or another East Coast port in packed and labeled sections and then assembled in England.[26]

Seizing on the question of the delivery dates, and on the apparent indifference of the British to the contract price, Schwab made a daring offer to Fisher and Churchill. Since time was of the essence, why not set up a system of bonuses and penalties on top of the basic $10 million price? Tossing out numbers seemingly at random, Schwab said that for every week that Bethlehem was ahead of the agreed delivery date, the Admiralty would pay a special bonus—£2,000 (about $10,000) per submarine. And if the company was late, Schwab said he would be willing to forfeit £1,000 per ship per week. Fisher and Churchill said yes, and these provisions were drawn up by F. W. Black, the Admiralty director of contracts.[27]

Sir John was convinced he had done fabulously with the American. Schwab was assuming the risks and the Admiralty was getting the needed war vessels. "We have made a wonderful coup (after you left) with someone abroad for very rapid delivery of submarines and small craft and guns and ammunitions," he boasted to Admiral John Jellicoe in a letter. "I must not put more on paper, but it's a gigantic deal done in five minutes! That's what I call war!"[28]

Equally delighted was Bethlehem's chairman and chief stockholder. In a cable to Snyder and Eugene Grace before he departed from London, Schwab waxed ecstatic: "This is an opportunity for energetic men to make good.... Will sail Wednesday. Would like to have meeting soon after returning as possible with all heads and in meantime assemble all material, make all contracts and start without one moment's delay. I am personally deeply interested and we must make good."[29]

*

When Schwab arrived in New York on November 20, reporters surrounded him at dockside and asked him if the rumor circulating on Wall Street that he had been on an arms mission to England was true. Absolutely not, he answered. "I went to London, where the financing of the fortifications I am constructing in Chile is being done. I am well satisfied with my trip abroad, but have brought back no new contracts."[30]

Within days, however, his explanation was discredited. The *Times* carried a story reporting the substance of his $40 million munitions contract with Lord Kitchener. Even worse, there were published reports that a "belligerent government" had ordered 20 submarines from a U.S. shipyard. While the government wasn't identified, the article identified the contract recipient as Bethlehem's Fore River yard.[31]

With his cover blown (by a German spy, according to one report, although leaks by commercial rivals appear equally likely), Schwab had to come to grips with reconciling his "opportunity to make good" with the government's policy of neutrality. President Wilson's declaration of neutrality called on all Americans to put a curb on "every transaction that might be construed as a preference of one party to the struggle before another." On October 15, 1914, the administration had issued a public circular on munitions traffic. The circular noted that in the absence of U.S. or international laws to the contrary, a private citizen was entitled to sell any article of commerce so long as it was a bona fide transaction. "It makes no difference whether the articles sold are exclusively for war purposes, such as firearms, explosives, etc., or are foodstuffs, clothing, horses, etc., for the use of the army or navy of the belligerent." The argument was hinged on a double negative—that private trade in firearms, explosives and other goods used exclusively for war, being private, did not constitute a non-neutral act by the U.S. government.[32]

There were many critics of the policy, including many of Wilson's fellow Democrats in Congress, but the president did not budge. As a committed free trader he believed that the marketplace should determine the course of commerce. An expression of the president's thinking was contained in a December 8, 1914, letter to Jacob H. Schiff, a prominent German-American businessman. Schiff had written to the president asking him to urge U.S. companies publicly not to add to the bloodshed in Europe by supplying any of the belligerents with arms. In his reply Wilson admitted that the question of munitions traffic was "one of the most perplexing things I have had to decide," but concluded after much soul-searching that private commerce was outside the purview of government. It was a matter, he said, that must "settle itself."[33]

In the case of Schwab's submarine order, though, the agreement unmistakably violated the same canon of U.S. law that Wilson had cited in permitting unrestricted munitions traffic. A key premise of maritime law was the absolute freedom of U.S. ports from "entanglements" with foreign powers. George Washington had first laid down restrictions on the arming of foreign vessels within U.S. waters in 1793 after Citizen Edmond Genet had attempted to organize French attacks on British shipping from American shores. Washington's rules were codified into federal law in 1818. Under Section 5, Article 283 of the revised statutes, it was a violation of U.S. neutrality laws for a private citizen to fit out or arm a ship in U.S. waters when the nation was at peace. Violations were punishable by $5,000 in fines and not more than three years in prison.[34]

Over the years the statute received a powerful assist from the *Alabama* case. During the Civil War the Confederacy had managed to purchase a ship in Britain that was later named the *Alabama*. Receiving arms in the Portuguese Azores, the ship went on to attack Union shipping during the war. Having protested futilely against the *Alabama*'s departure from England, the American government held the British government accountable for the ship's depredations. The dispute caused deep strain between the two nations until Britain finally agreed to the American demand for arbitration. In 1872 an international tribunal in Geneva took the American position that a neutral government was obliged to exercise "due diligence" in preventing the departure of vessels of war for a belligerent. The British were assessed $15.5 million in gold as indemnity.[35]

Thus, Wilson wrote in his letter to Schiff:

> In a single recent case I saw my way clear to act. When it came to the manufacture of constituent parts of submarines and their shipment abroad, complete, to be put together elsewhere, it seemed to me clearly my privilege, acting in the spirit of the *Alabama* case, to say that the government could not allow that, and the Fore River Ship Building Company which is said to have undertaken the contracts has cancelled them.[36]

Wilson's letter accurately summed up the administration's position in the submarine matter. On November 30, 1914, Wilson had written to State Department counselor Robert Lansing that Schwab's submarine contract was illegal under the *Alabama* case. On December 2 this message was conveyed directly to Schwab at a meeting in Washington with William Jennings Bryan, secretary of state.

Bryan told Schwab that he and the president regarded the submarine sale as a breach of the spirit of neutrality, and asked him to cancel the order voluntarily. Schwab asked for time to think the matter over. Two days later he placed a telephone call to Bryan. He said he would submit to the president's view and "would not build submarines for any belligerent country for delivery during the war." Delighted, Bryan issued a press release describing Schwab's agreement. The release ended: "This closes the submarine incident."[37]

The New York Times echoed Bryan's self-congratulatory remarks. Under the headline SCHWAB HEEDS PRESIDENT the paper noted, "The importance of the [Bryan] statement lies in the fact that it means the formation of a definite policy, the taking of a definite stand, by President Wilson, that the construction for belligerents of submarines or other war vessels in this country would violate the spirit of strict neutrality."[38]

*

The submarine incident, in fact, was far from closed. Between the time of his initial meeting with Bryan on December 2 and his telephone call on December 4, Schwab had been exceedingly busy. He and his entourage had taken a train from Washington to Montreal, where they were given a tour of the shipyard of Canadian Vickers, a subsidiary of the British munitions giant Vickers, Ltd. Then on December 5 Schwab had slipped aboard the *Lusitania* in New York. This time he was registered under the name "James E. Ward."[39]

Eight days later, on Sunday, December 13, Schwab reached London. In a top situation room he met with Churchill and Fisher. They had been informed of the administration's opposition to the submarine contract, and Churchill was said to be furious. He lashed out at Schwab, calling him an American bluffer who had fooled His Majesty's top command into believing he could deliver the goods. After Churchill paused to draw a breath, Schwab answered, "We'll deliver your submarines on time, though we will need a certain shipyard in Montreal."[40]

Then he outlined his plan. The submarines would be completed at the Vickers yard in Montreal, using the parts and materials made at Bethlehem plants in America. Relieved and convinced of the soundness of the scheme, Churchill and Fisher drew up a supplemental agreement dated December 15, 1914. It specified that 10 of the 20 vessels would be fabricated in Montreal and the rest, due to space limitations, would be completed when feasible at Bethlehem's Fore River yard.[41]

The contract included some fascinating provisions. The Admiralty agreed to lease the Montreal facility from Vickers in the name of the British government and then provide the facilities free of charge to Bethlehem; further, the British Government would pay a portion of the salaries of the Vickers officers displaced from the shipyard. All this suited Schwab fine. But then he said he couldn't go ahead with the contract. When a startled Churchill and Fisher asked why, Schwab said it was because he would lose money. He said the cost of doing work in Canada would be higher than if the job was done wholly in the U.S. So under his insistence that he could not proceed otherwise, the supplemental agreement raised the base price of the Canadian submarines from $500,000 to $600,000, or 20 percent. In addition, the delivery dates were pushed back by an average of four months with the first vessels due on August 31, 1915, rather than on April 24, 1915. Given the scale of bonuses and penalties provided by the amended contract, Schwab was in an even better position to reap a substantial lump-sum bonus for delivering the vessels ahead of time.[42]

Having reached an agreement that held substantially more profit, Schwab hurried back to America on the *Lusitania*. He arrived in New York two days before Christmas. Reporters again clustered around him on the gangplank. Why had he gone to England? they asked. He answered with a prepared statement that he and Johnston had written at sea:

> I was in London only two days, and that for the purpose of relinquishing several governmental contracts for the building of submarines which I had obtained on a previous visit to Europe.
>
> I was in doubt as to the status under the neutrality declaration of any American shipbuilding company in executing orders for submarines or other war vessels for belligerents.
>
> Accordingly I went to Washington to consult with the authorities after obtaining these contracts. After several consultations with Secretary Bryan . . . I was shown a law passed after the settlement of the *Alabama* claim to the effect that any ships in whole or in sections designed for belligerent countries could not be shipped from neutral countries without incurring a breach of neutrality.

> Secretary Bryan assured me that it was against the spirit and letter of the law for me to endeavor to execute any such contract, and accordingly there was nothing left for me to do but go back to Europe and cancel the contract. That was the principal part of my business abroad.[43]

In these few words Schwab put into play a bold deception that it was he, not the Wilson administration, who had had doubts as to the legality of the submarine contract under American neutrality; that it was he, not the administration, who had initiated the interviews with Secretary Bryan that led to the government's pronouncement that any vessels designed for belligerent countries could not be shipped from American shores; and, most remarkably, that it was for the purpose of canceling the contracts that he had returned to Europe.

As Schwab was portraying himself as an honest tradesman on good terms with his country's laws and his own conscience, his agents were in Canada making preparations for construction of the vessels at the Vickers yard. On December 21 the executive committee at Vickers huddled in secret session and pledged to exercise utmost discretion in connection with the Bethlehem submarines. Within days skilled workmen from the States began trickling into the shipyard. They were followed by two top Bethlehem officials. Eventually a force of over 500 men and supervisors was assembled—housed in private barracks, fed at a special canteen, and sworn to secrecy.[44]

In order to disguise the flow of materials to Canada (which was a belligerent in the war and an ally of Britain), American newspapers were told that the company was at work on "structural steel for bridges to replace those destroyed in Europe" and that shipment of the steel was being made by way of Canada. While the press accepted the company's statements at face value, the German government was not so easily deceived. Something was going on, that was clear, and the German embassy in Washington endeavored to discover the details. Acting on intelligence reports, Johann von Bernstorff, German ambassador to the U.S., confronted Secretary Bryan with evidence of Schwab's subterfuge. In a private communication to Bryan on January 27, 1915, Bernstorff wrote that he had obtained knowledge that "the Bethlehem Steel Works are secretly sending to Canada parts of submarines." Bernstorff complained that the arrangement was an unlawful evasion of American neutrality, and said that the bonuses for early delivery and other contractual arrangements demonstrated that the vessels were for use in the present war. He demanded an investigation.[45]

No evidence has come to light to indicate whether Bryan or other administration officials might have been aware of the Bethlehem ruse before the Bernstorff letter. However, State Department correspondence released in 1928 and other data compiled by historian Gaddis Smith point to the conclusion that, after being informed of the situation by the German ambassador, the Wilson administration lost its will to battle Schwab further. Having achieved a domestic public-relations victory in the name of strict neutrality, the administration was unwilling to let the Germans snatch it away.

According to Smith's findings, Secretary of State Bryan promised Ambassador Bernstorff that his complaint would receive a thorough investigation. Then he sent a letter to Schwab asking for an explanation. Replying that he would be delighted to come to Washington to talk over the matter, Schwab said he was unfortunately suffering from a cold and could not leave his New York residence. In his place was sent his Wall Street corporate counselor Paul D. Cravath. Cravath conferred with State Department lawyer Lansing on Saturday, February 6, 1915.

While no written record of this conversation survives, Smith makes a convincing case that the lawyers reached a tacit understanding that day. The key point of agreement was a redefinition of Bethlehem's materials sent to the Canadian yard. If the materials shipped over the border required further fabrication before the submarines could be launched—as Cravath apparently argued—then Lansing agreed that the U.S. government would consider the materials commercial articles permitted under the public circular of October 1914, and not prohibited "component parts" of a vessel of war. In the wake of the Cravath-Lansing interview the above distinction became the legal centerpiece of the administration's new position on the subs—"followed," notes Smith, "as faithfully as if it had been written into law."[46]

To put this disinformation into play, Schwab secured a letter from P. L. Miller, general manager of Canadian Vickers, which stated that it was *his* company that was building submarines and using materials purchased on the open market, including "comparatively insignificant" purchases from subsidiary companies of Bethlehem. Counselor Lansing then prepared a letter to Ambassador Bernstorff in response to the German protest. The president never received a copy of the letter, and Bryan, whose distaste for detail work was legendary, signed it apparently with little more than a cursory glance. The letter assured Bernstorff that "no component parts are being built by the Bethlehem Steel Works and being sent to Canada" and that the ten subs being built at Fore River were "not for delivery to any belligerent nation."[47]

Bernstorff refused to accept this explanation. He urged Bryan to look into the matter again, but his protests were to no avail. The Wilson administration stood firm. Within a few months the issue became moot. The first six Bethlehem submarines were launched from Montreal in July 1915 and the entire fleet was completed by September—ahead of schedule and netting Schwab about $4 million in profits and bonuses.[48]

*

While Bethlehem soared, Pennsylvania Steel floundered. Before the war the firms had been rough equals, the Pennsylvania combine larger in terms of overall steel capacity, and Bethlehem having the edge in shipbuilding contracts. In terms of earnings, too, the companies showed similar profiles. In 1912 Bethlehem's operating earnings were $4,846,814, or 6.26 percent on preferred stock and 3.43 percent on common, and Pennsylvania-Maryland's earnings were $3,113,819, or 7.80 percent on preferred and 1.53 percent on common.[49]

How fortunes changed: Bethlehem's net earnings rocketed 184 percent from 1913 to 1915, from $8,752,671 to $24,821,408, while Pennsylvania barely broke even, earning only $4.2 million over the entire 1913–1915 period. Underlying the woes of Pennsylvania Steel was a single fact—the precipitous drop in overseas rail sales. With the International Rail Association shattered and most colonial railroad projects stopped dead in their tracks, rail production at Sparrows Point plummeted to 147,048 tons in 1914, an 82-percent drop from the previous year and the lowest level since the recession of 1904. The last major order—16,000 tons for the Victorian Railways of Australia—was booked in June 1914 and completed on the very day that war was declared. In the wake of this overseas slump, combined with lackluster demand by domestic railroads, the company registered losses of over $600,000 in rail production, which was only marginally offset by an upswing in war-related merchant shipbuilding contracts.[50]

For the man who had spent a lifetime constructing Sparrows Point, the shrinkage of the rail market was a devastating blow. Swiveling in the chair of his office above the company store, Frederick Wood could see his handiwork—efficient docks at portside, modern skip cars climbing up blast furnaces with 2,500 tons of daily capacity—but on too many days he saw not a wisp of smoke on the far finishing end of his complex. Sticking to one specialty, Wood and his associates had steered the company through the competitive storms of the late '90s. They had recapitalized the concern in the conglomeration era of 1901, and had seemed to achieve safe harbor with the tapping of the Mayari deposit in 1909. But the world was a fickle place that could confound the logic of the engineer.

While the steel managers did not lack for practical expertise (nor were they innocents abroad), the fury unleashed on August 1, 1914, overwhelmed them. "Small orders that have come in during the past twenty-four hours will keep the works in operation a few days longer," Frederick wrote ruefully to his wife, Caroline, upon the onset of the crisis. "Beyond is blank and we have to muster up faith that a greater demand will spring up before long."[51] *Beyond is blank.... We have to muster up faith*—these were words of a businessman entangled in forces beyond his control.

So it was left to Wood to batten down what he had built. The rail mill was put on part time. It would run for a week or ten days, then was shut for as long as three weeks as new orders were slowly, painfully, accumulated. The other departments followed suit by necessity. Because blast furnaces could not be started and stopped at will, Wood ordered two of the four furnaces shut or "banked" semipermanently. Nor did it end there. What was happening at the mill knocked the props from the model village on the east side of the peninsula. At a place where independence and self-sufficiency were explicitly traded for a job, lack of work was the cruelest cut of all. The same men who had been overworked for years now were tossed aside.

As the mill hit rock bottom in late 1914, 2,000 of the plant's 4,000 workmen found themselves without work and facing the long winter months ahead. In the crisis even senior employees were handed the pink slip. Griffith M. Liebig, for example, had begun working in 1888 at the plant in the voucher department. He and his wife Emma had lived for 27 years in the company town, where they had raised their two sons. Suddenly, in November 1914, at age 53 and apparently ill, Liebig was terminated. In a rare letter that survives from this era, a distraught Mrs. Liebig poured out her plight to Frederick Wood:

Dear Mr. Wood.—

I don't know whether it is worth while for me to come to you in this my hour of great trouble & some distress. Nothing has pained me & broken my heart more then [sic] to think of the past 27 yrs. of hard & strenuous night and day work Mr. Liebig has toiled day after day using his eyes and his brain, now they have dismissed him & said his services were no longer needed, making no recompense, & I must bear all this trouble, we have not been able to tell him on account of his weakened condition.... True, Mr. Wood I know you have been kind to Mr. Liebig & I appreciate it very much, but oh my how hard it is to take every thing at once & he is so sick I can hardly realize it. Him working here all these years, they don't seem to consider, I am hardly able to work & will still do more to shoulder this burden if only some thing could be done for poor Mr. Liebig, I feel so badly because he is sick & has practically lost his eyesight working here night & day hoping if he improves something may turn up for him, for the boys to shoulder this burden with small wages is more than I can bear. Hoping this may bring brightness to this said [sic] heart.

Very Sincerely—
Mrs. Liebig.[52]

Wood's reply, if any, to Mrs. Liebig does not appear in his files, and the family was not heard from again.

At the start of 1915 Wood and his associates entered a period of concentrated activity to drum up sales. In February 1915 they completed experimental work on a vanadium steel rail which showed up to 40 percent more elastic strength than regular carbon steel rails. The rail yielded a much-needed order from the Lackawanna Railroad. Wood also continued tests on heat-treating Mayari ore for higher-strength rails. (Since 1911 "Mayari rails," with a natural alloy of chrome and nickel, had been sold for $5 a ton above standard carbon rails.) In July 1915, Wood and Edgar Felton oversaw another industry "first": the first commercial rolling of super-heavy 125-pound rails for the Pennsylvania Railroad.[53] The executives, in short, responded to the downturn with several strategies. They reduced the size of the manufacturing unit, banking several furnaces; they kept close tabs on labor, slashing whenever possible; and they strove to improve the quality and range of their product. Given the firm's long experience and established reputation, the strategy might have succeeded in normal times.

But 1915 was decidedly not a normal year. The New York Stock Exchange was the scene of the greatest boom ever experienced—the boom in "war bride" stocks. The deluge of munitions orders released to American companies from Britain and France was causing a wave of buying in munitions-related securities. Chief among these was the company that brought home the first big munitions contract. Before the war Bethlehem common stock never sold above 46⅝ on the exchange, and even at its normal price of $30 it had not been regarded with great favor. In 1915, propelled by the Kitchener contract, it rocketed up and up, finally to over $600 a share.[54]

Flush with cash and prestige, Schwab approached two fellow financiers on October 5, 1915. They were William Donner and Edward Stotesbury. Donner was a Midwesterner closely identified with the Mellon brothers (Richard B. and Andrew W.) of Pittsburgh and the National Tinplate combine. In 1912 he had gained entry into the upper ranks of Philadelphia through the purchase of a large block of stock in Cambria Steel at Johnstown. Stotesbury was the grandee of Philadelphia finance—head of Drexel & Company and partner of J. P. Morgan & Company, whose possessions included a 172-room mansion on the Main Line and the Wingwood House at Bar Harbor. (Later he would erect the fabulous El Mirasol villa in Palm Beach.) Donner and Stotesbury had been designated as financial agents of the Pennsylvania and Reading Railroads' stock holdings in the company. What Schwab proposed to Donner and Stotesbury was a mutually beneficial buyout.[55]

Thus began a series of negotiations in which Schwab put together a package that met the demands of the sellers. A tentative agreement among the parties was reported in *The New York Times* on October 16. But then Stotesbury rejected Schwab's offer, and a bid was tendered by Donner himself. He purchased 65,418 shares of preferred stock at $90 a share ($5 a share above Schwab's proposal) and 52,780 shares of common at $30 a share ($3 above Schwab). That his move was orchestrated to compel Schwab to raise his bid was underscored by the fact that Donner's stock purchases were not made in cash or securities, but rather in personal notes guaranteed by the Pennsylvania Railroad.[56]

The stock gambit worked because of the worth of the properties that the men were shopping. After a banking syndicate including Percy Rockefeller made a financial pass at the company, Schwab resubmitted his bid. He raised his tender for the preferred stock from $85 to $100 a share, while his bid for the common shares stayed at $27. Donner accepted the offer and reaped $495,840 in stock profits.[57]

The buyout was characterized by an absence of input from the company's two senior officers, Edgar Felton and Frederick Wood. They were shunted aside by Donner and Stotesbury, who dominated the board of directors' executive committee. When the deal was consummated on February 16, 1916, Schwab acquired all the assets of Pennsylvania Steel and its subsidiaries for $31,941,630, plus the assumption of $17.3 million in liens. No cash actually changed hands; instead 20-year Bethlehem gold notes bearing 5 percent annual interest were issued to cover the transaction.[58]

With no provisions having been made for him, Wood crash-landed as senior vice president. He had paid the ultimate price of not owning the company which he had influenced so profoundly. When he was asked by Schwab to stay on as general manager of Sparrows Point, Wood agreed, in part because he needed the money.

*

With the purchase of Pennsylvania Steel, Schwab's company was propelled into the front ranks of "independent" steelmakers. In one stroke, ingot capacity was more than doubled to 2,200,000 annual ingot tons, which moved Bethlehem above Lackawanna Steel, Jones & Laughlin, and Midvale-Cambria. It was still considerably below the size of the Steel Trust in terms of assets and steelmaking capacity, but the merger greatly strengthened Bethlehem's claim over Middle Atlantic markets. What's more, the merger gave the company access to the Mayari ores whose metallic properties were of great worth to arms manufacturers. Noted *The New York Times*, "Mayari ores are famous throughout the industry, the steel made from them being of great value in the manufacture of armor plate and ordnance. This steel is a natural chrome-nickel alloy, and when forged it has from 8,000 to 10,000 pounds per square inch more tensile strength and elastic limit than the ordinary carbon steel."[59]

Schwab's oft-stated ambition to match the empire of Elbert Gary did not seem so absurdly improbable any more. "A few short years ago," said *Moody's* reporter Arundel Cotter, "if it had been stated that

Bethlehem Steel was buying Pennsylvania it would have been regarded as a case of the tail wagging the dog, for Pennsylvania Steel has long been one of the most important steel producing organizations in the country.... But within the past eighteen months the financial world has come to regard Bethlehem with something even more than respect."[60]

In an unsettled time the steelman's aggressiveness and guile had worked wonders. He had found a creative solution to the president's strict neutrality, and in so doing gave lie to the conventional wisdom that arms trafficking and reputable commerce do not mix. In sanitized versions of his life and business practices, Schwab was lauded for personifying the hustle and bustle that was quintessentially American. His voice had a baritone heartiness that revealed his "manliness" and a smile so "hypnotic" that it was said to gain the confidence of a stranger immediately. But what put him on the map, of course, was his fantastic success. "Bethlehem has been the *magnum opus* of Schwab's life," wrote Cotter. "It has been, in so far as a corporation can be, an expression of the man himself. He nursed it through its infancy, guided it during its adolescence, and now it is, for its size, the richest steel company in the world."[61]

Now he cast his eye at the plant on the Chesapeake as the jewel of his crown. "For over 20 years I have had an ambition to own the steel plant at Sparrows Point," he confided to a reporter.[62] That ambition having been realized, the peninsula now lay before him like a great ship at anchor, brightly lit and latent with activity and power.

PART II

WAR AND PROSPERITY 1916-1931

CHAPTER 6

THE MUNITIONS MACHINE

"Ladies and gentlemen, I am here for a very brief and simple duty, a very delightful duty, that of welcoming Mr. Schwab to Baltimore."

Before James H. Preston, mayor of Baltimore, a throng of Eastern capitalists and politicians had come to the banquet hall of the Belvedere Hotel for a "Dinner of Welcome Tendered to Mr. Schwab by the City of Baltimore." The date was November 21, 1916, ten months after Bethlehem's takeover of Sparrows Point.

As the tinkle and clatter of dinnerware died down, the mayor began his oratory, welcoming Schwab as the "greatest individual employer of labor in the United States." In an effusive introduction, he praised the businessman "for your warm-hearted breadth of spirit, for your personality, for the higher traits over and above those things you are doing for Baltimore, its harbor and vicinity.... You are making a contribution not only to Maryland and Baltimore," Preston said, "but also to the greatness of our country, to American manhood, to American citizenship, to the broad humanitarianism of mankind, in which you exemplify in yourself the highest ideals. I present to you on behalf of the citizens of our city, this gold box."

The gold box was a cigar humidor. At the upper edge of the box was carved a skyline of Sparrows Point, on the lower edge a skyline of Baltimore City. Superimposed between the carvings were two medallions. One carried the coat of arms of Baltimore. The other had the steel mogul's personal monogram.[1]

Schwab, moved by the mayor's and the audience's reception, tossed aside the speech that had been written for him by Ivy L. Lee, the publicist.

"I am going to say what is in my heart tonight," he told the crowd.

"When I first entered the steel business 35 years ago the whole country consumed scarcely 1 million tons of steel. When the Steel Corporation was formed in 1901 the whole country only consumed 12 million. We then felt that we had reached the zenith of production. You will not be surprised, I am sure, when I tell you that this year the United States will produce about 40 million tons of steel. Now I don't think that is the zenith, or anywhere near the zenith. I think we are going onward and upward, and that this country is destined to be the greatest manufacturer in the world of iron and steel, as great as all the other countries put together in the years to come."

Schwab raised his hands to hush the audience. "People will say, 'Where is it going? What is going to be done with it?'"

The answer was consumption, he said. Consumption would solve the cycles of production as the United States and the world fed on "commercial steels"—plate steel for the shipbuilding and construction industries, tin plate for canners and meat processors, bar and sheet steel for the auto industry, rod steel for wire, springs, and nails of all descriptions, and, yes, steel for rails and railroads too.

"The production today has been much augmented by reason of the war," Schwab conceded. "People say, 'What is going to happen after the war is over?'"

The answer was simple: buildings would grow higher, ships would become heavier, automobiles more numerous and urban areas more sprawling—"the trend of the future must show ultimately greater and greater production of iron and steel."

But to capture the industrial and consumer markets of the future, Schwab cautioned, Bethlehem had to diversify from what he delicately referred to as "special lines of highly technical articles [i.e., guns and ordnance] and things of that kind." The war had shifted the center of gravity to the Atlantic Coast and it was one of Schwab's insights to secure his company's future by building up the East as the company's exclusive preserve. "I have expressed the opinion that the development of the iron and steel industry during the next decade would be in the East. Here is where I have placed my money and my fortune, and here is where I intend during the remainder of my days to develop."

This was where Sparrows Point fit in. At Bethlehem, Pennsylvania, ore had to be shipped from port to plant by rail, at a penalty of $1 per ton of pig iron. At Sparrows Point, of course, the freight charge was eliminated. Overall, the tidewater location of Sparrows Point could save the company $2 in manufacturing costs per ton of finished steel.[2] And as the years progressed Schwab estimated that Sparrows Point would increasingly tip the balance in favor of Bethlehem over U.S. Steel and other Midwest works.

"I am going to tell you why," he said. "The Middle West depends entirely for its supply of material upon the great ore deposits of the Northwest. They are large, it is true; but how many of you realize that they are now taking ores out of the Northwest at the rate of 53 million tons a year? It is not going to be many years before ores are going to be very scarce and valuable. In my time, in Pittsburgh, we would not use ore that did not have at least 58 or 59 percent of iron. Now we are glad in Pittsburgh to use ores that will run as low as 49 percent in iron. Now, the East is so located that by reason of our great water transportation facilities the great ore deposits of South America, Cuba, and other countries, of which there is a vast extent available, can be brought here. For 50, 100, nay, 200 years to come the supply of steel will be practically undiminished."

For these reasons, he said, Sparrows Point had been targeted as the centerpiece of the Bethlehem chain. "Now, boys, I am going to tell you something about what we contemplate here in Baltimore—*We contemplate here the largest steel plant in the United States!* The final estimate of what we will probably be obliged to expend on our plant here in Baltimore is about $50 million. Now, it is safe to say this is the largest undertaking ever contemplated by an industrial concern to my knowledge."

Speaking intensely, Schwab continued, "I glory and rejoice in the opportunity to expand. It is not a mere question of the amount of money we shall make, although that is a measure of success, but it is the satisfaction of conceiving and carrying to successful issue great enterprises. I am sure not many people believe when other people say, 'Oh, I don't work for money.' But not many people are apt to say that, 'I don't work for money.' I can assure you that, as far as our enterprises are concerned, the chief and propellant motives that urge us on and on is the sense of the satisfaction of accomplishment, and the close association with the boys who surround you, who do things, and who let the old man, who gets the credit for it, just look after the high spots, as Mr. Snyder so well expresses it.

"My greatest joy and delight during these last 15 years have been to see so many able men, some of whom sit here with us this evening, advanced in our enterprise to positions of dignity and importance, to positions of respect and confidence, to positions that are the envy of the people in the community in which they live. I love the boys that are associated with me; I love the people who do business with me; and I love above all to rejoice in their prosperity and success. And I express the hope that Baltimore, for her enterprises here, may furnish some of these *boys*, as I call them, some of these men who will add to the lustre and dignity and success of your local enterprises."

The speech had its intended effect. The locally prominent were riveted. Columns of ink were spilled over the steel promoter and his dream. A MARVELOUS MIDAS was the editorial headline in Felix Agnus's *American*, which commented, "Baltimore is realizing a dream of a lifetime, the industrial development which belongs to the city by right of location, by natural advantages and shipping facilities.... His words were eagerly grasped by men who make the wheels of a great city 'go around,' who have long looked forward to the day when Baltimore would hum with industries, when the blast furnace would ever be aglow, when the factory bustle would be music to their ears."[3]

The *Baltimore Sun* seconded the acclaim. "From a material standpoint the coming of a man like Schwab, with his dazzling plans of development for Sparrows Point, is of inestimable value to this city. It generates new energy and initiative in our local enterprises; is bound to attract other business and speed up the wheels of local progress." An editorial cartoon limned the paper's whoop and holler. It showed the chairman of the board astride a fireplace, carrying a bag of "good things" while smiling his winning smile. "Baltimore's Santa Claus Came Early This Year," said the caption.[4]

Yet where had Schwab's bag of "good things" come from? After floating $31.9 million in bonds to acquire Pennsylvania-Maryland Steel several months earlier, Bethlehem's chairman was signing requisition orders for rolling machines and heavy equipment that almost defied comprehension. The capital budget of Baltimore, the nation's eighth-largest city, was under $12 million in 1916. Where had the industrialist obtained double that amount ($25 million was slated for the first phase of mill building) to embark on a vast recrafting of Sparrows Point?

In the heady boosterism that accompanied Schwab's talk, with the attendant sense that the city was getting something wonderful for free, the source of his company's profits was not an "angle" that the press thought newsworthy. If Americans worshipped the making of money, most preferred not to know how it was really done. Nor was it something that Schwab volunteered as he departed from Baltimore the next morning, telling his local Boswells, "I had the time of my life!" as his train was about to pull out of Union Station.[5]

To answer this question one had to step back into the shadowy world of arms contracting and production, for the expansion of Sparrows Point was inextricably bound to the expansion of the European conflict in which, according to official U.S. policy, America still maintained neutrality "in thought as well as in action."[6]

*

> *When Mr. Schwab came to Bethlehem the town had 13,000 inhabitants. To-day the employees of Bethlehem Steel alone are sufficient to make a city of 150,000 people. Ten years ago the annual sales of Bethlehem were $10 million. Last year they were $230 million.*
>
> —*IRON AGE*, DEC. 28, 1916

During 1916 Schwab's company grew to become the largest munitions manufacturer in the world. With millions of men engaged in combat and battle lines drawn across hundreds of miles, the demand for arms swelled. Inundated with orders from Britain, France, and Russia, Schwab had mounted a corporation-wide push to increase munitions production. To this end a powder plant had been constructed at New Castle, Delaware, close at hand to explosives maker E. I. Du Pont de Nemours; a fuse and timer mill established at Redington, Pennsylvania; machine shops acquired at Reading and Lebanon, Pennsylvania; and 25,000 acres of land at May's Landing, New Jersey, for "the greatest proving ground in the world." Sparrows Point became part of the arms manufacturing matrix, its open-hearth furnaces supplying billets of "Mayari" nickel-chrome steel for shipment to the shellmaking division.[7]

Operations central was South Bethlehem, Pennsylvania. There on 600 acres of property along the Lehigh River shells were being forged, guns bored, and armor plate manufactured day and night by an army of 25,000 workmen. Shells were the prized object of the Allies—the high-tech instrument of combat—and from a prewar level of 3,000 shells a day Bethlehem had enlarged its capabilities to produce 25,000 shells a day, or 750,000 shells per month, in late 1916. Not even the famed Essen Works of Krupp could match the output of bombs shipped daily from the little town of Bethlehem. Not bad for a town that had been named by its Moravian founders after the birthplace of the Man of Peace.[8]

In its specialty of armor-piercing projectiles and explosive shells, Bethlehem was said to lead the world. Both types of ammunition were based on the use of shrapnel. A shrapnel projectile was a hollowed-out bomb which contained several hundred bullets as well as an explosive charge. There were two basic types. One was designed to explode on impact with the ground; the other to burst in the air before reaching the target and thus release its shower of bullets over a wider zone of combat. "Shrapnel is the greatest man killing power of all ammunition used in field artillery, and the Bethlehem product is said to be the most efficient," noted a contemporary report.[9]

While details of production were protected behind barbed-wire fencing and armed guards, a picture of the company's military wares could be gleaned from scattered data in technical journals as well as in the company's own privately published catalogs. The ammunition shells delivered to the Allies were 85-percent steel by weight, and they varied from 1½ to as much as 805 pounds. High precision metalworking was the key ingredient. By one count there were 38 machining steps in the making of a shrapnel shell.[10]

The finished value of a World War I shell ranged from $10 to $415 for the 14-inch Bethlehem armor-piercing navy shell. The 2.95-inch (75-mm) shell was a popular caliber for infantry. Its length was 9.55 inches from nose to base, it had a diameter of 3.285 inches at its widest point, and its loaded weight was 18 pounds. The shell was filled with about 260 lead bullets and an explosive charge of about 14 ounces of black powder, and propelled with a charge of nitrocellulose, smokeless powder, and primed with smokeless percussion primer.

Good practice demanded that the shrapnel be burst at a point about 150 yards short of the target fired at. The effect was to release the bullets, which then rushed on at the full remaining velocity of the shell and struck the ground within a rectangle about the size of an ordinary company formation. When the shell was burst over earthworks, the bullets formed a horizontal sheaf that "searched out" the troops sheltered behind. (In the first war using shrapnel, the Boers found English shrapnel so searching that they dug their trenches at a slope, the bottom sloping toward the enemy, to escape the barrage.) The bullets had a striking force of about 700 foot-pounds, enough to kill man or horse instantly.[11]

Multiply this shell by 25,000 and one had an approximation of the scene every morning on the banks of the Lehigh River, where an acre or so of shells stood upright in long rows, glistening with a fresh coat of paint, awaiting shipment to New York Harbor, where they were packed beneath the decks of outgoing steamers, all a day's work at the "American Krupps."

Supplementing shell production, Bethlehem's gun shops made every kind and size of field artillery imaginable. Bethlehem's line of guns started with the venerable naval landing gun, the 1-pound (37-mm) semiautomatic Mark C, shooting 1.7-pound shells and suitable for trench warfare, to the immense 16-inch, .50-caliber battleship guns that required boring machines 195 feet long to manufacture. In a 1916 catalog of wares, *Mobile Artillery Material,* distributed to the U.S. Army and Navy and war offices abroad, the company provided the dimensions and improvements of its major lines. A trench workhorse was the 3-inch semiautomatic Mark O field gun. Made of forged nickel steel, it was equipped with a modern breech block, improved sights, and variable recoil. Its principal characteristics were as follows:

Weight of projectile	15 pounds
Weight of complete round	18.9 pounds
Length of gun	87.8 inches
Rifling length	72.72 inches
Width of grooves	0.2927 inches
Recoil	Variable, 30 to 18 inches
Muzzle velocity	1,800 feet per second
Maximum range	10,800 yards
Diameter of wheels	56 inches
Elevation limits	−5 to +70 degrees
Height of axis of bore	40.875 inches
Weight in firing position	2,950 pounds[12]

The "surprise guns" of Germany and Austria spurred Schwab's organization to develop its own army siege mortars. Skoda of Austria-Hungary had brought out a 12-inch (305-mm) mortar early in the war; another, built by Krupp, was a 16.5-inch (420-mm) giant which was 24 feet long with its gun carriage, weighed 98 tons, fired a shell weighing 1,800 pounds at a range of nine miles, and required a crew of 200 attendants. Needless to say, the Krupp and Skoda guns were hard to move around, which limited their effectiveness.[13] Bethlehem's 11-inch siege gun and Mark A carriage were designed to meet the changed condition of the European war by stressing mobility. The 11,500-pound mortar was separated from the carriage, the carriage and limber forming one load and the mortar, carried on a transport wagon, the other. Again for purposes of speed, the siege mortar was equipped with a "quick-return mechanism" for returning the mortar to its loading position after the mortar and cradle were disengaged from the rocker. Sighting also was improved. Chief characteristics of this heavy-metal monster were:

Weights of projectiles	805 pounds, 604 pounds
Length of mortar	143 inches
Muzzle velocities	900 feet per second, 1,160 feet per second
Maximum ranges	7,600 yards, 11,000 yards
Elevation limits	+20 to +65 degrees
Weight in firing position	43,000 pounds[14]

The Allies were willing to pay extraordinary prices for Bethlehem's guns and shrapnel. By the end of 1915 the British War Office had more than doubled its order of $40 million negotiated earlier between Schwab and Lord Kitchener. Among the major transactions were $60,625,348.34 for shells (paid on July 24, 1915), $8,498,727.54 for more shells (paid on October 4, 1915), $1,200,000 for guns and sights (paid on July 7 and 20, 1915), and $3,800,000 for fuses (paid on November 30, 1915). On November 1, 1915, the French government paid the company $553,925 for guns, and on December 15, $70,570 for ammunition carts.[15]

Records of these remarkable purchases were subpoenaed 20 years later from the ledger books of J. P. Morgan & Company, general purchasing agent for the British and French governments in the United States, and published by the Senate Special Committee Investigating the Munitions Industry (the Nye committee). The committee conducted a wide-ranging investigation of American munitions trafficking in the period of American neutrality. It revealed an ongoing relationship between hundreds of private American companies and Allied war offices, with a corollary disregard for the president's call for neutrality as an inconvenience (or "technicality") to the pursuit of business trade and profits. The state of affairs was typified by the remarks of Thomas W. Lamont, senior partner of the Morgan company. When asked by Senator Gerald P. Nye (D., N.D.) how the financial tie-in between Morgan and the Allies squared with the question of neutrality, Lamont said in reply, "Bankers or nobody in the business world at that particular moment were thinking about the questions of neutrality very much. They were thinking about the danger and perils of our own situation. They were thinking about the heavy American maturities abroad. We were looking after our own troubles."[16]

In 1916 Bethlehem Steel's net earnings soared 149 percent over 1915 to $61,717,310. Having pulled off the feat of earning double its stock capitalization, the company declared a 200-percent stock dividend. By owning 60,000 shares of common stock, Chairman Schwab was entitled to 120,000 shares of new class-B common stock. When issued on February 17, 1917 the new stock sold on the exchange at $140. (Schwab kept his old common which had voting privileges and was quoted in the $400–$450 range.) It was further announced by the board of directors that the dividend rate would be increased to $10 a year for preferred stock, of which Schwab owned 90,000 shares. Thanks to the arms business, Schwab could have his cake and eat it too—could enjoy a cash dividend of $2,409,576 and new stock worth $16.8 million, and still have plenty left over for plant expansion.[17]

*

Granted, Bethlehem and its chairman waxed prosperous, but didn't the company's weapons save Britain and France from certain defeat at the hands of Germany? Here the record is highly debatable, given the false hopes and awful blunders of the Allies. Although Britain and France drew forth and mercilessly used their most deadly weapons of war, even the most lethal machinery did not guarantee victory. On the contrary, the buildup of arms, matched by the German command, led to escalating rounds of violence and the mass squandering of men, as well as a draining of the national treasuries of three great nations.

To begin with, there was the cautionary tale of the fate of the submarines Churchill and Fisher had so avidly sought from Schwab. Launched from Montreal in 1915, six of the vessels were assigned to patrol duties around England, where they proved of minimal utility because the blockade of German ports had deprived them of their raison d'être—there weren't any German merchant ships to attack. Compounding the problem was a technological deficiency that somehow had escaped the attention of Churchill and Fisher: submarines could not stalk their underwater counterparts. ("You cannot fight submarines with submarines. You cannot see under water. Hence you cannot fight under water," one expert noted astringently.) Therefore, they were useless against the real menace to British shipping—German U-boat attacks.[18]

None of the six vessels defending British waters managed to sink a U-boat; one of them became stranded on the Dutch coast and was impounded by the Royal Netherlands Navy. At Gallipoli, where the four other Montreal subs were assigned, shells were lobbed into Turkish positions and local sailing ships were destroyed, but regarding the main action, the assault of Australian, British, and New Zealand troops on Turkish machine-gun emplacements, the vessels were irrelevant. Churchill was dismissed as first lord of the Admiralty after the failure of the Gallipoli campaign, and the submarines began rather aimless patrols of the Mediterranean that would last out the war.[19]

Indeed, so quickly had circumstances diminished the usefulness of the vessels that the Admirality decided in 1915 to "write off" the ten other Bethlehem submarines that had been built at the Fore River shipyard under the original 20-sub pact. After paying the agreed-upon price of $6 million (plus bonuses), the Admiralty willingly let the vessels be placed under the custody of the commandant of the Boston Navy Yard. There the submarines stayed, gathering barnacles, for a year, until the British Foreign Office finally found a use for six of them: they were given to the government of Chile as a gift. Four remained. Two were offered to Canada as a gift and the final two were taken to Britain and scrapped.[20] None saw any action.*

Miscalculations of infinitely more tragic dimensions took place on the Western Front in 1916, the very year of Bethlehem's great profit surge. Desperate to crack the German trench lines, the Allied command conceived of a plan to smash German defenses around the Somme River. Over months of preparation some 1,400 artillery guns were moved into the 23-mile front, and several million rounds of ammunition shells accumulated. Bethlehem's contribution to the

*The story of the disposition of the subs is not in Churchill's well-known history of World War I. On pages 497–98 of the first volume, Churchill makes a glancing reference to the decision to order the vessels (without mention of the cat-and-mouse antics involved), then drops the subject, never to raise it again, leaving the reader to assume that the expenditure was of value. While highly insightful in many respects, Churchill's *The World Crisis* also had a reputation to protect.

Somme arsenal is hinted by the purchasing records of J. P. Morgan & Co. On May 1, 1916, Britain paid Bethlehem $11,809,054.73 through the Morgan Company for field gun shells and $4,199,672.40 for shell fuses. Based on prevailing prices, the payments would have provided about 600,000 6-inch, 90-pound shells as well as about 5 million fuses. Eight days later, on May 9, the French government paid Bethlehem $7,821,724.67 for "shell steel"—specially treated steel shipped to French foundries for artillery shells. A month later the French paid another $1,205,110 for powder.[21]

There is no way of knowing whether these items found their way to the Somme front or if they were used in other battles fought at the time. What proved critical, though, was that while the Allies prepared their offense, the Germans prepared their defense. They threw up second and third lines of trenches spaced several miles apart, reinforced the trenches with steel-beam framing, and provisioned them with guns supplied by their weapons makers, Krupps and Deutsche Waffen. Between these positions entanglements were strung, nailed to stakes and interlaced with barbed wire as thick as a man's finger.[22] Thus the battlefield of the Somme was drawn in steel: mountains of shells trained against underground fortresses of steel, the machines of Bethlehem versus the machines of Krupp. On June 24, 1916, the Allied bombardment began. For seven days and seven nights, the British and French guns roared, tearing paths through the wire entanglements and battering the trench parapets relentlessly. For seven days and seven nights the Germans hid in their steel-beamed trenches, "a belt full of hand grenades around them, gripping their rifles and listening for the bombardment to lift," a German soldier wrote after the war.[23]

At 7:30 on the morning of July 1 the British army rose from the trenches. Ordered to proceed at a steady gait, 150,000 troops began to file across no-man's-land. German troops scrambled from their dugouts and ran to the nearest shell holes. Crews dragged out their ammunition boxes and a rough firing line was formed. Within minutes the first line of British came into view. The troops walked right up to point-blank range before the Germans started firing. The line quivered, then fell in whole sections. Trapped without protection in no-man's-land, the men were mowed down mercilessly by German fire—at day's end the British army had suffered the worst losses in the history of warfare, 57,470 men killed, wounded, or captured by official count. Military censors withheld the magnitude of the losses. Lloyd

George, successor to Lord Kitchener as secretary of state for war, ordered a resumption of assaults. The mailed fist prevailed over common sense. Now, however, the objective was no longer conceived as a gallant breakthrough of enemy lines, only bulldogged attrition, of "wearing out the enemy," in official nomenclature. German General Fritz von Below responded in kind, ordering that every yard of trenches lost must be regained in counterattacks regardless of casualties.[24]

From July 1916 through the time of Charles Schwab's speech in Baltimore in November 1916 the battle raged back and forth, locked across the same parcels of land. Trenches were obliterated, barbed wire pulverized, assaults taken amidst a wilderness of shell craters. Even on days when nothing of significance happened, hundreds of casualties were produced from random shellings. Winston Churchill, a trenchant critic when evaluating military operations not of his own making, compared the Battle of the Somme to an immense machine —a war machine whose wheels grabbed, then ground up, the bones and flesh of whole armies. In *World Crisis* he wrote:

> The battlefield was completely encircled by thousands of guns of all sizes, and a wide oval space prepared in their midst. Through this awful arena all the divisions of each army, battered ceaselessly by the enveloping artillery, were made to pass in succession, as if they were the teeth of interlocking cog-wheels grinding each other. For month after month the ceaseless cannonade continued at its utmost intensity, and month after month the gallant divisions of heroic human beings were torn to pieces in this terrible rotation.[25]

Mud finally ended the stalemate. As the autumn rains fell on the battlefield, the land became a sea of mud. Churned up by thousands of vehicles, by hundreds of thousands of men, and by millions of shells, the mud bogged down man and gun. "Still," Churchill pointed out, "the remorseless wheels revolved. Still the auditorium of artillery roared. At last the legs of men could no longer move; they wallowed and floundered helplessly in the slime. Their food, their ammunition lagged behind them along the smashed and choked roadways."[26]

Five months after it started, the machine clicked off, destroyed by its own nihilistic power. The armies could give no more. British and French losses at the Somme later were put at 614,000 men, the vast bulk British, and German casualties at 440,000. Over the same period (June 24 to December 1, 1916) Bethlehem Steel sold arms totaling $23,267,135.18 to the British and French governments, according to the records of J. P. Morgan & Company.[27]

*

Money has a way of creating its own context. First with trepidation, then with growing confidence, other producers took the path blasted out by Schwab. By the start of 1917 every major U.S. steel company was delivering war-related materials to Britain and France. Five of them (U.S. Steel, Lackawanna, Inland, Midvale, and Brier Hill) had moved directly into the production of lethal war equipment, mostly shells and heavy guns. Although none of the companies sold anywhere near Bethlehem's volume ($297.6 million), cumulatively the shipments added up to about $150 million, based on the subpoenaed Morgan records.[28]

The story of the industry's shift to a war footing was told by Elbert Gary. The chairman of U.S. Steel had been a prominent critic of the conflict when it broke out. In his annual address to the Iron and Steel Institute in October 1914, for example, he said that there were no high ideals in this war, only the mercantile rivalry of three major countries. "I venture the opinion that the struggle for commercial supremacy was the underlying cause of the war—that the questions at issue largely relate to dollars and cents," Gary said. "We cannot think of this conflict without feelings of horror. It is impossible to realize the extent of the suffering and misery which it entails." He called on the warring nations to submit their differences to binding arbitration.[29] A year later, in his 1915 address to the AISI, he was no less critical. Accusing both the Allied and German governments of suppressing the horrible facts of the conflict from their citizens, Gary wondered whether, if told the truth, the civilian populations wouldn't rise up and demand peace. "We are at peace with all the world and it seems likely that the wise policy which has permitted this state of affairs will be continued. We sincerely hope and pray that the wars which are raging in Europe may soon be brought to a close."[30]

Behind Gary's pacifism was the philosophy that war was the quickest way to cause social unrest. Herbert Spencer of "social Darwinism" fame had been an influential spokesman of this position. He repeatedly expressed worry of the threat of military elites in the progress of industrial nations. "All future progress depends upon the increasing predominance of industrialism over militancy," he wrote, adding, "the antagonism between the industrial and the militant types practically determines everything in civilization."[31]

But the economic tonic of arms sales, together with longstanding cultural and political attachments between the U.S. and Britain, were altering the Spencerian formula. The presumed antagonism between the military and industry had given way to intimate cooperation between the House of Morgan and the British Ministry of War, a cooperation that was characterized by secret investments in military builders, monopolization of credit arrangements, and orchestration of war purchases which reached an incredible rate of $10 million a day. Even the world's largest manufacturing concern was not immune to these economic pressures. U.S. Steel could not live on pacifism alone. Its stockholders must have dividends. So while voicing public concern about the war, Gary made a fateful, if unpublicized, decision early in 1916—he had two rail mills at the Gary works outside of Chicago torn out and converted to shellmaking plants. On May 22, 1916, his company received the first payment for shell sales. It was for $7,909,895.03. Over the next 11 months the company was paid another $5,530,064 for shells from His Britannic Majesty's Government as well as $34,193,376 for general steel forgings.[32]

If Gary worried about the implications of war, and quoted from the Bible in speeches to affirm his faith in Christian brotherhood, many of his fellow moguls were not so morally constrained. Locomotive maker Samuel Vauclain had turned his Baldwin complex at Eddystone, Pennsylvania, into a bomb factory. In 1915 and 1916 Baldwin produced over 3 million rounds of shells, including the deadly 12-inch "high explosive" howitzer projectile, in its own name and through a subsidiary, Eddystone Ammunition Corporation, which was then handed over to the British Government in return for personal loans given to President Vauclain by J. P. Morgan & Company. The Morgan records show Baldwin-Eddystone got $60 million in foreign shell orders. Other shell contractors were American Locomotive with $95 million; General Electric, $25.7 million; American Can, $33.3 million; E. W. Bliss, $33 million; and Bartlett-Hayward, $14.5 million.[33] Du Pont shipped the explosives to fill them.*

*In terms of dollar value, Du Pont was the largest recipient of Allied orders during the neutrality period. The Morgan records reveal that the company sold $424.9 million in nitrocellulose powder, TNT, guncotton, and detonators. Number-two supplier was Bethlehem with $297.6 million. Guggen-

As shellmaking was building up a new economic infrastructure, the war also powerfully stimulated the creativity of engineers and inventors. One measure was the sheer volume of stories in *Iron Age* which described advances in metalworking machinery for the purpose of making shells and guns. In 1916 there were 18 major articles regarding shell manufacture alone. Black & Decker's boring head blades for roughing the side walls of shells were prominently discussed in one story. There were other articles giving helpful tips on how to maximize production in shell factories, for example, "Routine for Producing One Hundred 6-In. 100-Lb. Shells Per Hour from Forgings—Proper Varnishing and Marking."[34] GE, Ingersoll Rand, and other companies evolved improved methods for drawing and piercing steel, all of which in turn augmented and expanded the market for the metal. Thus when Schwab spoke in his Baltimore speech of the new demand for steel he wasn't gazing idly into a crystal ball; he was expounding on trends already apparent to industry insiders.

The munitions business, though, was more than a technological drive—it was an instrument of international relations. Because the arms merchants dealt largely with governments, and because their products were used against other governments, their business inevitably influenced the relations between America and the belligerents. In a war of machined goods, the goods produced in U.S. factories rippled under and around the diplomatic and political disputes of the period. Changes in industry were occurring at such a fast clip that few fathomed this—least of all, it would seem, President Wilson. In 1916 he

---

heim's Copper Trust, American Smelting & Refining, was number three, selling $242.6 million in copper and brass, an essential component of ordnance making. (In 1916 British munitions factories depended on America for 75 percent of their copper.) Rounding out the who's-who of major Allied contractors were the Yankee gunsmiths of Connecticut Valley. Colt Patent Arms bored out machine guns, Winchester specialized in .303-caliber Lee-Enfield rifles, Remington and Marlin made cartridges, and Scovill handled fuses and brass tubes. All told, the Connecticut Valley sold about $80 million in arms and ammunition.

The list would not be complete without a word about the bankers. Thomas Lamont's comments to the Nye committee have already been noted: J. P. Morgan, Jr., testified in 1936 that his firm ordered about $3 billion of materials for Britain and France, earning a commission of $30 million. The Allies also used the banks to secure private loans. During the neutrality period over $2 billion was extended privately in the United States to Allied governments, compared with only about $27 million to Germany. (From *Munitions* hearings, part 26, exhibits 2152–58, and testimony, Jan. 7, 1936.)

ran simultaneously on a platform of peace and a platform of prosperity. One cannot doubt his sincere hope for the former; yet his inability, or unwillingness, to acknowledge the link between prosperity and a $10-million-a-day Allied arms market would come to haunt him and his administration. Wilson saw the danger of the country moving willy-nilly into a state of war, but he defined the danger as a matter of German attacks on American shipping and of overemotionalism at home. In his first term he had won from Germany a promise to suspend U-boat attacks on American shipping. He had won political plaudits for his diplomacy of neutrality. His party's banner—HE KEPT US OUT OF WAR—was generally credited with bringing him victory in 1916 over a more hawkish Charles Evans Hughes.[35]

Yet war was encroaching nevertheless. The anomaly of a nation proclaiming neutrality while its factories were in a state of war with a power overseas could not last indefinitely. And shortly after the Battle of the Somme stalled out, war did wash to these shores. In January 1917 Germany announced the resumption of U-boat warfare on neutral shipping without warning or examination. The declaration caused a virtual blockade of American shipping: goods piled up in U.S. ports; shippers were panic-stricken by the announcement. For Wilson, the decision by Germany was an act of international lawlessness which violated the sanctity of the free highway of the seas. Diplomatic relations between the countries were severed. Relations deteriorated. Several ships were sunk. People once anxious not to fight in a European war were swept up in hatred for Germany, stoked, it was later demonstrated, by skillful British propaganda. What's more, the "ultimate" war exerted an undeniable fascination. In Baltimore, socialites promoted the British cause by holding an "Allied bazaar" at the Fifth Regiment Armory. On the lawn outside the bazaar a 200-foot replica of a Western Front trench had been dug. Visitors walking through the trench could view an officers' dugout and Red Cross station, put their hands on a machine gun, stare at wire entanglements strung along the lawn through a periscope.[36]

Did munitions makers force America into the war? No. Other factors were involved. But unrestricted munitions traffic certainly helped prompt Germany to declare unrestricted warfare on U.S. shipping, which finally made conditions intolerable. On April 2, 1917, the president, addressing a joint session of Congress, asked for a declaration of war on Germany. He framed his war speech in lofty terms. "We have no selfish ends to serve. We desire no conquest, no dominion. We ask no indemnities for ourselves, no material compensation for the sacrifices we shall freely make. We are but one of the champions of the rights of mankind." Concluded Wilson: "The world must be made safe for democracy."[37]

*

In the year that America went to war, Sparrows Point underwent its greatest transformation since the plant had been started 30 years earlier. The very shoreline of the peninsula was soon reshaped. Steel pilings were rammed 2,367 feet beyond the south shore, and huge amounts of earth, stone, slag, and gravel were dumped into the space between. When the fill had settled out, Sparrows Point was 100 acres larger. Near the "made" land the Army Corps of Engineers and the Baltimore Harbor Engineer authorized construction of an artificial harbor. A 30-foot-deep channel was dug to the high wharf and a turn-around basin established for ore steamers. Baltimore's Mayor Preston offered his support by providing city money for the harbor project and for temporary housing. In coordination with these projects, keels for four new boats for ore shipments were laid down in the shipyard. The first boat was launched on July 9. It was the S.S. *Cubore* (short for "Cuban Ore"), christened with a bottle of champagne by the chairman's wife, Rana Schwab. Down the ways glided the boat, 400 feet long and the largest of its class.[38]

Changes were underway also on the farthest tip of the peninsula. The swampy land around Shantytown was elevated and a coal-handling plant was installed on a pier extending westward into the harbor. Adjacent to the coal-handling plant preparations were made for a 240-battery coke oven facility to be built by the Koppers Company. Included in the complex was to be one of the largest plants for the extraction of benzol-benzene, a coke chemical used for explosives as well as many other manufactured items. All coal for the works was to be hauled by barge from the B&O and Western Maryland Railway wharfs in Baltimore as well as from Norfolk, Virginia, where the company had signed a contract for Pocahontas coal with the Norfolk & Western Railway.[39]

On the north side of the peninsula the last tracts not owned by the steelmaker were acquired. This included the 533-acre Numsen tract. With these purchases all land washed by the waters of Humphrey's Creek was company owned, which "gives an unlimited area for any improvements that might be planned and an added acreage for the economical dumping of slag," according to a company report.[40]

At the same time that the Point was being remolded and its perimeter expanded, tremendous activity was going on within. On the cattail-fringed flatlands platoons of structural workers could be seen setting massive concrete footings in place and piecing together the skeletons of new rolling mills. A visitor could not help but be impressed with the number and size of the buildings that were sprouting up. One example was the tin-plate mill. Rising in great sweeps of steel north of town, the mill doubled the roofed space of the original rail mills. The hot-mill department, for example, consisted of a massive steel shed 980 feet long, 147 feet wide, and 41 feet from floor to upper vent stacks. The annealing and cold-rolling departments were only slightly less huge: 756 feet long by 115 feet wide. There was a large bar shed, a tin-coating house, assorting shop, machine shop for handling in-house repairs, a blacksmith shop, carpenter room, and storehouse.[41]

Twelve hot-mill units were installed. They were driven by 1,200-horsepower motors. There were 36 furnaces for heating the steel before it was charged into the hot mills and squeezed down to sheets by crews. In the cold-rolling department, a special rope-drive mechanism was installed to improve the heavy-pressure rolling needed for the surface of a finished sheet. Elsewhere in the building, workmen bolted into place the sulfuric acid tanks needed for cleaning the sheets. Next were 21 tinning stacks where the sheets were to be dipped in a bath of hot tin. A coating of tin of about 1/40th the width of a human hair produced tin plate. For specialty rolling, a bar and jobbing mill were also established. Initial net output: 3,500 tons of tin plate a month, plus 2,000 tons of specialty bar.[42]

A mile west of the tin-plate mill rose the 110-inch sheared-plate mill. The idea here was to make the large plates used at the shipyard. Up to that time the shipyard had purchased its plates mostly from the Central Iron & Steel Company in Harrisburg, which obtained the slabs from the Steelton plant. The Sparrows Point plate mill had the then enormous capacity of 12,000 tons per month. It was carefully located for the economical delivery of hot slabs from the furnaces. A large crane runway was extended through the mill for handling the slabs, and a narrow-gauge railroad constructed to connect the mill's innards with the steel department and a future bloomer mill.[43]

On July 26, 1917, the sheet and tin-plate mills were opened. It was the only fully integrated tin plant east of Pittsburgh. Immediately all of its production was consigned to "defense work," since the U.S. Army urgently needed tin-plate stock for the canning of rations and other supplies for the forces to be transported overseas. On September 10, 1917, the 110-inch plate mill was placed in service. Immediately all of its output was absorbed by orders for merchant ships. As luck would have it, although entrance into the war by the U.S. stimulated sales for all Bethlehem products, nowhere did business balloon as sharply as in shipbuilding. In a matter of months Bethlehem's order books for ships had zoomed to $273 million. Having a shipyard adjacent to a steelmaking mill gave Bethlehem a competitive edge over Newport News, New York Shipbuilding, and other rivals, a situation that accounted in part for its leapfrogging growth.[44]

Even rail sales raced upward in that fateful year. The army had developed plans for railroad requirements in Europe which eventually called for 305,000 tons of rail to rebuild destroyed track. The army allocated 212,450 tons of rails for the French forces and 64,484 tons for Italy. In fact, by early 1918 Sparrows Point and the other Bethlehem plants were so clogged with orders that backlogs of as much as 25 weeks were reported for certain items.[45] Although the same problem was occurring at several other Eastern companies, at less well situated plants in the Middle West excess capacity was reported.

The War Industries Board was very unhappy about the development. A meeting was held in Washington at which Schwab was asked to reallocate some of his contracts to the companies that were not so heavily burdened. According to a reporter, the discussion was not long underway when, despite the presence of Secretary of War Newton Baker and Board Chairman Bernard M. Baruch, Schwab lost his composure:

> Tears welled into his big brown eyes. He said that he didn't mind giving up the contracts, but that taking them away would be a reflection upon his company, and that it would be excessively painful to him, for he had no children, and the company was the only thing he had in life, and that it would break his heart to have anything said against it. To the further amazement of the assembled war lords, he then turned to his friend Barney Baruch and said, "Let me have a minute alone with you, Barney." The two men left the room. All this puzzled the beholders, and one of them turned to his neighbor and said, "I wonder what he's doing out there with Baruch?"

> "Probably selling him stock in the Bethlehem Corporation," was the unfeeling reply. A few minutes later Schwab and Baruch returned. The steelman launched into a new and impassioned plea against having any of his contracts taken away. Bethlehem kept the contracts.[46]

Early in 1918, Frederick Wood resigned from the company. His two years as Schwab's general manager had been difficult ones. He had carried out the heavy responsibilities of the expansion, working with experts from around the country and spending more capital in the space of two years ($25 million) than he had over the course of three decades. Still it wasn't the same. Wages were set by South Bethlehem and in matters of personnel he had to consult with (and in some cases defer to) Schwab or his coterie of subordinates.

Wood's aide, Quincy Bent, had been transferred; paperwork got heavier; his superintendents were presented with lectures on the "Bethlehem way" of doing business. Running Sparrows Point always had been an intensely personal and private matter to him. So when the interference got to be too much, he quit. Wood joined Stone & Webster's American International Shipbuilding Corporation as a vice president and took an active part in the construction of the wartime shipyard at Hog Island near Philadelphia. Returning to Baltimore in 1921, he started a second career as a mining and steel consultant. He chaired the American Iron Institute's Committee on Rail Manufacture and served on the boards of a number of Baltimore companies. In 1933, he was elected chairman of the board of Eastern Rolling Mills, a specialty sheet roller. He retained this position until his death in 1943 at the age of 86.[47]

Wood was replaced by W. Frank Roberts, age 38. He had started out at age 12 as an office boy for a coal company near Bethlehem, later entering the steelmaker's steam department and working his way up the job ladder to assistant superintendent. His big break came in 1915 when Schwab picked him as superintendent of the shrapnel and gun projectile departments. In the vast upsweep of business that ensued he gained rank and was promoted to general superintendent of the works, responsible for the 25,000-a-day shell assembly operation. The *Baltimore Sun* described him as "short, thick-set, powerfully built," with a deliberate manner. "Like many big men, he is more given to listening than talking, but what he says is direct and to the point, and his mild blue eyes are keenly, minutely observant."[48]

Motivating working men to work harder was the greatest challenge facing industry, he told the paper. "Mr. Roberts does not believe that the meaning of the war has yet stirred the industrial classes very deeply." He was sponsoring preparedness drills and "100 percent Americanism" rallies at Sparrows Point and the results were encouraging. He said another top priority was to build more working men's houses.

"Last year at Sparrows Point we just broke even, but it is a good industrial stabilizer. Men who have settled down in homes of their own, bringing their wives and families with them, take root. They become interested in their homes and gardens, and they reach the point where a few additional cents elsewhere would not repay them for the trouble of moving. Good men like these exercise great influence in the plant. We use them as industrial balance wheels, so to speak. Whenever a number of new men are taken into the plant the old men are spread out—helpers are made machinists, machinists subforemen. We have a number of these subforemen, and their moral effect is splendid."[49]

Four thousand workmen were added to the rolls in 1917 and another 3,500 were needed in the early months of 1918. For the duration of the war, daily employment averaged 12,500—triple the prewar total. Twice a day a swollen tide of exhausted men ebbed from the mill, while a new tide rushed forward. Because of the emergency, the 24-hour "swing shift" had been reinstated after an absence of several years.[50] So after toiling for a week of 11-hour day shifts, then a week of 13-hour night shifts, the men faced the grueling test of 24 hours straight in the line of duty.

*

As the war effort grew, so did the stature of Charlie Schwab. He was a sought-after speaker at war bond rallies and was named Salesman of the Year by the World Conference of Salesmen. But nothing was to elevate his reputation more than his summons, in April 1918, to serve as Wilson's wartime "shipbuilding czar."

Schwab's hour of destiny was a result of presidential desperation. Handed $2.63 billion by Congress to build a "bridge to France," Wilson's Shipping Board was bogged down and under attack. Its shipbuilding arm, the Emergency Fleet Corporation, had called on Major General George W. Goethals, builder of the Panama Canal, to put the program in order. But Goethals had a temper tantrum and resigned.

The output of troopships from private builders was disappointingly low. With the press clamoring for action, Edward N. Hurley, chairman of the Shipping Board, proposed to President Wilson the appointment of a prominent businessman to inspire and direct private production.[51]

Henry Ford, Hurley's top choice, refused the job. Hurley's second choice was Schwab. Wilson authorized Hurley to telegraph Schwab and ask him to come to Washington on urgent business. When Hurley and his associates, Charles Piez and Bainbridge Colby, made the proposal to Schwab, he protested that he could not leave Bethlehem. He also worried that the appointment might cause complications for the shipbuilding contracts his company had with the Emergency Fleet Corporation. The three men assured him that practical accommodations could be made. Schwab returned to Washington on April 16. This time he was swept into the White House for an appointment with the president. "I vividly recall the interview," Bainbridge Colby wrote later. "The President, with whom I had fully discussed the question, and who was entirely in sympathy with the effort to 'requisition' Mr. Schwab, came out of an inner room assuming that the matter was settled and that Mr. Schwab was willing to undertake the work. He put out both hands to Mr. Schwab and spoke in acknowledgment of his sacrifices and his patriotism in a way that would have moved any man."

"Will you stand back of me?" Schwab asked Wilson.

"To the last resources of the United States of America," Wilson replied.[52]

Schwab agreed to take the post, but only on two conditions. First, he wanted total autonomy from Washington so he could run the program without being boxed in by government procedures. Hurley, Piez, and Colby agreed. Second, he insisted on a written agreement that excused him from dealings with his own company and cleared him from potential conflict-of-interest charges. This condition was also met, and a special committee of the Shipping Board was established to handle negotiations with Bethlehem. In this way Schwab was not required to sever his financial connections with the company while in government service, nor place his holdings in a blind trust. He could retain his title of chairman of the board of Bethlehem and was free to exercise his rights as a shareholder. He would receive a nominal government salary of $1 a year, thus entering the ranks of "dollar-a-year" businessmen supporting the war effort.[53]

These matters settled privately, a statement was prepared for the press. Mr. Schwab, it stated, had been "requisitioned" to serve in the new position of director general of the Fleet Corporation and given complete direction of the work of shipbuilding. "Mr. Schwab has agreed to accept this position in answer to the call of the nation. He agreed to take up the work at the sacrifice of his personal wishes in the matter. His services were virtually commandeered. His great experience as a steel maker and builder of ships has been drafted for the nation."[54]

With the good offices of the government behind him, Schwab caught the fancy of the nation. With a superb mind and an undeniable command of the ship business, he was successful in speeding up production. Shipbuilders responded to his fast-paced awarding of contracts and the dispersal of awards for yards that exceeded their production quotas. While a few grumbled about the unassailable place of his company in government contracting, most realized his value to their business. Moving the offices of the Fleet Corporation from Washington to Philadelphia, he insulated the program from both congressional carping and government auditors, the former a definite plus in terms of accelerating production, the latter causing years of litigation after the war.

What attracted the nation was his barnstorming stump style. On June 17 he kicked off a campaign to tour the nation's shipyards and put it straight to the workers. He began in his bailiwick, in Baltimore, where he dedicated the steel refrigerator ship S.S. *South Pole.*

"I hope the ladies present will pardon me for saying it, but the Kaiser is in the for d——dest licking of his life! He is going to make us lose more ships—that's why we're building so many—but, by God, we are not going to lose this war!" Making a sweeping gesture to Bainbridge Colby, he proclaimed, "These are the type of men, Mr. Colby, who make up America! These are the type of men who never lose! Stand by me and we will win! Now go to it, boys, and give 'em h——!"[55]

Up and down the Atlantic, across the Gulf states, and on the Pacific Coast, Schwab brought his high-decibel message to all and sundry. In one celebrated incident, he made an address to workers at a shipyard in Seattle noted as a hotbed for Wobblies, the radical group opposed to American involvement in the war. Knowing that his reception would be chilly, if not even hostile, Schwab had an opener on hand to disarm the audience. What was the one thing socialist workmen would scarcely suspect a famous capitalist to talk about openly? "Now, boys," he began, "I'm a very rich man..." The startled men laughed and applauded, and Schwab went on to make his pitch. When he was done a delegation of men offered him an honorary membership in the IWW. He accepted.[56]

The tour reached its patriotic climax on July Fourth when Schwab presided from San Francisco over a gala launching of 95 ships nationwide. Brass bands played Sousa marches, Pastor Eaton tossed out prayers like confetti, and workers cheered the dollar-a-year man like a returning military hero. And he cheered them back.

"You men who swing the cranes are in charge of the big guns," he declared. "You who drive the rivets are operating the machine guns of the shipyard. Every man who does a full day's work is doing his share to win the war!" In its next-day editions *The New York Times* proclaimed Schwab the "Field Marshal of Industry," saying, "No vocabulary could do justice to the zeal, energy, driving power, magnetism, and inevitable success of this remarkable man."[57]

Schwab had learned a technique. Whenever he wanted to stir up a crowd, he would sing the praise of the common man, pillory the kaiser, and invoke the name of the "beloved president" he had shown no hesitation in defying four years earlier.

> WASHINGTON, July 23.—Charles M. Schwab, Director-General of the Emergency Fleet Corporation, transmitted today to the men working with him a letter from President Wilson expressing the "keen interest" with which the President is following the progress of the ship building program.
>
> "I am greatly pleased to be able to send you with this a copy of a letter of July 10, addressed to the Chairman, Mr. E. N. Hurley, by our beloved President, Woodrow Wilson....
>
> "I know that I may count on you to see that the President is given no cause for disappointment in the future, but, on the contrary, will be more and more delighted at what we can do when spurred by a deep sense of patriotism."
>
> WASHINGTON, Sept. 1.—In a Labor Day greeting cabled

to American troops in France, Director General Schwab of the Emergency Fleet Corporation says:

"We are with you in every possible way. We are devoting all of our energies and making all of the sacrifices, no matter how great, so that you may be supplied with food, ammunition, and other necessities. The ships are coming out at a splendid rate, and this shows what our industrial workers can do when inspired by patriotic enthusiasm. God be with you all, and may you return covered with honor and the glory of magnificent deeds in keeping with American traditions."

SCHWAB READY TO LOSE WEALTH TO END WAR

NEW YORK [SEPT. 19]—Charles M. Schwab, Director General of the Emergency Fleet Corporation, visited the Standard Shipyards on Staten Island and took part in a ceremony at the plate works of raising a flag.

"Boys, I have a plant in which I employ 150,000 [*sic*] men. I have been away from there about five months and I do not intend to return until I have finished this job and ended the war.

"I may lose all my money there—I am pretty rich—but even if I do lose all my money, if I can help build ships that will end the war, I will be satisfied."[58]

Schwab's oratory about losing his money in order to end the war was grandiloquent. But neither his company nor he had much of a chance to do so when they straddled the "narrows" of the war effort. Bethlehem continued on its trajectory of high earnings as company plants churned out over half of the U.S. Army's big guns in 1918 as well as 42 destroyers and submarines for the navy. The war years were ones of extraordinary bounty for Schwab personally. Altogether he earned $21 million from his holdings in Bethlehem, according to available data, $11 million in cash dividends and $10 million in new common stock distributed in a special stock dividend. Such returns would be equal to $204 million in 1987 dollars.[59]

What changed in 1918 was not the dynamics of making money on munitions, but the mechanics of warfare. On the battlefield the trench stalemate had been broken by new offensive machinery that triumphed over the power of defense. The British "caterpillar" (tank) afforded the first means of advance against trench artillery and wire entanglements without much risk of loss of life. It pushed back the once-impregnable German line. The airplane helped break the deadlock and accelerate the Allied drive to the Rhine.

War ended on November 11, 1918. A month later, President Wilson received a telegraph from Director General Schwab while sailing across the Atlantic to the Versailles Peace Conference. "The changed industrial conditions brought about by the signing of the armistice have brought with them many serious problems affecting industrial operations. I enlisted for the emergency, and now that the emergency is over I desire most earnestly to be released for the great problem of bringing the industries in which I am interested back to normal conditions. May I have your consent to retire as Director-General of the Emergency Fleet Corporation?"

Accepting Schwab's resignation, the president said in a wireless message, "You have been exceedingly generous in giving your services, and they have been invaluable. Let me thank you very cordially indeed for all that you have done. I shall always remember it, as I am sure your associates in the government will, as a service of unusual value and distinction."[60]

CHAPTER 7

CHASING DOLLARS

I don't care how much a man earns. The more he earns the better I like him.

—CHARLES SCHWAB, ADDRESSING PRINCETON UNIVERSITY STUDENTS, MARCH 16, 1920[1]

In 1914, when the World War began, the U.S. steel industry produced 23,513,030 tons of steel. In 1918, when the armistice was signed, the industry produced 44,462,432 tons. The trade had rolled up its highest profit margins in history—28.9 percent before taxes in 1917. Notwithstanding the War Revenue Act, sizable amounts of this income escaped federal taxation under an IRS provision which allowed steelmakers to place part of their income in special tax-free equipment accounts. U.S. Steel, for example, put into the account $191 million in profits which it claimed would be spent after the war for replacement of worn equipment. The Nye Senate Committee later found that not a single penny of this fund was ever used for its intended purpose.[2]

Bethlehem Steel had used its earnings of $111,168,129 in 1917–18 to continue the expansion of Sparrows Point and for the retirement of outstanding debt. On February 1, 1919, the company received $37.6 million in British Treasury Notes in payment for munitions supplied to Russia and guaranteed by the British government. Release of these funds enabled Bethlehem to practically wipe out its outstanding indebtedness. Over the course of the year $25,988,364 was used to retire bonds, including the 5-percent gold notes for purchase of Pennsylvania-Maryland Steel. On December 31, 1919 long-term debt stood at a negligible $3,607,189. The steelmaker's balance sheet bulged at $357.2 million, with $47.8 million in cash and marketable securities. (On December 31, 1914, the balance sheet had been one-third of that—$106.4 million—and cash and liquid securities had been only $5.7 million.) "Phenomenal" was the assessment of business writer B. C. Forbes.[3]

With an awesome war chest at his disposal, Charlie Schwab was ready to embark on a campaign of industrial growth that would keep him fully before the public eye. A war hero and a very wealthy man, "Smiling Charlie" was keenly sought out by the press and he freely dispensed advice on all kinds of matters. He became a regular fixture in *The New York Times,* taking a place among businessmen like Walter Chrysler, Otto M. Kahn, Harvey Firestone, and Henry Ford, whose accomplishments enraptured a money-conscious nation. If postwar men and women had become cynical about politics and religion, they still had faith that "at the end of the rainbow there was at least a pot of negotiable legal tender consisting of the profits of American industry and American salesmanship," wrote Frederick Lewis Allen.[4]

Underlining the enormous confidence in business was the enormous progress made in the manufacture of new things. From GE toasters to Easy Electric washing machines, Americans found an array of manufactured goods entering their daily lives. Nearly all of these goods used steel as a component part. The automakers' need for steel for the 2.2 million cars they built in 1920 (up from 573,000 units in 1914) represented a large potential market for Bethlehem Steel. Soon the drop forges that were used to make machine-gun pistons at South Bethlehem were manufacturing pistons for automobiles. The gun-projectile furnaces were making rear wheel housings and brake drums as well as electrical generators, locomotive engine parts, and rivet sets.[5] The Bethlehem munitions machine had turned quiet.

But quite aside from the multiplication of steel-based "implements of peace," the '20s were a proving ground for new factory methods and management. With little enlargement of personnel, a vast enlargement in labor output took place. Leading the productivity charge was the manufacturing sector. Output per individual worker from 1919 to 1925 rose by 15 percent in transportation, 18 percent in agriculture, 33 percent in mining, and fully 40 percent in manufacturing. In part, this stunning performance was a result of the increased use of machinery and power to replace human labor. But a large part stemmed from a better organized and harder-working factory force.[6]

The war had presented tangible evidence of the productive capacities that were released when the patriotic spirit and competitive energies of the common man were fully engaged. As a memo to Frederick Wood in 1918 had stated, "With a cause of the highest patriotic potential as a background, it is thought possible to make the building of ships a National Game just as baseball has been for many years."[7]

Why not make factory work a national pastime? While the impulse to "make every man a capitalist" had been present in corporate America for a long time, in the 1920s the methodology was to undergo major changes. The traditional value structure lost its luster. The old Calvinist producer ethic with its stress on savings and self-denial was troublesome when companies found they could sell more if workmen and their families were consumers of their products.

Soon a number of books appeared in Britain, France, and Germany on the new industrial order in the United States, praising its accomplishments, condemning its materialism, but always seeking its secret. If Communists had their Marxism-Leninism, Americans had their "New Capitalism," a term popularized by the U.S. Chamber of Commerce to fit into the laissez-faire policies of Presidents Harding and Coolidge. Soon the business world was throbbing with concepts like "scientific management," "Fordismus," and "human engineering" to capitalize on the cooperative spirit of labor.[8]

It was here in the realm of factory productivity that Schwab again demonstrated his Midas touch. "Winning Workmen to a Right Spirit in Their Tasks" was his professed goal, and the idiosyncratic wage plan he instituted at his mills helped bridge the gap between the rigid ideas of wages and profits of the past and modern ideas about worker motivation.[9]

*

On March 16, 1920, Charlie Schwab and his publicist, Ivy Ledbetter Lee, went to Princeton University, where the chairman gave a lecture on "what it took" for a young man to lead a successful life. After laying down a few rules as old as Horatio Alger, if not Ben Franklin, the chairman got down to business. "You will be interested to know something about how we do things at Bethlehem," he asserted.

"I pay the managers of our works practically no salary. I make them partners in the business, only I don't let them share in the efforts of any other men. For example, if a man is manager of a blast-furnace department he makes profit out of the successful conduct of his department, but I don't allow him to share in the prosperity of some other able man in some other department of the establishment. I give him a percentage of what he saves or makes in the department immediately under his own control and management.

"For example, if it takes a dollar a ton to make pig iron, and if it takes him a dollar a ton to make pig iron, I say to him, 'Well, you are no better than the average manager over the country. Therefore, you are entitled to only the usual wages. But if you make pig iron at 90 cents a ton you are entitled to share with me in a large part of the profits. And if you make it for 40 or 50 cents a ton you share to a very large degree.' "[10]

In most departments, a dollar was used as the base for calculating incentives. To use Schwab's example: if the superintendent cut the cost of pig production from $1 to 90 cents, he would receive an additional 1 cent to his salary for each ton made. If he reduced the cost from 90 to 80 cents, his bonus per ton would jump to 2 cents. And if he lowered costs from 80 to 70 cents his bonus would not be 2 cents per ton, but 3.[11]

Ingeniously, for every reduction in unit cost there was a greater incentive for further effort, and for every further effort there was greater profit to the company. But the opposite was also true. If a superintendent's unit costs rose above a predetermined level, "he knows something is wrong and I don't have to say anything," Schwab pointed out. "We get rid of him."[12]

"Now I do the same with the working people," he declared. At Sparrows Point and the other plants bonuses were meted out to skilled and semiskilled workers according to the job assignment. A mechanic, for example, was permitted a certain time for repairs. If he did the task in the time permitted he received his regular pay. But if he got the work done ahead of schedule, he got his regular pay and was assigned to another job whose pay was added onto his hourly wage. In the tin-plate mill incentives were based on the weight of "prime plate" coming off a mill—if output showed a 40-percent gain, wages would rise by approximately 20 percent.[13]

Schwab's faith in the common man's urge to make money (and to work harder to make more) had inspired a system of wages unique in the steel trade. How unique may be observed by comparing the plan with the policies of his predecessor at Sparrows Point. To Frederick Wood, wages were a consequence of the costs of mill production, not a spur to lower the costs. Wood wanted no part of flexible wages; like most industrialists he believed profits belonged exclusively to those who took the investment risks. Moreover, the question of workman discipline and communal order entered the picture. Wouldn't bonuses, especially cash bonuses, cause a loosening of discipline among the workforce? Wood only had to look at the trouble his brother had had with liquor consumption on payday to be convinced of how easily a workman could be led astray with too much cash in his pocket. So Wood set employee compensation on a fixed basis (day rates for laborers, set tonnage rates for skilled men), then relied on close supervision to keep everyone in line. It should be pointed out, though, that Wood practiced the same sort of wage discipline on himself. In his years as president of Maryland Steel his income averaged $18,000 to $20,000 a year, a comfortable but scarcely extravagant sum by contemporary standards, and he seemed little inclined to seek more.[14]

Not only did Schwab turn Wood's labor principles on their head, perceiving money to be the inciter, not the weakener, of the famous American work ethic, but he actively sought to institutionalize quick gratification into his bonus program. Bonuses were paid once a month and were timed to come out after regular checks for maximum psychological effect—to cause "each employee to be constantly keyed up to the highest possible pitch of efficiency." While conceding that the bonus system required a lot of recordkeeping, Schwab said it "pays for itself a hundred times over."[15]

While Schwab rejected straight pay as old-fashioned, he had little use for the "scientific management" school founded by Frederick W. Taylor. Taylor advocated "time-and-motion studies" in order to determine the time a job would require if all wasteful time and motion were eliminated. He had developed what he considered optimal ways for a workman to handle his factory duties, including an exactly prescribed sequence of body movements for shoveling coal into a furnace. Many of Taylor's experiments had been conducted at the South Bethlehem plant after he was hired in 1898 by Joseph Wharton. When Schwab assumed control of the company he threw out Taylor's procedures. He said there was no need to have slide-rule men and stopwatch men—"supernumeraries," he called them—getting in the way of the work. Schwab's policy was simple: he didn't care how a job was done, only that it was done. His check wasn't the stopwatch but the cost sheet.[16]

A book of advice that Schwab presented was a valuable text to his incentive system. Called *Succeeding with What You Have,* the book was mostly a compilation of first-person anecdotes and autobiographical tidbits. Most of these were drawn from his speeches, since he intensely disliked writing. *Succeeding* began with young Charlie's start at Carnegie Steel's Braddock works in 1879 (where he fled from his rural hometown of Loretto, Pennsylvania), and traced his success in the world of steel, first as superintendent of the Edgar Thomson works and then as U.S. Steel's first president at age 39. The book was filled with advice whose themes were suggested by the chapter headings: "Seizing Your Opportunities," "How Men Are Appraised," "Thinking Beyond Your Job," "Women's Part in Man's Success." Predictably, one found stories of young men who had started out as lowly laborers. Predictably, too, one followed the rise of these ex-laborers from wages to riches.[17]

Compared to Frederick Taylor's theories of scientific management, this was threadbare stuff, but then Schwab wasn't selling science. He was selling sweat: sweat to workmen who were doing jobs still largely classified as "heavy labor and hazardous." To the 11,000 wage earners at Sparrows Point as well as to the 35,000 other men who labored at the company's mines, mills, shipyards, and railroad short lines, Schwab offered these words of encouragement in *Succeeding:*

> Don't be afraid of imperilling your health by giving a few extra hours to the company that pays your salary.... Be thorough in all things, no matter how small or distasteful.... Don't be reluctant about putting on overalls.... Bare hands grip success better than kid gloves.... Achievement is the only test. The fellow who does the most is going to get the most pay, provided he shows equal intelligence.[18]

Under Schwab's regime, wages did increase. This is important to remember. In absolute terms the increase was quite large. By 1925, average earnings were about $2,000 a year. This compared to approximately $1,100 at the end of the Wood years. Inflation, though, had accounted for a large part of this increase. Between 1915 and 1925 the Consumer Price Index went from 30.4 to 52.5, thus taking 74 percent of the 81-percent hike in earnings over the period and leaving an increase in real earnings of about 7 percent.[19]

However, this did not account for the discrepancy in pay between unskilled and skilled workmen. Schwab's incentive system had the effect of widening the "spread" between the lowest- and best-paying jobs at the plant. General laborers were paid a flat 37 cents an hour, and significantly were not part of the incentive plan. This made for parsimonious living. Moreover, earnings were undercut by seasonal layoffs. When orders from automakers and railroads slacked off in winter and late summer, so did the need for men in the labor gangs. Given the prospect of three to eight weeks' unemployment even in a "boom" year on top of a low hourly rate to begin with, it scarcely comes as a surprise that the labor pool at Sparrows Point was black. As in the Wood years, laborers were chiefly recruits from elsewhere—from rural Maryland and the South, where limited employment opportunities made $900 to $1,200 a year at the Point look good.[20]

On the other hand, skilled and semiskilled workers were doing quite well. Their gains under the Schwab system were real and in some cases substantial. Head tin-mill rollers, for instance, could earn $10 to $12 a day. Theirs was a job that required great technical skill on top of legendary sweat to handle cherry-red bars with tongs and feed them back and forth through the rollers that pressed them into sheets of specified thickness. At least ten years of apprenticeship were needed. They were the "aristocrats" of the mill. Other members of the tin crew also fared well—$6 to 8 a day for catchers and doublers, $4 to 5 for heaters and roughers, and $3 for the screw boy. At the open hearths there were other high-paying jobs for men who could withstand the heat and strenuous labor. In brief, in assessing the sources of stability in the mills during the '20s, money ranked high. Under Schwab's rule the skilled ranks had passed through the threshold of the middle class. They could afford that Easy Electric washing machine that sold for only $189.99 at the company store. They were "making money," and therefore had a powerful disincentive to buck the system that was adding dollars to their paycheck.[21]

*

When Schwab took charge of Bethlehem, he made a number of management decisions of far-reaching consequences. One was the principle of a central board of directors to monitor affairs closely. Under Schwab's organizational scheme, all cost and wage data from the mills were to flow upward to the supreme command through orderly written reports. This was opposed to the old ways of doing things—"each department going pretty much its own way."[22]

Another cornerstone of company policy was the doctrine of promotion from within. In seeking men for the supreme command he made a point not to bring in executives from other companies, but rather to select from those inside the organization who had proven themselves. He believed strongly that there were quite a few young men at Bethlehem who could eventually assume high-level positions under his tutelage.

A final Schwab tenet was derived from the first two—absolute loyalty to the organization. He demanded a level of fidelity to the company extreme even in the close-knit steel industry. A Bethlehem executive was required to live in Bethlehem. He was forbidden from engaging in any business activities outside the company unless directed to do so for the higher good of the company. That meant no sitting on the board of the local bank. Officers were not to be active in politics. Schwab wanted no competing loyalties, not even from after-dinner theater which he scorned for taking management acolytes away from "study that will add to their business knowledge."[23]

From these organizing principles sprang the elite corps that ran the steel company. He called them the "boys" and he never tired of expressing his love for them. ("I always drive by encouraging words of cheer and commendation," he told Clarence W. Barron.) Limited to 16 in number, the "boys" were not only operating officers of the company, but most of them also were on the board of directors. This was highly significant. Most companies as large as Bethlehem had seats reserved for "outside" directors selected from banking houses and major stockholders. These directors were responsible for reviewing management decisions and helping decide general policy. But Schwab ran Bethlehem like a small company. The *corporation* did not operate any iron or steel works; the Bethlehem Steel *company* did this, and all members of the company's board of directors were Schwab's "boys." The officers were also on the corporation board, thus holding the majority of seats under Schwab's control.[24]

On weekdays the "boys" gathered in a large room on the top floor of the Bethlehem Steel Office building in South Bethlehem for a luncheon. (Lesser officials met for conference luncheons in adjoining rooms.) After the luncheon, over cigars and coffee, they reviewed cost data from the mills and in general discussed company business said to include "every particular from the hiring of a man for a vacant position to the purchase of properties involving millions."[25] Because the executives were also directors of the steelmaker, their daily congregation was the equivalent of a board meeting.

The luncheons permitted the officers to make decisions rapidly, which was the point, after all, for a man in a hurry. And to make sure that the competitive spirit in the executive offices was at the highest possible pitch (for he rarely attended the daily meetings, letting the participants take the initiative in daily matters), Schwab devised a special reporting system to keep him abreast of financial results. The moment of truth occurred on the third of each month. On that day Chairman Schwab found on his desk a report from each division itemizing the production, costs, and profits of the units under their jurisdiction for the past month. Schwab studied the cost sheets thoroughly. One of his lesser-known talents was a photographic memory that enabled him with a personal mnemonic system to memorize a hundred pages of cost data in an hour. It was said he could spit back an obscure price years later to the astonishment of his staff who, rushing back to their files, would find the chairman to be right to the decimal point.[26]

Armed with the cost data, Schwab calculated the monthly bonuses for his officers. When he said he paid "practically no" salaries he spoke the truth. The big men at Bethlehem seldom got more than a few hundred dollars a month in salaries. But their bonus envelopes bulged. If the corporation cleared $1 million before depreciation in a month, Schwab divided about $50,000 in bonuses among his 16 assistants. But if the monthly earnings jumped to $2 million, executive bonuses tripled to about $150,000. And if $4 million was garnered, then the cash bonuses were upped not to $300,000 but to nearly $350,000. Schwab was shrewd enough to realize that if you paid your subordinates more than they were innately worth, you had them under your power. Bonus pay began at 3.43 percent of net earnings, and this percentage increased with increasing earnings to a maximum of 8 percent.[27]

Although no one knew it at the time, the Bethlehem officers were the best-paid executives in America. After a few years of his six-figure bonuses they were millionaires too.[28]

*

Of all the "boys," the one who rose to greatest prominence and wealth was Eugene Gifford Grace. In the hands of business reporter B. C. Forbes his story became the stuff of dreams. His success, Forbes proclaimed in a 1920 profile in *American Magazine,* was "a narrative of typical American diligence, vision, forethought, pluck, and unremitting application based on searching mental analysis at every stage of the journey. It is a narrative rich in suggestion for every striving youth and every progressive businessman in America. It reveals a pathway which each one of us, be our work what it may, can seek to tread, for it is a pathway not of miracles, but of common sense applied at every turn of the way and every hour of the day." The article was headlined, "'GENE" GRACE—WHOSE STORY READS LIKE A FAIRY TALE."[29]

Like a good fairy-tale hero, Grace had started out in obscurity. He was a small-town boy steeped in the Bible-thumping Methodism of his father, John Wesley Grace. He was born in Goshen, New Jersey, on the south Jersey coast, on August 27, 1876, 15 years the junior of Schwab. Early on, what encouraged him to compete was his family's modest financial status. His father was a New Jersey sea captain in coastal trade who had retired to run a general store. (The family was of no relation to the W. R. Grace shipping family.) At Pennington, New Jersey, where he attended a preparatory school with his older brother John, Eugene competed aggressively in the classroom and athletic field. He entered Lehigh University in Bethlehem in 1895, and soon was at the top of his class in grade point average and was captain of the baseball team.[30]

Following his graduation from Lehigh, Grace entered Bethlehem as an electric crane operator. He was transferred to the open hearth and became superintendent of yards and transportation in 1902. He first glimpsed Schwab in 1904 while hanging on the back of the businessman's private railroad car. "Even on the lower rungs [of management] I could feel the breeze of a new personality," he wrote. After Schwab took over active management of the company, he called on the young Grace to go to Cuba. Vexed by the high operating costs of the Juragua mines, Schwab asked him to "clean up the situation." He would double his pay to $500 a month for the trouble. "He asked if I thought I could do it. Of course, I told him I thought I could, but I must put a condition in it. He asked what it was. I told him that when I got the job done in Cuba I wanted to come back here and continue, I wasn't looking forward to my life's work being in Cuba. He said that was fair, go down and get things in shape and come back."[31]

After completing his work in Cuba, he was named by Schwab assistant to the general superintendent at Bethlehem. Less than four months later, in June 1906, he had the general superintendent's job. Nothing stopped him from then on. In 1908 he was named general manager; in 1911 he was made vice president; and in 1916, at age 39, he was appointed president, succeeding Schwab who had previously handled the post.[32]

Schwab had found a subordinate who pursued the techniques of making money with almost Messianic fervor. It was not happenstance that Grace founded the corporation's internal slogan, "Always More Production," and told journalist Forbes, "The earning of profits is a matter of supreme moment to every executive." Lanky, with a small face pierced by deepset eyes, Grace was shy in public, hot tempered in private, and had the power of concentration to put out of his mind any stray thoughts until a problem was solved. And in contrast to his boss (whose habits of spending were uncompromising), Grace seemed compelled to scrutinize every cent that passed within his view. He built himself a large house on "Bonus Hill" overlooking the mill at Bethlehem, but furnished it sparingly. He did not indulge in alcohol, and imposed a reign of strict accounting on his wife, the former Marion Brown, a daughter of a local family, and their three children. With pride he told Forbes how he trained his eldest son about the ways of money. Sent at age 10 to a boarding school, the boy lived in a small, bare-walled room and was allowed no more than 5 cents of spending money a week. In this way, Grace reported, the boy was thrifty and didn't take money for granted.[33]

Given his penny-pinching ways, it was of small wonder that Grace perfected the practice of the "cost underrun" that was to engulf the company in controversy over wartime shipbuilding costs. Six months after President Wilson had declared war on Germany, Grace had set up the Bethlehem Shipbuilding Corporation to unify the company's shipyards under his leadership. Vice president Joseph W. Powell, an ex-navy officer, was authorized by Grace to enter negotiations with the Emergency Fleet Corporation for building 86 troop and tanker ships, a large portion of which were to be built at Sparrows Point. The contract arrived at with the government was a so-called "bonus-for-savings" agreement which permitted the company to receive half of any savings over the estimated cost of the ships. The job was estimated to cost about $140 million, including a fixed profit not to exceed 10 percent. When the ship job wound up costing only $93 million, Bethlehem was entitled to get $24 million in extra profits above the fixed profit margin.[34]

The bonus-for-savings contract came under intense criticism when the Shipping Board filed suit in 1925 accusing Bethlehem of deriving "excessive, unreasonable and unconscionable profits" on the job. The government asked for the return of $14,789,708. Most of this involved alleged excess profits, but there was a supplemental claim that the company fraudulently billed the government for wage bonuses. (For every $1 in bonuses the company gave to its shipyard workers at Sparrows Point, $2.625 was allegedly billed to the Fleet Corporation.) The case threw into sharp relief the question of the relative strength of the private and public sectors during the war. Evidence presented before the Special Master showed that Bethlehem had insisted on the bonus-for-savings contract and then strove, on pain of withdrawing from negotiations, to get the highest estimated price possible. On January 5, 1918, the two government negotiators had formally complained to Charles Piez, vice president of the Fleet Corporation, of the company's attitude: "We wish to place on record the fact that [Grace and Powell] have insisted on comparatively high prices for these vessels; that they have only with difficulty been persuaded to quote us on the types of ships referred to; and, that their attitude has been characterized by an arbitrary refusal to guarantee or stand behind delivery dates."[35]

What particularly irked the members of the Shipping Board was that Schwab had not ended Bethlehem's bonus-for-savings contract when he became director general of the Fleet Corporation. In one of his first decisions Schwab ordered shipbuilders negotiating contracts to comply with a 10-percent profit restriction. But he failed to apply this stricture on his own company. "They were the first contracts he should have assured himself were within the earnings limitation of the profit he set," said Shipping Board member Frederick Thompson, who was not involved in the original negotiations.[36]

But Grace refused to buckle under. Handling the case for his boss, he filed a countersuit against the Fleet Corporation and announced publicly, "Bethlehem has been advised that its contracts are valid and that it will not only defeat the Government's claim but will sustain its claim for the balance remaining unpaid."[37] Eventually he and lawyer Paul Cravath, who was quite indispensable, prevailed. In February 1936 Special Court Master William Mason ruled that, despite the nearly 30-percent profit on the contract, it was valid. According to the Special Master, the Fleet Corporation was fully aware of the company's profitmaking intentions and had signed the contract in order to induce the company to undertake the work. "It was Bethlehem's organization that was necessary to insure success to the shipbuilding program of the Fleet Corporation," a key passage of the decision noted, "and as the Government did not have power to compel performance by an unwilling organization, if Bethlehem demanded its price on the basis of substantial commercial profits rather than contribute such services on a patriotic basis, the Government was obliged to take the contract on such basis or not at all."[38]

The Master further exonerated Schwab fully, saying that he could not be held responsible for the failure of the Fleet Corporation to amend its contract with Bethlehem. After disallowing a few erroneous billings to the government, the Master ruled that the steelmaker was entitled to an additional $5.4 million in compensation on the ships.[39]

*

Steaming for Europe in February 1923, for his annual restorative at the roulette tables of Monte Carlo, Chairman Schwab had a long chat with Clarence Barron. As diarist of the American dream (Barron owned *The Wall Street Journal*), he sought the answer to what was then on the minds of many prospective moguls: What accounted for the towering success of Bethlehem?

The chairman answered with a story. "This is the damndest thing I ever knew or heard of. At 17 I lived at Loretto and had never worn a suit of clothes that was not made by my mother. As I swung around on the railroad curve I saw the great Edgar Thomson steel works for the first time in my life. It was the biggest thing I had ever seen, employing 10,000 men. Yet in seven years, or at the age of 24, I was president of the Edgar Thomson steel works and a few years later had built the Homestead plant."[40]

Ten thousand steelworkers offered 10,000 opportunities to bend the men to his will. Soon Schwab had his method down. "One of our best furnace men was never able to get his heats above 15 when I knew there was a possibility of 30, so I arranged with him without disturbing his organization that I would take a hand in it. With a piece of chalk I marked '15' on the floor in big figures, the number of heat turns that day. When the men returned they asked what it meant and were informed that the big boss had been down there and chalked up '15' on the floor. When I came around the next day the men had crossed out that '15' and chalked up '16,' and they continued chalking it up until I had to stop them for they would have wrecked the plant. They were turning out more than 30 heats."

Barron was fascinated. Whenever he had a problem with an employee, he said, he bulldogged it. "I believe in shaking that man up, giving him a jolt and a new start. Every man, of course, must work by his own policy. The men who love me the most I sometimes think are the men to whom I have given the biggest shake-up."

Schwab replied, "For 22 years I never spoke a harsh word to anybody in the entire organization of the Carnegie Steel works. My feeling is that the average working man is a better man than the rest of us. He is a good, loyal citizen and family man.... You can't express yourself to working men and get the best out of them unless you feel in your heart the sympathy for them you express."

So was that it? Barron asked—the deep sympathy that rested in his heart for the workingman?

Not quite, for Schwab had another story to tell. Six years after he became president of Edgar Thomson, when he was 30, he said, he faced the greatest challenge in his career. He had been assigned by Mr. Carnegie to restore order at Homestead during the bitter strike and lockout with the Amalgamated union. "I never left the works from the 6th of July until Christmas," Schwab recalled, but he accomplished his task.

"Since that day I have never had labor unions in any of my concerns. We make our own labor unions. We organize our labor into units of 300 and then the representatives of these 300 meet together every week. Then every fortnight they meet with the head men.... We discuss matters, but we never vote. I will not permit myself to be in the position of having the labor dictate to the management."[41]

Schwab was describing to Barron the Bethlehem Employee Representation Plan, a company union which had been praised in many quarters for offering employees "democracy on the job" without the ill effects of trade unions or strikes. Dr. Charles W. Eliot, for example, the former president of Harvard, championed the Bethlehem Plan as an answer to Bolsheviks and others who might try to foment discord among the working class. If America was going to preach democracy abroad, it couldn't forget to implement it at home. "The Bethlehem Plan is not supposed to be perfect but perfecting," wrote Dr. Eliot in *The New York Times*. The bottom line was that it "has won a place in the esteem of the employees in every plant."[42]

Dr. Eliot's fulsome praise notwithstanding, the Bethlehem plan owed less to the desire for "perfecting" steelworkers than to a need to get Schwab and his cohorts out of a tight spot following a critical government ruling in August 1918. The historical context of the Bethlehem Plan was important: American doughboys were in France, fighting in the Saint-Mihiel salient; public passions in the United States were in full flood; and Grace, youthful president of Bethlehem Steel, was faced with an acutely embarrassing situation. At the very moment that his boss, as director general of the Fleet Corporation, was exhorting the working man to aim his rivet gun at the kaiser, the National War Labor Board had come out with a report condemning Bethlehem's own labor practices.

"It appears beyond doubt," read the document, issued on August 1 and signed by the board's cochairmen, former President William Howard Taft and Frank P. Walsh, "that the dissatisfaction among the employees of the company has had a serious detrimental effect upon the production of war materials absolutely necessary to the success of the American Expeditionary Forces. This was clearly developed in the testimony of the officials of the Ordnance Department." Grace and his associates were faulted for complicity in various antiunion tactics. One involved raids on suspected union members by Bethlehem city police. The police chief had admitted that he ordered city police to enter the shops and threaten men, or arrest them on misdemeanor charges.[43]

The agency also had a considerable difference of opinion from Grace and Schwab as to the fairness of the incentive wage system. According to the agency's summary of findings, the plan was capricious and arbitrary; not even company officials were able to explain to the satisfaction of the board why wages fluctuated so greatly between pay periods. The agency directed Bethlehem to revise or eliminate the bonus plan, saying it was an unacceptable substitute for overtime and Sunday pay mandated for companies holding war contracts. It also ordered the company to collectively bargain with elected shop representatives to resolve the serious morale and pay problems revealed by the investigation.[44]

If there was any question about how Eugene Grace would respond to the board's award, it was answered over the next month. He did nothing. His defiance was a major breach of government-industry etiquette in a war emergency. With tensions mounting, Grace was summoned to Washington by the labor board to explain his defiance of its order. In answer to questions from Taft, he said the company could not afford the overtime wage scales ordered, since it "would result in wiping out our entire earnings." Expressing skepticism of his answer in light of the company's recent earnings, the board nevertheless agreed to forward Grace's concerns to the War Industries Board. (Robert S. Brookings, chairman of the WIB Price Fixing Committee, rejected the plea of poverty as absurd.)[45]

Although they sparred some more, the differences between Taft and Grace didn't seem insurmountable. When Taft emphasized the absolute need for collective bargaining to forestall further employee dissatisfaction and turnover, Grace agreed. Amost meekly he promised that shop elections would be held in the very near future. The reason for Grace's surprising acquiescence soon became apparent. Two weeks after his appearance before the board, the company announced that its employees had caught "collective bargaining" fever. In conference with management they had "adopted" an independent scheme of collective bargaining called the "Bethlehem Employee Representation Plan." According to the press release, worker representatives had become so inspired by the ERP that they had traveled all the way to New York where the plan was finalized in the New York suite of Schwab's publicity agent, Ivy Ledbetter Lee.[46]

Lee was not a novice in "employee representation." He had hatched a similar plan for another client, John D. Rockefeller, Jr., following a brutal labor confrontation. In 1914 coal miners at the Rockefellers' Colorado Fuel & Iron went on strike over the company's refusal to recognize the United Mine Workers. Evicted from their company houses, some strikers and their families were living in tents and starving. The climax came when 20 of them were killed in a brawl with armed guards at the "Ludlow Massacre." In Congressional testimony it was acknowledged that Rockefeller money had paid the wages of the guards involved in bloody Ludlow.[47]

In response to the public outcry Lee had prepared a plan of employee representation with the advice and assistance of W. L. Mackenzie King, later prime minister of Canada. King devised a mechanism which would avoid recognition of a union while allowing for bargaining of a different sort. The plan called for regular conferences between management and employee representatives in a kind of unicameral legislative setting. Because channels of communication were opened between the employer and employed, the program was billed as a "square deal" for labor that fostered self-improvement among the toilers. The Colorado Plan did wonders in repairing the tattered reputation of the Rockefellers in Congress and in the mainstream press.*

Would the Bethlehem Plan work the same magic for Grace and Schwab? Lee left little to chance. Promoting the Bethlehem plan tirelessly in the influential New York press, he picked up positive blurbs from the *Times* and other papers as well as the trade press.[48] A few blocks from Lee's quarters, at Cravath & Henderson, Grace huddled with Paul D. Cravath. The lawyer, who had worked some magic of his own in 1915 with the British sub deal, reviewed the employee plan to make sure that it would meet the legal requirements of the War Labor Board's award. He and Grace had seized upon the wording of the board's award, which called for recognition of employee shop representatives, but not specifically union employee representatives. In other words, the decree could be interpreted as giving sanction not to the Amalgamated Association of Iron, Steel and Tin Workers, but to a conference relationship between employer and employed. So long as the company could dominate the conference relationship it would be free of the "taint" of unionism.[49]

Between November 1918 and February 1919, the Bethlehem Plan was instituted at the mills, winning reluctant acquiescence from the

*Lee's resourcefulness was not lost on other businessmen. Soon he was considered the premier publicist for Wall Street and represented a consortium of millionaires and corporations, including Walter Chrysler, Harry Guggenheim, Thomas Fortune Ryan, Benjamin Javits, American Tobacco Company, Armour & Company, General Mills, Standard Oil, Pennsylvania Railroad, and United States Rubber as well as Charlie Schwab and Bethlehem Steel. Lee's critics were also legion. In *The Brass Check* (1919) Upton Sinclair anointed Lee "Poison Ivy," a nickname that stuck through his career.

labor board. (The board's skepticism diminished in direct proportion to the pressure it received from Congress to clear its docket and disband after the armistice.)[50] Like the Colorado plan, the Bethlehem ERP bore a superficial resemblance to a governmental legislative body. At Sparrows Point, where the workforce numbered in excess of 10,000, one representative was to be elected for every 300 employees. Nominations for office were made with the assistance of management, with three persons to be nominated for every person to be elected. Departmental elections were held for the persons nominated, and these were followed by plant-wide elections involving a narrower range of candidates who had previously won a plurality in their departments. Once elected, the employee representatives had two functions: to participate in a general conference with management once a year, and to sit on one of 13 plant committees that met on alternate months with management representatives in joint conference.[51]

J. Howard Burns was elected chairman of the ERP at Sparrows Point. He was a machinist with 30 years seniority at the mill, whose family rented a prized company house on E Street. His committee involved itself in athletics and recreation sponsored by the company; other topics of discussion taken up by the ERP committees were employees' transportation, bonus and tonnage schedules, "practice, methods and economy," housing and domestic economies, safety and prevention of accidents, and "condition of industry," later renamed "education and publications."[52]

The program was wholly advisory. Matters voted on by the panels were to be discussed first with company representatives in joint session and then passed to higher management as a recommendation. A similar advisory system encompassed employee grievances, called "adjustments" in ERP lingo. Article X of the bylaws said, "Any matter which in the opinion of any employee requires adjustment, and which such employee has been unable to adjust with the foreman of the work on which he is engaged, may be taken up by such employee, either in person or through any representative of his department." There was a catch, however. While any matter could be grieved, management retained the right to terminate all grievances. Final authority was vested in the office of the president.[53]

The significance of the plan lay not in what it did, but in what it didn't. The employee representatives, even if they had the desire, lacked any power to bargain as equals with the company or to sign legally enforceable agreements. Committee meetings were held at the general manager's office and the company determined the compensatory pay for the representatives. Needless to say, the plan did not recognize the right of employees to call for a strike. That was outside the parameters of the debating-club atmosphere. It went without saying that the panelists could not negotiate with management on wages, hours, and working conditions, although, as noted, there were advisory committees on these subjects.[54]

The essence of the ERP was to avoid the appearance of antiunionism while keeping, as much as possible, the reality. And the policy worked. For a multitude of reasons the Amalgamated Association failed to attract the loyalty of the workforce. This was especially true after the "Great Steel Strike" of 1919 collapsed and Amalgamated supporters were routed from steel country. The strike was centered in the Pittsburgh-Youngstown and Chicago districts, where it was hard fought for several weeks. At Sparrows Point, however, no more than 500 men responded to the strike call. They were principally laborers and tin-mill workers, some of whom had been involved in an earlier walkout. Most returned to work a few days later when it became apparent that they couldn't shut down the works. Plant manager Roberts applied the screws to the renegades, saying he would fire those absent from work since they were violating the "contract" with the company ERP which barred strikes. It was a no-win situation.[55]

"You kept your thoughts to yourself," remembers Frank Amann, a bricklayer. "Back then there was guys around and they wanted to start a union. But all of them lived in Sparrows Point. So the company calls them in and says, 'Look, we want this union stuff cut out or we'll bring the moving truck over tomorrow.' And the guys said okay 'cause the company had everything sewn up. Hell, in them days Sparrows Point was Siberia for the union."[56]

CHAPTER 8

TIN CANS AND TAKEOVERS

The 1920s proved to be the perfect setting for the organization that Schwab built. Bethlehem had better raw material resources than its competitors; it could deliver steel at lower prices in the Northeast and on the Pacific coast; it was better financed and was constantly on the prowl for new conquests. It played the postwar game of acquisitions with the same consummate skill it showed in eliciting record production from the men inside the mills.

"Don't waste time talking about what you have already done," said President Grace when asked his business philosophy by B. C. Forbes. "Use the time in planning and executing something new. Dividends are not earned by reminiscing on the past."[1]

Sparrows Point in particular was growing fast because it was supplying basic products that were part of the revolution in consumer goods. The empire of Charlie Schwab increasingly became interlocked with the rising empires of other big thinkers, men who had found new ways to help Americans spend their money. One big thinker was James D. Dole, the "Pineapple King." A New Englander by birth with family ties in Hawaii, young Jim Dole had set sail for the islands after his graduation from Harvard with a nest-egg of $1,200. He had thought of going into the coffee business, but changed his mind after he secured some acreage in a homesteading district opened up on Oahu Island northwest of Pearl Harbor. He decided he would raise pineapples, but not for the fresh food market. He would grow them and put them in cans. In 1903 his total pack was 1,893 cases; in 1918 his company was producing over 1 million cases a year. By 1922 he had so outgrown his original plantations on Oahu that he purchased the entire island of Lanai and turned it into a lush agricultural community of towns, roads, and 2,000 acres of land suitable for pineapple cultivation.[2]

Dole's Hawaiian Pineapple Company became a major consumer of tin plate from steel mills. A lot of it was ordered from Sparrows Point. The economic underpinning of the enterprise—the supply of tin plate for the canning of Dole's popular pineapples—was a case study of the global mechanics of mass-volume production. To make tin plate, Bethlehem hauled iron ore 1,200 miles from Cuba, and purchased tin from the Straits Settlements near Singapore, 13,650 miles distant by boat. These raw materials, together with coal from Bethlehem mines in West Virginia and Pennsylvania, and limestone from southeast Pennsylvania, were dumped into the furnaces and worked into steel at Sparrows Point. Two weeks were required to make a sheet of tin plate—three weeks if one went back to the pig iron, which had to be refined into steel and then rolled into bars before the tin mill even started its work. Despite its paper-thinness, tin plate had to be "malleable," able to withstand the stress of being bent, folded and stamped into complicated shapes without breaking, yet strong enough to absorb the pressure that built up inside a container during sterilization. The coating of tin on the steel was excellent for preservation of most vegetables and fruits. It resisted the attack of food acids as well as prevented oxidation (rust) on the outside of the can. Being inert, tin inhibited the metallic taste from leaching into the contents.[3]

Shipped 6,700 miles from Sparrows Point to Honolulu through the Panama Canal, the tin sheets were fabricated into cans and taken next door to the Dole packing factory. An incredible piece of machinery had been invented by Dole engineer Henry Ginaca. In a series of clattering movements out of a Rube Goldberg fantasy, the Ginaca grabbed a pineapple, peeled it, sized it, cored it, cut off its ends, and delivered the cylinder to the packing tables. As refinements were added, the output of the Ginacas increased until the machines could handle 100 fruits per minute. After further processing and inspection, the fruit was packed into sterilized cans with a dollop of syrup for flavoring.[4]

Shipped back to the United States, Dole's pineapples found their way onto the shelves of grocery stores, where they were purchased by housewives for desserts, fruit cocktails, or "just plain snacks," as Dole advertisements suggested. As for the tin plate, its fate was inglorious. After the fruit contents were emptied from the can, it was thrown into the wastebasket. The garbage man then came to collect it and cart it away to the dump.

*

Dole's was one of many success stories which helped the mills of Sparrows Point grow. Prior to John T. Dorrance, for example, the Joseph Campbell Preserve Company had been just another South Jersey packer of all kinds of soups, jams, fruits, ketchups, preserves, and vegetables. But Dorrance had a better idea. A chemist by trade, he wondered why the company should pay freight rates and storage charges mostly on the water in soup. He discontinued all product lines except for extra thick concentrated soups, and focused his firm's attention on making those items an object of necessity in every home. With the help of the cherubic "Campbell Kids," the soupmaker became a household word—and a huge consumer of Sparrows Point tin plate. In 1923 Campbell purchased 500 million cans through an arrangement with Continental Can Company, its sole source supplier, which in turn purchased a majority of its plate from Bethlehem Steel.[5]

The surge of tin-canned foods was an outgrowth of many factors that, combined with longer-term economic trends, were altering the daily life and outlook of the average citizen. Most profound was the acceleration of the 25-year drift of population from rural to urban regions. The "passing of the family farm" with its grow-it-yourself ethic swung the door open to a ripe market for canned foods. In the cities, where fresh produce was hard to come by or out of season, canned fruits and vegetables became acceptable, and inexpensive, substitutes.

The apartment buildings constructed in the decade were notable for their shrunken cooking areas, optimistically labeled "kitchenettes." As people moved into apartments, they prepared less food from scratch. Between 1914 and 1929, consumption of canned fruits and vegetables more than doubled, and consumption of canned milk nearly tripled. In 1925 Hormel & Company introduced a new canned meat mixture which was called "luncheon meat"; it was not sterilized, but preserved with sodium nitrate and nitrate salts and could be kept in refrigerators for long periods. "In many city homes," William Leuchtenburg wrote, "the family sat down to a meal that started with canned soup, proceeded to canned meat and vegetables, and ended with canned peaches." Many found that pineapples reposing in a Dole Diamond Head or Outrigger can added a touch of exotica to dinner in a cramped city apartment.[6]

Thanks to the advances in canning, foods that had been unknown or available only to the rich became part of everyday life. The average American consumed a more varied diet of meats, vegetables, and fruits than ever before, as food entered the factory and became "processed" rather than grown. If some aspects of the canning revolution were less than desirable, including the heavy use of chemicals and inferior taste of vegetables and meats sterilized before they were canned, the business had a beneficial effect on grocery stores and the emerging network of national food chains. The humble tin can made sense to food merchandizers on the following grounds. At about half the weight of standard glass bottles, cans reduced shipping costs. They were easier to shelve, store, and display, did not break like glass, and did away with the bother of returning empty bottles to suppliers.[7]

As tin cans were used more widely, the industry expanded into the packaging of nonfood items. A distinction was made in the trade between "packers' cans" and "general-line cans." Packers' cans were sealed cylindrical containers used for cooked food (tomato products dominated this market, filling over 200 million cans in the mid-'20s). General-line cans were developed for everything else. Improvements in the reclosure of metal cans were instrumental in increasing general-line sales. Paint in cans, for example, had been around since 1906, yet had the drawback of poor resealing. In 1916 the American Can Company introduced the "double-tite" friction cover for effective reclosure. Thereafter sales of canned paint picked up greatly. Canmakers became sensitive to the issue of "convenience" and "quickness" in improving the structure and opening of their containers. The key-opening coffee can was brought out in 1920. Screw-top or double-tite friction covers were developed to enclose products like floor waxes, putty, and polishes that heretofore had been sold in bulk. At the same time, canmakers stamped out bigger and bigger drums for industrial users and shippers.[8]

Still another market for canmakers was the "fancy" retail trade. Advances in lithography had made possible the printing of designs and multiple colors on metal cans with the facility of paper printing. Medicinal articles, spice cans, and tobacco boxes were sold in striking cans in the '20s, while even more pretentious work was reserved for lipstick, perfume, shampoo, and other signature items of the "flapper look." (There was an interesting tie-in here: nearly all women's cosmetics then were manufactured from benzol and benzene, by-product coke chemicals extracted from coke ovens. With one of the country's largest extraction plants, Sparrows Point was a leading supplier of the chemicals for cosmetics wrapped in its tin plate.)[9]

Bethlehem Steel's place in the tin-plate business was overwhelming in the Northeast for the simple reason that the Sparrows Point mill was the only facility east of Pittsburgh which produced all grades of tin plate, lead-coated terne plate, and non-tinned black plate directly from sheet bars. Schwab had managed this feat by buying out the single competitor that had challenged his position. In January 1916 John E. Aldred, the Baltimore financier, had organized a tin-plate company with John M. Dennis, president of Union Trust Bank and state treasurer of Maryland. However, before the plant was erected on a 12-acre site in East Baltimore, Schwab had bought Aldred out, paying him about $2 million for the company. Schwab then closed the facility, sold off the land, and moved the equipment to Sparrows Point. Once freed of the Aldred threat, Schwab made certain that no other company moved into the Baltimore district by signing contracts for the supply of tin plate with Southern Can, Metal Packaging Corporation, and American Can; these concerns collectively canned the greatest volume of tomatoes in the country as well as many other vegetables, seafood, and general-line items.[10]

Between 1919 and 1924 tin-plate production at the Point was tripled from 1 to 3 million "base boxes" a year. Production was then raised to 4 million base boxes in 1926. A base box consisted of 112 sheets of 14 by 20 inches; the plant had the equipment available to manufacture some 448 million sheets annually. Each of these sheets in turn could be made into four to six cans, depending on the container size. Adding up the output, the number of cans that could be produced out of Sparrows Point tin plate was approximately 2 billion a year. In all, tin plate accounted for $20 million in annual sales.[11]

The relationship between steelmakers and automakers underwent a similar dynamic in the '20s. Not only did the volume of individual vehicles produced race forward, but the use of steel per vehicle increased. Here a key milepost was the stamped-steel body, replacing the wood frame, that made its commercial debut in 1923. The car body was the achievement of Edward G. Budd, a brilliant young Philadelphia engineer. When approaching the problem of constructing a car body, Budd conceived of steel not as a surface covering laid on a wooden frame, but as a structural material which would provide both the framework and "skin" of the vehicle. Such a unified frame had tremendous resilience. The flexibility of steel enabled a vehicle to absorb shocks from the chassis that would crack the varnished wood frame of the typical car. With steel, greater strength could be achieved with less overall weight. This increased the speed at which a car could be driven using the same motor. Budd's all-steel frame was an important breakthrough in American auto manufacture, as significant in terms of consumer safety and vehicle design as Ford's assembly line had been for expediting the production of cars. But Budd's invention met resistance in Detroit. Established body makers denounced the steel car as cheap and flimsy, and insisted that thin sheet steel construction was not strong enough. In spite of its many advantages, the Budd frame did not become universally accepted by the industry until the middle 1930s.[12]

The sheet-jobbing mill at the Point manufactured extra-wide (up to 72 inches) sheets for the Chevrolet assembly plant in East Baltimore. Opened in 1920 and expanded in 1922 and again in 1924, the sheet-jobbing mill also handled corrugated and galvanized sheets. The plate-flanging mill sold parts for stamping machinery required for building crankcases and power trains. In still wider circles of trade, the automobile business stimulated sales of tin plate (for oil cans and gasoline cans) and pumped public dollars into big-ticket road projects (highway bridges, tunnels, and the like). Bethlehem partook in many of these road projects in its sales zone, winning contracts for the supply of 83,000 of the 110,000 tons of cast-iron segments needed for the tubing of the New York–New Jersey Hudson river tunnel (1922), and for 20,000 tons of plate and girder for the approaches of Camden-Philadelphia bridge over the Delaware (1924).[13]

*

The chain of processes that ended with the sale of tin plate, sheet steel, and heavy plates, bars, and rails on the market began with the effective control and transport of the base elements of steelmaking: limestone, coal, and iron ore. Keenly aware of the economies of backward integration, Schwab and Grace pursued the prewar "trust" policy of acquiring ownership of the best supplies of raw material available. In 1918 Bethlehem Mines purchased the Hanover limestone quarries in southeast Pennsylvania and entered into a long-term contract with the Western Maryland Railway for shipment of the crushed rock to Sparrows Point 65 miles away.[14] (The railroads continued their "special relationship" with Bethlehem and the other major steelmakers. While no longer the financial bankrollers of the past, the railroads purchased large amounts of steel for rails, locomotives and cars, and the steelmakers made long-term contracts with the railroads for hauling raw materials to the mills and transporting finished goods out.)[15]

Schwab and Grace also took action to enlarge coal holdings. During the war they purchased the Gray mines in Somerset County, Pennsylvania, 232 rail miles from Baltimore. The Western Maryland Railway handled the contract, hauling the coal in single shipments exclusive of other cargo. For years to come the "Gray train" would be known to the locals as it rumbled past the little mountain hamlets of Sand Patch and Mount Savage, down the graceful Potomac Valley and over the Blue Ridge to Baltimore, where the mineral was transshipped to the Point by barge. Less fabled as an institution but equally important were the cars of coal handled by the B&O over its West Virginia routes. In 1919 Bethlehem acquired the mining interests of former U.S. Senator Henry Gassaway Davis—46,000 acres of coal-bearing lands around Morgantown, West Virginia. Together with mines in Indiana County, Pennsylvania, and the Pocahontas district of West Virginia, these new holdings provided the Sparrows Point plant with numerous sources of good metallurgical coal.[16]

Five hundred pounds of limestone and 1½ tons of coal were needed to make a ton of steel—along with 2 tons of iron ore. Schwab's ex-employer, the Steel Trust, controlled the richest domestic source of ore at Mesabi in the north country, while Schwab pushed his organization south to work the best supplies of foreign ore. During the war and thereafter the Cuban mines, especially Mayari, had served their new master well. Shuttling between the island and Sparrows Point had been the S.S. *Cubore* and its three sister ships. They handled 11,000 tons per trip; more than 2 million tons landed at the man-made harbor immediately south of the blast furnaces between 1918 and 1922.[17]

At the same time the company was busy developing a new open-pit mine in Chile called El Tofo. Located four miles from the Pacific coast north of Coquimbo, the Tofo deposit had been known for more than 30 years. In January 1913, Schwab had completed negotiations to lease Tofo from the Schneider-Creusot interests of France. Bethlehem agreed to pay 12.5 cents in royalties to Schneider for each ton removed from the ground, plus a yearly fee of $200,000. It had been Schwab's intention to develop the property with $5 to $7 million in financing provided by French bankers, but the war had upended these plans. The fields lay fallow during the war as Bethlehem relied on Cuba as well as shipments from Sweden.[18]

The Chilean ore showed great promise because of its unusually high iron content. As Schwab pointed out in an interview with *Iron Age*, "Our Chilean ore runs 67 percent in iron, whereas Lake ores that once averaged 60 percent in iron are now down between 50 and 55 percent. Chilean ore hauled 4,400 miles can be laid down at Sparrows Point at $6 for the ore required to make a ton of pig iron, or about 6 cents per unit. While Pittsburgh has its advantage on fuel, the ore advantage of the Eastern seaboard plant is such that pig iron can be made more cheaply at Sparrows Point than at any point in the Middle West."[19]

Like the Mayari excavation, the most difficult aspect of the Chilean project involved transportation. The tops of the mountains were 2,000 feet above sea level. The French had built a bucket cableway from the top of Tofo to the loading docks at Cruz Grande. Bethlehem removed the cableway and blasted a standard-gauge railroad up the hillside in tiered layers. Over ten miles of track were used to scale the peak. Ore started coming down the mountainside in trains of stubby ore cars late in 1921. Over the following year the mines began their first steady output—222,947 metric tons.[20]

The successful tapping of Tofo was yet another example of Bethlehem's formidable presence in the constellation of steel. From the viewpoint of other producers it was in the enviable position of having both low-cost foreign ores all to itself, and a secure market for the sale of its goods in the great steel-consuming areas of the East.

By all statistical measures Bethlehem was heading up the ladder of power while the Steel Trust was slowly descending. U.S. Steel still held the predominant position with 45 percent of national ingot capacity, but this market share reflected a steady decline from 68 percent when the Trust was formed in 1901.

There were two factors that explained the Trust's loss of market share. One was its lack of technological inventiveness. The giant was slow to take advantage of new technology, and lost business to its more adventuresome competitors. Having so much capital tied up in the status quo, U.S. Steel was loath to undercut its investment. A number of Midwest firms thrived with small modern mills. The second factor was political. As the world's biggest corporation, the enterprise raised the hackles of politicians, including President Taft, whose Justice Department had filed suit in 1911 to break up the conglomerate into its original parts. Chairman Gary decided as a matter of policy to refrain from further acquisitions and to content himself with internally generated growth.[21] The fact that the Trust wasn't looking for new properties was a crucial piece of intelligence for Charlie Schwab.

In 1921 there were 12 important steelmakers independent of U.S. Steel. The larger ones owed their origins to the rail-rolling fraternity of the prior century. There were also several newer companies that served the "light-rolled" sheet, tin plate, and tubular markets. In order of steelmaking capacity, they were:

1. *Bethlehem:* 3.1 million annual tons at Sparrows Point, Bethlehem, Lebanon, and Steelton
2. *Midvale Steel & Ordnance:* 2.9 million annual tons chiefly at Johnstown and Nicetown, Pennsylvania
3. *Jones & Laughlin:* 2.6 million annual tons at Pittsburgh and Aliquippa
4. *Lackawanna Steel:* 1.8 million annual tons near Buffalo, New York (Earlier operations at Scranton, Pennsylvania, were transferred there in 1902)
5. *Youngstown Sheet & Tube:* 1.5 million annual tons at Youngstown, Ohio
6. *Republic Iron Company:* 1.4 million annual tons at Cleveland and Youngstown, Ohio
7. *Inland Steel:* 1.2 million annual tons at Indiana Harbor, Indiana, near Chicago
8. *Colorado Fuel & Iron:* 1.1 million annual tons at Pueblo, Colorado
9. *Wheeling Steel:* 1 million annual tons at Wheeling, West Virginia
10. *Steel & Tube Co. of America:* 900,000 annual tons at Chicago
11. *Brier Hill Steel:* 600,000 annual tons at Youngstown, Niles, Warren, and Girard, Ohio
12. *Weirton Steel:* 350,000 annual tons at Weirton, West Virginia[22]

There was a definable geographic separation of the companies. In the Northeast, Bethlehem was snugly anchored on or near the coast, and Midvale was concentrated at Johnstown. The zone between Pittsburgh and Chicago had Jones & Laughlin, Youngstown, Lackawanna, Republic, Inland, Steel & Tube, Wheeling, Brier Hill, and Weirton. West of the Mississippi, CF&I was peripheral, serving a limited market, while the Birmingham district was controlled by U.S. Steel, which had acquired Tennessee Coal and Iron a few years prior to the Taft antitrust suit.

In December 1921 came the startling news that Theodore Chadbourne was preparing to merge seven of the independents. His idea was to make a Midwest conglomerate rivaling U.S. Steel in size. The members of his "North American Steel" were presented as Midvale, Lackawanna, Youngstown, Republic, the Steel & Tube Company of America, Inland, and Brier Hill. Chadbourne had started off as the lawyer to financier George Gould, son of Jay, and had built up a strong reputation as a merger specialist, with recent credits that included the organization of Mack Trucks. The Wall Street house of Kuhn, Loeb was to handle the financing.[23]

In May 1922, just as negotiations appeared to be reaching a conclusion, Eugene Grace shocked Wall Street with the announcement that Bethlehem had arranged for the immediate purchase of Lackawanna. Bethlehem "intended to grow as fast as opportunity permitted," Grace said evenly. His news on behalf of the boards of the two companies was made at the closing bell of the May 11 session of the exchange.

When the market reopened the next morning, Lackawanna and Bethlehem were the subject of wild speculation. Lackawanna jumped up $9\frac{1}{4}$ points in the first few minutes, making one of the biggest leaps to date on the Big Board. Bethlehem common also made "violent opening advances." It was not to be the only violent outburst of the day. A few hours later Robert LaFollette—"Fighting Bob"—took to the floor of the U.S. Senate to address his colleagues. LaFollette said a few moguls were taking the opportunity to stamp out competitors like Morgan had 20 years before. "Iron and steel today lie at the base of every human activity," he declared. "The men who control this basic commodity control the nation."

By unanimous voice vote the Senate passed LaFollette's resolution directing the Federal Trade Commission to inform the Senate on the probable effects of the mergers and requesting the attorney general to study the advisability of proceeding against the proposals under the provisions of the Sherman Anti-Trust Act. When reports of the Senate resolution reached Wall Street, the stocks of Bethlehem and Lackawanna dropped, but only temporarily—by the end of the day they had posted a combined gain of $14\frac{5}{8}$ points.[24]

Bethlehem's raid on Lackawanna threw the industry into an uproar. The next Tuesday Grace and Moses Taylor, chairman of Lackawanna, disclosed the terms of the sale—Bethlehem would give $\frac{41}{100}$ of a share of its 7-percent preferred stock and $\frac{61}{100}$ of its class-B common stock for each share of Lackawanna stock, aggregating $35.1 million. Bethlehem also would assume $21 million in funded Lackawanna debt. On the same day (May 16) Chadbourne answered with a curse the rumor that Schwab was preparing to have himself installed as chairman of the North American combine. On Friday (May 19) Youngstown announced it was withdrawing from the Chadbourne talks, but "would reserve its options" in seeking other firms.

On Monday an unprecedented number of steel shares changed hands on the exchange; Consolidated Steel of St. Louis and Hoosier Rolling Mill of Terre Haute, Indiana, were reported to have drawn up defensive merger papers; Republic stock had taken a nosedive; and Inland was looking down the Calumet River at Steel & Tube. "STEEL MERGER TALK KEEPS STOCKS BUSY. Prices Fluctuate Widely as the Result of Many Rumors on Wall Street. MIDVALE NOW A FEATURE," was the headline in the *Times*.[25]

A week later the paper reported another tempest—a series of "secret" talks initiated by Schwab in Pittsburgh at the home of Emil Winter, president of Pittsburgh Steel Products Co. "SCHWAB SEEKING PLANTS?" wondered the *Times*.[26]

On June 5 the Federal Trade Commission gave its report on the probable effects of the mergers to the Senate. In regards to the Chadbourne grouping, the agency said it could not give a detailed analysis since amended plans following the Lackawanna defection had not been announced. However, regarding the consolidation of Bethlehem and Lackawanna, the agency warned of a "dangerous tendency" for monopoly in certain steel products. The combined firm would control in excess of $33\frac{1}{2}$ percent of steel rails produced east of the Ohio River and north of the Potomac, and over 39 percent of rail splice bars in the same territory. Bethlehem's already formidable position in structural shapes would be enhanced, posing a threat of price manipulation in the construction trade in New York, Philadelphia, and other population centers. The commission told the Senate that it had issued a complaint under the Clayton Act requesting Bethlehem to appear before the panel on July 24 and show cause why a cease-and-desist order should not be filed against the merger.[27]

Four days before the July 24 FTC hearing, Harry Micajah Daugherty, the attorney general, stepped forward on behalf of the merger. He said there was not the slightest reason to assert that the public would suffer if the consolidation were consummated. As a matter of fact, he said, the merger would increase rather than restrain competition in the industry. "I am persuaded that the motive which prompts Bethlehem to acquire the Lackawanna plant is the sole desire to secure greater efficiency and economy in the production, handling and distribution of steel products, and that the thought of acquiring a monopoly or of enhancing the price was never present. The whole transaction from beginning to end impresses me as being thoroughly clean, honest and straightforward." He refused to take antitrust action, saying he was obeying the recent resolution of the Steel case before the Supreme Court.[28] "I need not stop to point out that in *United States* v. *United States Steel Corporation* the Supreme Court refused to declare illegal a combination of much greater magnitude.... The merger now under consideration will be neither an actual monopoly nor even an attempt to monopolize; and, of course, the decision just referred to is controlling."*

While Senator LaFollette and other congressmen howled loudly at Daugherty and the Justice Department, Schwab and Grace gathered their subordinates together for a meeting of the board of directors. The board approved higher capitalization to cover the acquisition. Total authorized stock was raised by 10 percent to $170,108,500.[29] Thus fortified, Bethlehem absorbed Lackawanna in September 1922, without a dissenting vote cast by any stockholders. In acquiring the company, Bethlehem acquired important material holdings in the Northeast and on Lake Superior; this was one of the main attractions of the company to Schwab and Grace. Lackawanna's ownership of the Witherbee-Sherman mines on Lake Champlain, New York, provided access to ore that could be delivered to South Bethlehem at low cost over the Delaware & Hudson's route from Albany. The purchase eased the shortcomings of the South Bethlehem plant's location, which included a $2-a-ton penalty on foreign ores shipped from tidewater. Lackawanna also operated mines independently or on joint account in the Lake Superior district, including at the Mesabi and Marquette ranges. Coking coal was supplied by the Ellsworth collieries near Monongahela City, Pennsylvania, south of Pittsburgh, and hauled by rail to Buffalo. There was limestone available from company quarries

*In one of the most controversial cases of the era, the Supreme Court denied on March 1, 1920, the government's request to break up U.S. Steel into its original 190 companies. Speaking for the majority, Justice Joseph McKenna said the government had failed to prove that the combination had attained the monopoly status it had originally sought to obtain, and went on to declare that bigness, of and by itself, was not illegal under the Sherman Act. (This portion of the opinion was often called the "Magna Carta of Big Business.") Three justices dissented, arguing that U.S. Steel's intent to monopolize put it in violation of the law. Perhaps the oddest twist of all was that the two leading proponents of the government's case abstained from voting on grounds of judicial ethics. Louis D. Brandeis had been a vocal critic of U.S. Steel before going on the bench, and James C. McReynolds had been part of the prosecution as Wilson's attorney general. As a consequence, the view of four judges became the deciding opinion in the 4-to-3 decision. (*U.S.* v. *U.S. Steel Corp.*, 251 U.S. 417, 1920.)

at Pekin, New York, and a fleet of Lake steamers thrown into the bargain. The steamers offered economical distribution of the plant's steel products at Cleveland, Detroit, and Chicago, the first time these markets could be reached by Bethlehem-controlled transportation. And because the consolidation involved the purchase of physical assets rather than capital stock (a technicality of legal significance), Grace boasted to New York newspapermen of the merger—"it will be beyond the pale of any Federal Trade Commission action."[30]

Meanwhile, the plans of the Chadbourne grouping were falling apart. Over the summer three more steelmakers had jumped ship. This left Midvale, Republic, and Inland seeking to merge. Although Attorney General Daugherty once again gave his seal of approval, on September 28 the merger collapsed; the difficulties in obtaining financing had proved insurmountable.[31] From this failure sprang the most audacious move to date on the chessboard of steel. Despite the still unsettled FTC complaint against the Lackawanna purchase, Schwab and Grace began negotiations with the Midvale group. The implications were tremendous. Nearly as large as Bethlehem in terms of ingot capacity, Midvale was among the 12 largest industrial concerns in the nation. Its assets of $270 million dwarfed such giants as Westinghouse Electric, American Smelting & Refining, and American Can.[32]

Midvale was an outgrowth of the earlier set of steel mergers that had resulted in the Bethlehem-Pennsylvania Steel combination. The current company had been capitalized in October 1915 by Percy Rockefeller, William E. Corey, Frank Vanderlip, Samuel Vauclain of Baldwin Locomotives, and other important Wall Street figures. Mimicking Bethlehem's success, the Midvale organization had built a munitions works at Nicetown, Pennsylvania, and, securing large war orders from Britain and France, purchased the Cambria Steel facilities in Johnstown. As Eugene Grace wrote, "The Midvale Company had been formed to develop as a Bethlehem competitor in the East."[33]

Under tight security, Grace began negotiating directly with William Corey, chairman of the board and intimate of the Rockefellers. The negotiations involved the sale of all Midvale properties with the exception of the Nicetown munitions works, which would be spun off as a separate company. The numbers bandied about were some of the biggest in corporate America since the war. Grace agreed to a $97-million purchase price with the assumption of another $41 million in bonds and indebtedness. Schwab never questioned the price, saying that the property would repay many times over. Wrote Grace:

"Mr. Schwab was out at his home in Loretto during the latter stage of these negotiations. Late one afternoon Mr. Corey and I reached an agreement for the transaction. It occurred to me that it might be well to advise Mr. Schwab where we were at that time. He had a private wire in his home and I sent a message to him telling him what the conditions were.

"I got an immediate reply from him—'Get it.' I answered him at once—'Got it.' "[34]

*

After swallowing the number-5 steelmaker, Schwab and Grace had gone on to consume number 3. Announced on November 24, 1922, the Midvale purchase flung Bethlehem's bulk far beyond the Atlantic shore. Beginning at salt water at Sparrows Point, the empire now reached 200 miles inland to Johnstown, at the very doorstep of Pittsburgh, then, veering north, dropped down to fresh water at Lackawanna Harbor on Lake Erie. Owning every major integrated plant east of Pittsburgh, Bethlehem's dominance in the steel-consuming areas of the East was increased at the same time that it could enter the markets of the Middle West on a more favorable basis. An example of the overall market concentration afforded by the Midvale properties was the Gautier division at Johnstown. It turned out a large variety of products for automobiles, including special sections for wheel rims, window sashes, and crescent bars, tapping the Detroit market. Johnstown greatly increased Bethlehem's share of heavy-rolled products for railroads and the mining industry, including steel wheels and axles, mine hoisting equipment, hydraulic presses, and steel hopper and boxcar fabrication.[35]

The acquisitions gave the company an ingot producing capacity of 7,600,000 tons annually, or about 15 percent of the country's total. This placed it at one-third the size of U.S. Steel (excluding Bethlehem's shipbuilding and repair yards), and nearly triple the size of Jones & Laughlin, the third-largest steel producer. Weighed on the scales of big business, Bethlehem had assets of $637 million, more than either General Motors or Ford, and ranked only below U.S. Steel and Standard Oil of New Jersey in terms of overall size, according to *Moody's Analysis of Investments.*[36]

In carrying out the Midvale acquisition, Schwab and Grace were given unconditional clearance by the Harding administration. Even before the proposed merger was a day old, Attorney General Daugherty let it be known that his office would offer no opposition. In an interview published in *Nation's Business*, the magazine of the National Chamber of Commerce, Daugherty said it was not his intention to prosecute businessmen who may have broken one of the "technical" provisions of the Sherman Act. "So long as I am Attorney General I am not going unnecessarily to harass men who have unwittingly run counter with the statutes." Not long after, Daugherty was forced out of his post after he refused to testify in connection with his involvement in the Teapot Dome scandal.[37]

The Federal Trade Commission tried not to be so smugly satisfied. On January 26, 1923, the commission issued a rewritten finding directed at both the Midvale and Lackawanna acquisitions under the anticombination provisions of the Clayton Act. The agency expressed the hope that its action would at least delay the Midvale buyout until hearings were held. But Grace and Schwab brushed the complaint aside. Citing "competitive conditions," they proceeded with the merger. On March 30, 1923, Bethlehem officially took possession of Midvale's assets upon the delivery of 976,814 shares of newly issued Bethlehem common worth $97,681,400 for distribution to Midvale stockholders.[38]

Over the next four years the FTC amassed thousands of pages of testimony in an attempt to develop the facts concerning the competitive impact of the consolidation. The proceedings were an uphill battle with minimal cooperation from the steelmaker. The case, though, was halted in 1927 after the Supreme Court ruled in *U.S.* v. *Eastman Kodak* that the FTC did not have the authority to order divestiture of physical assets under the antitrust laws.[39]

CHAPTER 9

CHARLIE AND HIS WORKMEN

Old woman, we have had a good year and you can have $2,000,000 to do what you please with."

—CHARLES SCHWAB, TO HIS WIFE RANA IN 1926[1]

As Bethlehem's holdings accumulated, so did the accumulations of the chairman and chief stockholder. Inventorying Schwab's personal property and plant was an exercise in listmaking on a grandiose scale. Ever since culling his first millions at Carnegie, he had pursued possessions with a vengeance that showed little sign of abatement with age. Money, the cornerstone of his management policy, was similarly the key to his private life.

The first splurge was his residence on Riverside Drive in New York. Modeled after elements of three French châteaux (Chenonceaux, Elois, and Azay-le-Rideau), "On the Hudson" outdid in legal tender what the Astors, Charles Tyson Yerkes, and other nabobs had piled up on Fifth Avenue; when completed it was described as both the biggest and most expensive private residence in Manhattan. It occupied a full city block on Riverside Drive between 73rd and 74th Streets and West End Avenue. The main structure had 75 rooms, cost $3.5 million in land and building ($44 million in 1987 dollars), including the price of opening up a quarry near Poughkeepsie for the granite exterior. The same stone was used later to build the Cathedral of St. John the Divine on Amsterdam Avenue.[2]

Furnishings and artwork set Schwab back another $4 million ($50 million in today's coinage) as his agents scoured Europe for objets d'art and his architect, Maurice Hebert, imported craftsmen to duplicate what could not be acquired. On the main floor there was a Louis Seize drawing room, a Louis Quatorze dining salon, and a Louis Treize breakfast room, each of heroic proportions (the dining salon was said to seat 250 guests). Like any self-respecting tycoon, Schwab had a private gallery and filled it with Old Masters, among them canvases by Hals, Corot, and Innes, Titian's portrait of Cardinal Bembo, and Turner's *Rockets and Blue Lights*, purchased in 1917 at the then-exorbitant price of $250,000. There were some uniquely Schwabian touches, such as the ten-ton Carrara marble altar installed in his chapel room and the bronze sculptures of master rollers and other steel mill personnel lining his grand hallway.[3]

Whether one was a foreign diplomat or a New York building contractor, a recipient of Schwab's brand of hospitality was not likely to forget the experience soon. When he retained Caruso to sing arias to his guests at a dinner party, he forked over $10,000 to the singer—half a dozen years' wages for one of his employees. He kept three French chefs on around-the-clock call in his kitchen. Masseurs were available night and day. Some wags wondered if Schwab was operating a home or a hotel: The château's accoutrements included a full-size gymnasium, bowling lanes, and a basement swimming pool set amidst columns of Tuscan marble. This was quite an outlay by a man who was childless in his marriage, and who spent many months a year traveling. "Ostentation, it was generally admitted, was part of the Schwab scheme of things," remarked Lucius Beebe dryly in his chronicle of the very rich, *The Big Spenders*.[4]

While indisputably pricey, the Riverside palace paled by comparison to the country compound Schwab completed after the war on the hills that surrounded his boyhood home at Loretto, Pennsylvania. If the New York property displayed his early wealth from Carnegie Steel, "Immergrun" ("Evergreen" in German) constituted a sprawling monument to the success of Bethlehem Steel. Based on his known holdings of Bethlehem stock, Schwab had drawn about $21 million in cash and stock dividends between 1914 and 1918. A goodly portion of this windfall was spent in the religious little town he had fled from so many years before.[5]

Immergrun was a bid for immortality by a man disinterested in the stewardship of wealth as championed by his mentor, Andrew Carnegie. Openly he reveled in wealth. "I disagreed with Carnegie's ideas on how best to distribute his wealth," he told a journalist. "I spent mine! Spending creates more wealth for everybody. I say that with due regard for Carnegie's theories and the unquestioned hope in his mind that he was being a benefactor of mankind. But I believe that, let alone, wealth will distribute itself."[6] His message struck a responsive chord. Americans enjoyed the Steel King's excess; they took comfort in a life of apparently abundant pleasures.

*

Loretto lies high on the crest of the Alleghenies, 25 miles west of Altoona and not far from the Pennsylvania Railroad's Horseshoe Curve. Elevations of 2,200 to 2,500 feet give the district a brisk and changeable climate. In spring and summer there are swift turns from sunlight to clouds, with squalls playing across the woodlands and in the pastures. The winters are severe. Schwab's parents, John and Pauline, were Americans of German roots who moved to the community in the 1870s because of its adherence to strict Catholic values. Charlie, the oldest of five Schwab children, helped his postman father with the mail before dropping out of St. Francis, the local school run by friars, in 1879 and heading to Braddock and his rendezvous with history.[7]

Twenty years later, not long after he had become president of Carnegie Steel, he returned to Loretto and picked the highest hill for a summer retreat. In 1900, a pleasant Victorian cottage was completed on the site, at a cost of $35,000—spare change compared to what was to ensue. The original Immergrun stood intact until the war years, when Schwab was seized by a vision of grandeur that would demonstrate to his satisfaction the esteemed place he held among American moguls. As work began, he enlarged his land holdings from 600 to 1,000 acres. Roads around Loretto were moved and new electrical transmission lines strung. In one project, a whole hillside was cut down and an "Italian Renaissance Garden" placed on the leveled site. Teamsters with horse-drawn drays hauled the earth down to Loretto Station and dragged up huge blocks of limestone and other building materials. The original house was raised on tackle and pulleys and carried over the tops of a stand of elms and an orchard that were of sentimental value to the owner and placed at a new location 1,500 feet away.[8]

Then there was the "Normandy Village project." Schwab, it seems, had developed a liking for a farming village he had visited in France—so he had a copy of it made. When finished, the community occupied a 66-acre site on the northern edge of the estate. The village contained a woodworking shop, farmer's cottage, sheep folds, piggery, ice cream room, forge room, stables, smoke room, barn with a bell tower—everything a thriving French village should have, except people.[9]

In all, five years and hundreds of workmen were required to complete the main residence, gardens, grounds, dairy, and gentleman's farm to his specifications. The farm buildings and formal gardens were done by Charles Wellford Leavitt, the landscape architect. Under Leavitt's guidance Schwab's chickens were not housed in conventional chicken coops, but in full-scale replicas of French cottages. The water tower was a facsimile of an English coastal fortress. A reservoir with 1.5 million gallons of spring water was prepared for the estate and village. In accordance with his doctor's orders that he take up golf for exercise, a 133-acre private course was built. There was a stone building to house his fleet of cars, a horse stable, and a private business getaway called "The House in the Woods." Deeper in the forest an open-air theater was used for watching plays. Sometimes Schwab and his guests acted out charades on the stage.[10]

Extravagant, overpowering, whimsical, egocentric—Immergrun was part of the grand self-portrait that Schwab wished to present to the world. In the quiet of tiny Loretto the force of his personality burned as incessantly as in the round-the-clock blazings of Sparrows Point. When the iron gates were opened for the first public viewing shortly before 1920, the 400 residents of the town and representatives of the media saw an idealized baronial life fashioned by and for a lord of industry. The tour started at the stone obelisk built at the village crossroad. Passing a stone gatehouse, the road wound up the hillside through a profusion of copper beeches and white pines in a greensward expanse. To the right was the royal staff of the estate, an 80-foot-high flagpole flying the Stars and Stripes. The plaque at the base of the flagpole reminded viewers of Schwab's patriotism in the war. It quoted Wilson's thank-you letter in full.[11]

The road peaked and swung around an oval forecourt with a spectacular view of the 1,000-acre spread. The main residence was immediately to the south—a three-story Renaissance castle designed by Murphy and Dana and built of yellow limestone blocks with marble trim and a red tile roof. In the main rooms, the floorboards did not creak: They were constructed of the same limestone that formed the two-foot-thick walls. The rooms themselves—44 in all—were reprises of the style of the Riverside estate. Schwab's bedroom was finished in dark "masculine" woods of an Italian motif. Rana's room was French with white wood and gold trim. His room had four floor-to-wall wardrobes; hers boasted full-length mirrors in the dressing room.[12]

Leading up to the house on the south acreage were formal gardens laid out by landscape architect Leavitt: long lines of manicured boxwoods from Long Island, red cedars from New England, and other ornamental shrubbery not native to the region. The gardens were centered around three sunken lagoons that were flanked by life-size statuary, including eight voluptuous female figures. At the base of the lagoons a double stairway rose to the house 50 feet above. Between the double stairs an elaborate waterfall of 14 cascades had been constructed. The water dropped down to an underground tank where it was pumped back to the crest of the stairway by electrical motors.[13]

Seventy men were put permanently on the payroll to maintain Schwab's showplace. They were organized much like the labor in a steel mill. "Company" cottages were rented out to personnel responsible for the gardens, farm, cellars, and general maintenance. The superintendent of vegetables and flowers lived in the West cottage, close at hand to the propagating houses which kept the estate stocked with fresh vegetables, flowers, and herbs year round. (When the Schwabs were in New York a boxcar full of fresh meat, produce, milk and butter, herbs, and tobacco was shipped every Thursday, coupled to a Pennsylvania express train at Cresson.) Gardener's Cottage was located below the mansion. Among his duties was to oversee the hothouse cultivation of 52 varieties of roses. At the gatehouse superintendent Blair Seeds handled general administration and the payroll.[14]

Father Demetrius Schank, T.O.R., was born on the estate in 1917. He was raised in a farmhouse near Poultryman's Cottage, the oldest of eight children of the estate's master carpenter. He scarcely remembers Schwab at all in the flesh. During the summer months (late May to October) when Schwab and Rana lived at Immergrun, he was rarely seen around the estate. He did not like to putter around the flower gardens or even walk through them except when guests or photographers were present. Although he owned the most beautiful gardens in all of western Pennsylvania, whether he enjoyed them was open to question. In the 20 years that carpenter Schank worked for the steelman, he never once spoke to him. All orders came down through the hierarchy from superintendent Seeds. On some mornings Father Demetrius remembers watching a bustle of activity at the oval forecourt far up at the mansion, followed by a Packard speeding down the drive, the royal ensign, C.M.S., emblazoned on the car's doors. That was it.[15]

Mrs. Schwab was more accessible. Several times a summer she

stopped by the Schank farmhouse and chatted with Father Demetrius's mother. The two women had family connections that dated back to before she had married Charlie. Rana was the daughter of a Carnegie Steel chemist. She and Charlie were married in 1883. Suffering from rheumatism and overweight, she spent most of her time at Immergrun in the company of her two sisters and her nieces and nephews. "She was always very kind to us children who lived on the property," the priest remembers.*

There was obviously a story to be told about this venture in palatial living, and a good many magazines of the day found it: Charlie Schwab had built himself something as wonderfully big and ambitious as his company. On the pages of *Colliers* and *National Magazine* Americans could read all about it: about the rise of a boy from driving stakes at $1 a day to driving one of the nation's biggest corporations. But equally important, due to advances in photo processing, Americans could *see* the Steel King and take comfort in the life of enormous riches he enjoyed. They could see him sallying through his Renaissance gardens in breeches and English riding boots, a devoted collie at his side. Or watch him teeing off at the golf course knowing that only the most important sports personalities and business figures were honored with invitations to play there. Or observe him observing his herd of blooded cattle, which were said to be as much a source of pride to him as his famous thoroughbreds.[16]

In the pantheon of Business and Success he had truly arrived in the decade after the war. When he turned 65 on February 18, 1927, he was the subject of adoring interviews coordinated by publicist Ivy Lee. Cognizant that an international conference on naval disarmament was about to begin in Geneva among the U.S., Britain, and Japan, Lee and Schwab used the interviews as a platform to alter the image of Bethlehem Steel as a munitions maker. Lee and his assistants took the opportunity to review the past 20 years of the company to show that its devotion was to Vulcan, not Mars, as a manufacturer of peacetime products. It was further (and falsely) stated by Lee that "this [policy] was not merely a post-war development but a policy that had been

*Yes, Schwab also collected women, although his need for them was not so compelling as for steel mills and objets d'art. For brief flings, Broadway show girls would do. A lengthy liaison with Myrtle Hayes, a handsome businesswoman from Boston, ended in grief when he accused her of forging his name to $24,102 in promissory notes. She eventually pleaded guilty to second-degree forgery and was given a suspended sentence in 1924.

instituted 20 years before, when Schwab first took over." As a result of the publicity, Schwab received letters of congratulation from hundreds of peace-loving people from around the country.[17]

Bethlehem's chairman also attracted the attention of the "daring" journal of the time, *The American Mercury.* Edited by Baltimore's own H. L. Mencken, the magazine hailed Schwab as America's best-loved businessman. What the magazine liked about the steelman was his pumped-up, vernacular style. No dour Scotsman like Carnegie, nor preachy Jeremiah like Elbert Gary, Schwab sent a message of plenty, not piety. "Yust laugh, yust grin," was his advice, mimicking the Pennsylvania Dutch accent of his workmen. His dismissal of politics as trivial and his faith in business prosperity were perfect nostrums to the Puritan ethic that Mencken so rebelled against. Saluted the *Mercury,* "Parents with tears of hopefulness in their eyes have been known to recite the Schwab saga to their children, and awed biographers have written about it a thousand times in newspapers and magazines.... [His career] has done quite as much as any other career to establish the great American tradition that this is a Land of Opportunity, and that a poor boy can succeed gloriously among us, even if he is honest."[18]

*

Not until one encountered Sparrows Point could one appreciate the heights that Chairman Schwab and his associates had scaled as the '20s roared to a conclusion. Following the purchase of Lackawanna and Midvale, the corporation embarked on the biggest internal improvement program in history: Fully $157 million in cash was laid out between January 1, 1923, and December 31, 1928. Of the company's five major works (Sparrows Point, Steelton, Bethlehem, Johnstown, and Lackawanna), none received as high a proportion of the largesse as Sparrows Point.[19]

On the made land where slag dumping had continued to increase the plant's perimeter, the salient of steel pushed onward and upward. From the coke ovens on the southern flats to Numsen's Point on Bear Creek, construction workers continued the massive buildup of new mills and additions to old ones to capitalize on the plant's unique tidewater location through a greater diversification of its products.

A rod and wire mill, opened in 1926, completed the process whereby semi-finished bars were rolled down and stretched longer and longer until they emerged as rods and wire for bridge cable, industrial springs, bale ties, spikes, staples, and nails of all descriptions. Expansion of the tin department increased the capacity to produce tin plate and black plate to a fabulous 4.25 million base boxes a year. A new pipe mill permitted the manufacture of butt-weld and lap-weld steel pipe ranging in size from ¼ inch to 16 inches diameter, and new 17-inch and 21-inch sheet bar mills filled a gap in the demand for flat-rolled products by appliance makers and auto assembly plants.[20]

Twelve years after Bethlehem's takeover of Sparrows Point, the basic steelmaking facilities of the peninsula had been totally rebuilt and finishing capacity boosted to record levels. Whereas back in 1916 the plant never produced more than 35,000 tons of steel rails per month, in 1928 a month never passed at the Point when less than 115,000 tons of steel were made. A greater than threefold increase in physical output under Bethlehem ownership—no other major U.S. mill had expanded so fast. The yearly breakdown of capacity was:

Coke by-products:	
Tar	12.96 million gallons
Sulphate	18,720 tons
Benzol	3.36 million gallons
By-product gas	24 million cu. ft./day
Fabricating and construction steel:	
Sheared plates	150,000 tons
Universal plates	110,000 tons
Flanging department	20,000 tons
Rails	300,000 tons
Pipe:	
Butt-weld	100,000 tons
Lap-weld	116,000 tons
Sheets:	
Black (plain)	72,000 tons
Galvanized	36,000 tons
Blue annealed	20,000 tons
Jobbing (heavy gauge)	42,000 tons
Tin plate:	
Black plate for tinning	225,000 tons
Tin plate	4.25 million base boxes
Rod and wire:	
Rods	40,000 tons
Nails (all types)	30,000 tons
Wire, cable, bale ties, etc.	60,000 tons
Galvanized, blued, cement coated, etc.	14,000 tons

Brass foundry:
Propeller blades and castings 6,000 tons[21]

Big numbers were kicked around Sparrows Point, and the large-scale production yielded earnings of similarly mammoth proportions. For each ingot ton produced at the Point—and on average 2½ tons were produced every minute of the day—an average of $6.75 in income was flowing back into the company coffers. Blessed with the lowest iron ore costs of any mill in the country thanks to the mines in Cuba and at Tofo, Chile, the Maryland mill registered the best earnings ratio of all Bethlehem mills. With such felicitous results Schwab told the Baltimore Association of Commerce that he saw no end to the expansion of the plant.[22]

From the main harbor channel the works was shrouded by its own effluvia of smoke. Closer up the half-open sheds revealed tongues of white flame, rafts of slag-blackened metal twisting and roaring through sanded channels to "sub" or "hot-bottle" cars waiting below, ruby lights flickering hither and yon in sets of mill rolls blurred by motion. Amid the physical sensations of this machine culture—the fire, smoke, steam, and noise—a horde of humanity tramped three times daily through the clockhouses. At shift changes one encountered on the streets leading back to town and on the trolley cars bound for Baltimore young and old, black and white, all coated with the same grimy specks of ore dust. In the final two years of the decade, daily employment averaged 18,000, or 6,000 more than at the war's peak. If one counted wives, children, and other dependents, not less than 80,000 persons in the Baltimore region were bound to the iron molecule and the Bethlehem way of making it into steel.[23]

The world of Sparrows Point was made up of order givers and order takers, strong-willed bosses, demanding work, and often decent pay, according to those who lived out their lives in the mills:

"In them days the foreman had all the power. You went to him for everything. He'd take you out of a gang and put you on the floor or he'd break you if he didn't like ya looks."

"Where I was at the sheet-jobbing mill you just had to work it. Because you was workin' two bars at a time, each one weighed 304 pounds to the bar, and you had tongs and you'd catch the bars and then push 'em back over the top. It was red-hot work. We got $3 to $4 a day."

"If you could produce, you could make it."

"The heat was the worst part. You'd never see a guy bare-chested like they show in paintings 'cause he'd be toasted it was so hot."

"I know it was his hard work but I'll give the Point credit, we had a good livin'. My father raised us—he was from the old country—and he said, 'America didn't owe me anything and America gave me a good livin'.'"

"I led the buggies in the tin mill and you'd hit a bump on the floor and the handles would shake you and sometimes lift you right off the floor. I'd come home and my thighs would be just as hard as this table."

"You were really tied to the company then—your shelter, your food, your job."

"There was no such thing as pollution back in them days, 'cause nobody even thought about it."

"We had these bosses' pets. The cousin of the boss, he always got the pick of the jobs. They would push him up first."

"There was some union talk, yeah, but it was subversive. They had the company ERP around and you had to give lip service to it."

"Bethlehem Steel was so big, somehow you didn't see the owners, you saw the foreman who you knew and liked and was related to your buddy and you got along."

"You had to work till you dropped dead. You never had no chance to stop because you didn't make enough money to accumulate while you were working. Retirement—hell, there were no such thing."

"The thing about being a child in a steelworker's family—you had to keep quiet when he was on night shift and slept during the day. But when he was working 3 to 11 you could make a lot of noise. And the day shift was nice too."

"I remember I went with my mother down to the Point on the streetcar. We had a basket of food for my father, a whole roasted chicken and fruit and bread in a nice picnic basket—and we waited. All of a sudden there comes a man out of the mill. I didn't recognize him as my father he was so dirty."

"When a man fell into a ladle of hot metal, the ladle, the metal, everything was buried in the ground 'cause the men refused to work with it."

"It is only now that I can understand how rough it really was. When you are young things breeze by you a lot. You can't think about it, if you want to stay alive. Because if you was working down [in] the pits where there are cranes carrying tons and tons of hot metal over your head, you can't think about it. People will talk about this to a degree, but they would rather forget."[24]

*

The hiring pattern established during the Wood years continued with some variations through the '20s. By and large the skilled and semi-skilled labor for the mill migrated from the traditional steel capitals around Pittsburgh and eastern Ohio. Steel work was highly specialized and also tended to be self-selecting; as a result, a machinist was more likely to hail from a steel town in the Middle West than from the skilled industrial classes of nearby Baltimore. Although skilled jobs were still dominated by American-born workers, there was much less outright discrimination against Eastern Europeans. Bethlehem had no hesitancy to put "foreigners" on skilled jobs if they were efficient and could meet the company's tonnage quotas. For example, when a Steubenville, Ohio, rolling mill shut down in 1926, the Finnish crews there were recruited en masse to Sparrows Point, where they continued to converse in their native tongue and were bossed by an Italian foreman named Louie. Production, not ethnic origins, was the key to getting jobs.[25]

By the same token, though, the foreign-born could expect to go only so far in the mill hierarchy. By all accounts it was impossible for a man whose name ended in *-sky* or *-ini* to be appointed above the job of general foreman. And the management ranks were as much a preserve of the native-born WASPs as in the Wood days. "It damn near would take an act of God," remembers one worker, for a non-native applicant to be admitted to the company's "Looper" management trainee program.[26]

The Southern black man continued to fill the ranks of unskilled labor at the Point. The close of unrestricted immigration in 1921 increased this employment trend. In 1920 there were 1,527 black males employed in the two largest "steel side" departments, the blast furnace and open hearth; ten years later the number had increased to 3,887. With 6,000 black employees overall, Sparrows Point was the largest employer of black steelworkers outside of Birmingham. The company's need for blacks kept several recruiters busy, including Edward Watkins, who went through the backwater towns of Virginia and

North Carolina to encourage local men to come to the mill and, not coincidentally, to board at the rooming house he ran in the company town.[27]

Blacks could be found in nearly every department at the Point, but almost always in one occupation, that of general laborer. Records from 1930 indicate that 74 percent of the 6,000 blacks employed at the mill were general laborers while the remainder attained limited advancement as "dinkey-train" conductors, riggers, and in other semiskilled posts. "The black man's job was everything the white man don't want to do," said James Allen, a laborer in the tin department.[28]

Black or white, the men shared long hours. Until 1923 the 11-hour day shift, the 13-hour night shift, and the 24-hour swing shift on alternating Sundays remained in effect. Although such workweeks were the cause of controversy and complaints since before the war, it was not until the Interchurch World Movement report in 1920 on the causes of the 1919 strike that press and public attention were refocused on the matter. President Harding appealed to Elbert Gary in 1922 to reduce steel hours as a humanitarian gesture. Gary refused. Public censure increased, and in late 1923 Gary, Schwab, and other steel figures grudgingly agreed to end the 24-hour swing shift and to institute the eight-hour day and six-day week "as soon as the labor supply permitted." (Significantly, during the hours controversy the ERP employee representatives sidestepped the issue and issued no call for shorter workweeks.)[29]

Progress along these lines, though, was halting at best, as mill managers tried to squeeze the most hours out of the president's guidelines. For the most part hours were trimmed back partially—from 84 a week to between 56 and 70. While the 12-hour day was eliminated in the open-hearth and steel-rolling departments per Harding's wishes, the number of men working ten or more hours a day in other departments was not. And while the 24-hour swing shift was gone, as late as 1929 one-fourth of the workforce at Sparrows Point still worked a full seven days a week.[30]

Moreover, the men granted a "six-day week" found their newfound leisure reduced by the way in which the company calculated weekend layoff time. By most reckonings a six-day week meant a weekend between Saturday afternoon and Monday morning. But in the world of steel, when weekly changes of shift were scheduled on Sunday, each shift was set ahead by eight hours. As a result, weekends were of 24-hour duration. After finishing six days of daylight work at 3:00 P.M.

Saturday, a steelman was called back on Sunday at 3:00 P.M. to begin six days of 3:00-11:00 shift, and so on. Bethlehem called this arrangement a six-day week; workmen called it, with more accuracy, "one day off in 19"—meaning it took approximately three weeks of daily shiftwork to accumulate a full calendar day off work. Alluding to these hours, a report by an East Baltimore church organization noted ruefully: "The kind of work which most of the people do does not seem to be conducive to the cause of religion. At present most of the people are employed in the steel mills work. The irregularity of employment and hours of labor, in particular the latter, make it difficult to interest the great number of men in church attendance."[31]

The custom of "past practices" prevailed in other areas. There were no paid vacations for hourly employees at Sparrows Point, but there were two *unpaid* holidays a year. They were Christmas Day and the Fourth of July, when the machinery paused from midnight to midnight and a steelworker could sleep as late as he wished.

*

William Walk Womer accepted these grinding hours because he knew of nothing else. Of German stock, he was from people who had been ironworkers since the Civil War, beginning with his grandfather who had worked at the old Greenwood Iron Furnace in south-central Pennsylvania. William Womer was born in Mifflin County, Pennsylvania (not far from Schwab's Loretto), in the shadow of the Standard Axle Works, where he donned work denims and joined the payroll as a 15-year-old. Slowly before and during the war he advanced to "first helper," or head of the furnace crew, in the open-hearth department, which honed his skills and prepared him for the intersection of his ambitions and the grand designs of Charlie Schwab. When the 1922 recession put Standard Axle on part-time, word came through the grapevine that Sparrows Point was hiring. So William Womer went down to the plant and landed a job as first helper in No. 1 open hearth.[32]

"He came for the tonnage incentive," his son Ben says. By making fast heats he found he could earn up to $10 a day in wages and incentive pay. And while not all days were that good (when the furnace broke down his daily pay could recede to $4 or less), Mr. Womer still figured on a better income—about $3,200 a year—than in Mifflin County. So when Ben was 12 his father broke with 60 years of family tradition and moved his family to Sparrows Point. His skills were such that the company assigned the Womers a corner row house on H Street with six rooms and a bath. Ben's father then made the house a home by placing a big marble brick engraved with his initials, W. W. W., at the front door.

"I knew from way back what the open hearth was. I was no stranger to the open hearth," Ben says. "When my dad first went to work down there, they still worked those tremendous long shifts. Well, on Sundays during the long turns my mother would pack him his meal in a big aluminum bucket and I'd take it over. I'd get right up on the floor and see my dad." His mother never went along because "women didn't go up on the floor. Women were taboo."

Ben remembers a boyhood at Sparrows Point filled with good times. Afternoons at Pennwood Park soaking up the sun before taking a plunge in the harbor at the company beach. Saturday mornings at the company store on C Street munching on hot sesame buns. Saturday afternoons caddying for foremen and superintendents at the new company country club. Like his parents, young Ben took comfort at the cleanliness of the town, was proud of the country club and athletic fields established by Bethlehem for use by local residents, and accepted the social distinction between the families of workmen and bosses as a given.

In 1926, he entered Sparrows Point High. Because it handled all students from the whole Patapsco Neck and Dundalk area, the school was large, and because it demonstrated the steel company's commitment to education, its appearance was formidable. Above the front vestibule there were three concrete minarets surmounted by a flag of Maryland carved into stone panels and embroidered with a scroll bearing the state motto, *Fatti maschii parole femine* ("Deeds are manly, words womanly"). The school was the preserve of Principal Joseph Blair and his wife Nellie. The couple lived in a company house on Bosses' Row on C Street and nearly every summer took a vacation in Paris. In order that the sons and daughters of steelworkers could share their enthusiasm for Paris, Mr. Blair required that all students take two years of French.

Ben hated French. He found, he says, he could not pronounce it, could not understand it, and wanted nothing to do with it. His rebellion culminated on the day that Mrs. Blair came to the school to discuss her recent trip to Paris. "We got assembled down in the auditorium, the two classes of French, and she got up there on the stage and gave a talk. All in French. Well, I didn't understand a word she was saying. So when it come time for us to go back to our rooms, our teacher said, 'Now your homework for tonight is to write a thesis on what Mrs. Blair talked about.' So I turned my paper in the next day and I told 'em. I wrote on my paper, 'I don't know what the lady was talking about because she was talking in a foreign language.' Well, immediately, I got expelled."

Accepted back to school the following fall, Ben again found French class so painful that he decided to quit. "I got thoroughly disgusted—I needed that French like I needed a hole in the head—and I figured, well, I just might as well quit and go work in the mills. It was in April in 1928 and my mother had to take me up to City Hall to get a work permit. I wasn't 18 years old."

In the summer of 1928, still a few months shy of his 18th birthday, he started working in No. 2 open hearth. Due to his dad's influence he was put on the hot metal mixer as a skimmer boy. The mixer was a large enclosed tank where hot pig iron from the blast furnaces was mixed before the liquid was dumped into the open hearth. Ben was stationed on a platform on top of the mixer. When the iron was poured into the mixer, "you shoved this rod with an endplate up against the slag floating on top and tried to hold it back while the pig went into the mixer. Now the only thing between the ladle and you, protecting you from the heat, was this steel plate. I used to take wooden boards and lay 'em down on the platform to protect my legs from burns. Well, by the time they'd pour a ladle of metal, damn if them boards wouldn't be on fire. I'd hop off real fast."

Straining beside a cascade of molten iron, balanced on boards so hot they would catch on fire—"I did get a little tired of it," Ben says of skimming ladles. "But," he says, "I took it. I didn't say much to nobody 'cause I figured I had no choice but to stick it out for a while."

*

Charlie Parrish entered the mills because he figured it was the best opportunity available to him. To be raised in Goochland County, Virginia, was to be convinced that life elsewhere had to be better. Born in 1910 (the same year as Ben Womer) and orphaned in 1912, Charlie was brought up by his grandfather, an ex-slave who farmed property owned by his second wife on Little Creek. "I loved my granddaddy. He was a powerful man and he was a good man. But my step-grandmother, he was livin' on her place and he had to do what she say. And she didn't even allow my sister and me to eat at the table. She give you two pieces of corn bread and one biscuit at dinner, and a child he loves biscuits, but you only got one. So you was hungry. I was a big boy, strong, and I used to mind the cows. My granddaddy take a piece of bread from the table like he was gonna eat it. He'd take it outside and I'd eat that bread and it taste like cake."[33]

After his step-grandmother died in the flu epidemic of 1919, Charlie's grandfather moved to Dogtown with his two grandchildren. Charlie tended his aunt's hogs and began sharecropping on a white man's farm. In 1922, when he was 12, he was put in charge of all the farming. A year later he dropped out of the two-room schoolhouse down the lane. A routine settled into place. During the summer a quarter of all the corn he had harvested went to the white man. The wheat he took to the grain mill paid off the seed he had purchased in the spring. He had one pair of shoes, one suit for Sunday, and a strong desire to get out.

But where? On a spring day in 1926 the answer came rolling into Dogtown on the state road. Inside a Ford roadster sat his girlfriend's older brother. "See what you get working up North?" the boy said to Charlie. Charlie could see. "That car cost $600! And the boy drives the car to church and the best I can do is ride on the back of my uncle's truck. The boy made me want to come to Sparrows Point 'cause that's where he was workin'." After harvesting the corn that summer Charlie asked Carbon Carter, the storekeeper in Dogtown, for a loan. "Mr. Carter was a white man and he was a good man. He loaned me enough to buy a train fare to Baltimore. He gave me a pair of working clothes, two changes of underpants, and a good pair of shoes. And he told my aunt, 'Charlie's a smart boy. There's nothing here on the farm. He oughta go away.' He told me to pay him back when I gets the money."

In the morning light, Sparrows Point reared up in a maze of shapes and sizes. Bottle-shaped furnaces stood 100 feet in the air, trains whistled, and buildings crowded close before the whole scene flattened out and the company town came into view. Charlie stepped off the trolley car at 4th and D and made his way to the employment office past the throngs of men coming off the overnight shift. He remembered the date clearly: September 9, 1926.

"You had about 50 men in the employment room and a lot were laying down asleep. I guess they been there all night. Soon the employment man comes in and he says, 'Who's gonna work today?' and I hold up my hand and step right over the man that was sleepin' before me. The employment man wrote me up on a card and gave me a ticket to go to the dispensary. And they examined you and see [if] you got any disease. They asked me my age. I was but 16, but I said 18, and they said, 'Fine. You go on a gang.'"

Charlie's gang was No. 5 C.L. ("colored labor"), assigned to the blast furnaces. The gang dug ditches and shoveled coke and cleaned up the slag that slopped over on the casting platform. Because of the intense heat, work on the casting platform was intermittent, an hour of "slag-cracking," then a period of rest and recuperation. The work wasn't much different from the farm—"no-brain bullwork," Charlie called it, with pick and wheelbarrow substituting for rake and horse-drawn plow. Six feet tall and 200 pounds, Charlie could more than hold his own. What was different from Goochland County was the other men. The laborers were a rough lot, surly and profane, who played pranks on one another and hooted it up when a new fellow fainted from the heat. "They'd look at you and say, 'We'll have this boy down in a few days.'" Out-roughing them all, says Charlie, was the "straw boss," Big Jeff.

"Big from the feet up, near 300 pounds big—Big Jeff done hollered at me so when I started. 'Boy, where you come from?' he said. Oh, he cussed me out and yelled at me and made me bend down and tie his shoelaces. And I never been cussed at before. Didn't know what it was. And he'd go through our lunchpails in the morning when we was out workin' and he used to eat my lunch up."

Pure terror was Charlie's initial reaction to Big Jeff. Over the months, though, he came to see the man as "something of a bluffer." "Once he thought he had you he laid back. When nobody be around he'd set there on his bottom and look sorry tired. See, he was mean but he had no power. He wasn't nothin'. He just wanted to drive us black boys so he looked good to the white man." The white man was a tall ex-shipyard riveter named George Martin who was foreman of the mechanical crews. One day when the gang was cleaning up a spill on the casting floor, George Martin came over to Big Jeff and pointed at Charlie. "I want that boy," Martin said. With those words Charlie was taken out of the labor gang and made a "burner helper" in the mechanical gang.

It was an incredible promotion in his eyes. "I was real happy because with the race prejudice it was hard for a black man to learn a trade anywheres." Over the next year Charlie was one of a handful of blacks trained to use oxygen torches and burner equipment. It was the first time in his life he felt any real sense of stability, and it was a liberating feeling. "My wages were knocked up from 37 cents [an hour] to 41 cents. And I thought, well, the Bethlehem Steel Company was okay, it give you a chance."

*

A steelworker was no stranger to injury and he knew what it was to take risks. There was no way, though, he could prepare himself for the sheer number of perils he faced in the mills. With machine operations so encompassing, mishaps to life and limb were accepted as inevitable. Dangerous levels of carbon monoxide could (and did) leak through the fuel lines; powder charges could (and did) misfire; heavy objects fell; steel boiled out from furnaces and sparks shot across the walk aisles from the rolling presses. More than the hours and the heat, accidents brought home the vulnerability of the human organism to the requirements of the iron molecule.

Not long after he became a burner helper, Charlie Parrish lost a finger joint—a copper plate he was helping to set in place on a furnace bosh slipped off its rigging and smashed his hand. Ben Womer also won his stripes. He got a burn singe on his foot, having not hopped off the metal mixer quick enough. Bethlehem had erected a new dispensary in town to treat the injured, and retained two specialists in traumatic surgery. Although "lost-time injuries" were reported to have been reduced by half from the prewar rate, accidents and injuries were still a daily occurrence. Disabling, lost-time, and fatal injuries were covered by the 1914 Maryland Workmen's Compensation Law, which required modest insurance paybacks to the victims, and thereby gave a financial incentive to improve safety procedures.[34]

The most dangerous department was the open hearth, whose reported accident rate was 77 percent higher than the average for the industry. According to one report, "Accidents caused by falling objects are relatively more frequent in the open-hearth department than in any other except fabricating. Machinery is a special hazard in the open hearth and fabricating departments, and the open-hearth workers are also exposed to burns, perhaps especially the pitmen who clean up after tapping." The life expectancy of a worker who entered the steel mills at age 20 was seven years less than that of a person who entered a white-collar or professional job.[35] Joe Lewis, a door puller in No. 1 open hearth, relates an incident that shook him up:

"I was working 11:00-to-7:00 turn, it was in 1927, and I passed a form on a stretcher under a blanket. It didn't look right. It didn't look like a person there and yet again it did in a way. When I got to my job I saw they were sweeping a man up with a broom and a shovel and putting it in a shopping bag. So what I'd seen was half the form." The man, Joe says, had been torn up by a hot ladle car. "He had gone to dodge the charging machine on the platform and he ran right into the hot ladle car. At that time they had the motor out in front of the car and that's what ground him up. The worst thing was I knew the guy. His name was Jim Jacobs and he lived at Jones Creek. Quiet guy, around 40 years old. Well, I thought I was tough, but this made me real sick."[36]

Although the death of Jim Jacobs did not merit attention in the local press, several accidents in 1927 did. On December 1, 1927, the *Baltimore Sun* reported, "SIX MEN HURLED TO GROUND AS SCAFFOLD FALLS. Snapping Rope Causes Accident in Steel Plant at Sparrows Point. Injured Are Sent to Mercy Hospital. Several Still Unconscious, Half a Dozen Other Workmen Escape Unhurt." Four weeks later, on New Year's Eve 1927, there was an explosion at the power plant: "STEEL PLANT BLAST DARKENS ENTIRE TOWN. Transformer Explodes, Halting Mill Work at Sparrows Point. Power Unit Fire Fought Two Hours. Bethlehem Accident Laid to Short Circuit in Motor."[37]

A trolley car named *Dolores* normally carried the deceased to cemeteries in Baltimore. Owned by the transit company, the trolley was equipped with a special compartment to hold the casket and had seats for a mourning party. After a funeral service in a local church, the body would be placed aboard the car and people would gather along the tracks and view the coffin as the trolley moved slowly through the community. Since the company did not as a practice release the names of employees killed in accidents, *Dolores* let the townsfolk know that something had happened. "Nobody liked seeing it, but it was locally famous," says Ben Womer.[38]

CHAPTER 10
SMASH-UP

Maybe it was during the golfing party that he hosted on September 12, 1929, to celebrate the birthday of James Anson Campbell, president of Youngstown Sheet & Tube, who was speaking openly of retirement. Or perhaps it was a month later, when Bethlehem's banking syndicate successfully completed the sale of 800,000 shares of new common stock, raising $88 million in capitalization just days prior to "Black Thursday." Whatever the specific impetus, Charlie Schwab was again on the prowl, ready to clinch the title of master builder of steel.[1]

Like many businessmen of the nation, he was convinced that the collapse of the market in October 1929 was a temporary "adjustment" before the next spurt of growth, and solid evidence seemed to be on his side. Still clogged with orders, Sparrows Point and other company mills ran at near capacity levels through November and December 1929, with total income for the year reaching a high of $67,469,245, capping 60 consecutive quarters of profits since the chairman had made his munitions trip to London in October 1914. As of December 31, 1929, Bethlehem had a surplus of $117,546,496 in cash and liquid securities, again a record. "We have established such concentrated industrial strength in this country that we can all proceed with a high degree of rational self-confidence during 1930," Schwab instructed the Illinois Manufacturers' Association gathered in Chicago on December 10, 1929. "Never before has business been as firmly entrenched for prosperity as it is today."[2]

The plan he had conceived with Grace was gargantuan in scope. Targeting Mahoning Valley as the next step of Bethlehem's march westward, Schwab and Grace selected Youngstown Sheet & Tube as the best candidate for a buyout. Sheet & Tube had started out at the turn of the century as a combination of several furnaces around Youngstown, Ohio, for the purpose of rolling gas pipe for Rockefeller's Standard Oil group in Cleveland. Under the able leadership of James Campbell the company had expanded into sheet production, thus fulfilling the promise of its name. Close proximity to Detroit had guaranteed the company strong orders after the war. In 1922–23 Youngstown had grown to its current size. Originally part of Chadbourne's "Seven Steel" group, Youngstown had broken away from the proposed merger to buy up its two Middle West rivals: Brier Hill Steel, the most important sheet producer in the Ohio district, and the Steel and Tube Company of America, an Illinois company with an integrated steel plant on Chicago's south side.[3]

These purchases had made Youngstown the number-3 steelmaker in the nation. It was slightly under half the size of Bethlehem, controlling nearly 3 million ingot tons of capacity. Its emphasis on light-rolled products was a good balance to Bethlehem's overall concentration in heavy-rolled products. With its established mills and customers in the Middle West, Youngstown could give Bethlehem a strong market position in Detroit and direct access to Chicago markets for the first time. While Youngstown made sense as a buyout, so did several smaller steel companies. Inland Steel, for example, was as well located as Youngstown, on Lake Michigan at Indiana Harbor, and had a better earnings ratio; American Rolling Mills of Middletown, Ohio, and the National Steel group also were lean, high-profit producers. But they were handicapped in Schwab's mind by the comparative modesty of their size and, in the case of National Steel, the irascible nature of its prime mover, Ernest Tener Weir.[4]

Bethlehem's assets stood at $801,631,362 at the end of 1929, over five times what they had been at the end of 1915. By adding Youngstown to the fold ($235,740,721 in assets), Schwab could push Bethlehem over the $1 billion mark. That number, with its many trailing zeroes, was irresistible. Only three industrial concerns had attained such a summit in 1929. They were, in order: U.S. Steel ($2,286,000,000), Standard Oil of New Jersey ($1,767,000,000) and General Motors ($1,325,000,000). Of course, the biggest of the three bore Schwab's imprint as former chief selling agent and president of Carnegie Steel. Now only one more deal separated him from having his name attached to a second billion-dollar corporation.[5]

Several days before Christmas 1929, Eugene Grace found himself sitting in the suite he maintained at the Plaza. Across the coffee table was James Campbell. "Grace asked me if I was committed to anyone else," Campbell said in court testimony. The Sheet & Tube CEO answered to the negative. He had just advised L. E. Block, of Inland, that he would not go through with a merger discussed by the two companies. Thus began two months of private negotiations between Grace, Campbell, and other insiders which culminated in the surprise announcement on March 9, 1930, that senior management had contracted with Bethlehem for the purchase of all property and assets for $177 million in stock and cash.[6]

*

With his eyes fixed on Sheet & Tube ("the deal of the decade," marveled *Business Week*), Schwab had discounted the power of a rival suitor, Cyrus Stephen Eaton, the largest single stockholder of Sheet & Tube. At age 46, Eaton had carved out an impressive career in Midwest finance. An ex-divinity student from Pugwash, Nova Scotia, blond and good-looking, Eaton had come to Cleveland back in 1907 to work as a personal assistant for fellow Baptist John D. Rockefeller, Sr., who knew his minister father. Learning the ways of money at Rockefeller's knee, young Cyrus had gone into banking for himself in 1914. Soon he was merging small utilities into holding companies under the auspices of the Cleveland banking house of Otis & Company. Eaton moved into steel in 1925, and soon turned an $18 million loan to Trumbull Steel into board control of Cleveland's venerable Republic Iron Company. Then he piled onto Republic's shoulders other Middle West processors, including Interstate Central Alloy and Union Drawn Steel, to create a large holding company rechristened Republic Steel.[7]

Late in 1927, Eaton had begun accumulating shares of Youngstown Sheet. By the fall of 1929 he controlled about 20 percent of the common stock through the Cliffs Corporation, in which he held half interest with the prominent Mather family; Continental Shares, Commonwealth Securities, and International Shares, all investment trusts he controlled; and, of course, the banking house of Otis & Company. Precisely what he proposed to do with his Youngstown shares was open to speculation. He had urged Campbell to explore the possibility of merging with Inland Steel, and there were rumors that Eaton's eventual aim was to fold Sheet & Tube into the Republic company. On November 5, 1929, the banker was given what he thought was a strong assurance from Grace that Bethlehem was not interested in the Middle West. Eaton had invited Grace for lunch at the suite he rented at the Biltmore in New York. A major point the two men discussed was possible mergers in steel. In later testimony Eaton quoted Grace as saying that "Bethlehem's policy was definitely to stay where they were in the East, except for the property that they were then negotiating for on the Pacific coast, and that they had no intention of muddying the water of the Middle West."[8]

Eaton had interpreted Grace's comments at face value. As a consequence, he decided not to go through with a plan to buy a portion of Bethlehem's stock. Grace had voiced opposition to the proposal, saying it might be misinterpreted by the public. Preoccupied with the affairs of Republic Steel, Eaton was by all accounts unaware of the negotiations that took place between Grace and Campbell in January and February 1930, abetted by Price Waterhouse, the national accounting firm that the negotiators retained to compute a fair basis for the exchange of stock, and whose carefully edited report to stockholders would figure prominently in the legal fireworks to come. Eaton was shocked when he received a visit from Henry G. Dalton, vice president and director of Youngstown, on March 7, 1930, and was informed of Bethlehem's proposed contract to buy the company. After the conference Dalton telephoned Campbell and Grace to inform them of the ill success of the meeting. It was then decided that the principals should go to Cleveland on March 11 and confer with Eaton.[9]

"This 'Tuesday Conference' was one of the most curious in industrial history," *Fortune* magazine wrote later in a chronology of the dispute. Ostensibly aimed at gaining Eaton's cooperation, the meeting turned into an ill-conceived power play that antagonized both parties. In a sense the mission was doomed by the personalities involved. Campbell had developed a dislike of Eaton that his avuncular manner could not hide, and Grace was as clumsy at diplomacy as he was adroit at drilling steel numbers across a cost sheet. Surrounded by aides and advisors, the two laid out the financial advantages of the offer, but Eaton wasn't interested. Before entering the meeting, he had made up his mind to fight the merger. So without even a glance at the reports drawn up in support of the plan, he undertook a speech that denounced Grace and Campbell for their deception and personal treachery, and ended with a vow to do everything in his power to derail the plan unless the contract was withdrawn. This Grace refused to do. The meeting broke up with Campbell rejecting Eaton's request for a delay of the directors' meeting where the Bethlehem offer would be acted on. At 7:00 P.M. the conference ended and the war began.[10]

On the following day the Youngstown board met and, as expected, the directors approved the Bethlehem plan by a vote of six to three, the minority reflecting the Eaton appointees to the board. Accordingly, Campbell telephoned the financier and informed him that a stockholders' meeting was scheduled for April 8. Eaton organized a proxy committee that included a number of prominent businessmen. Within days they fired their opening salvo. It took a form that the proponents of the merger scarcely would have imagined. In every major newspaper in Ohio the public was invited by the proxy committee to assess the impact of the merger not only on shareholders but on the financial health of Ohio. "YOU SHOULD KNOW THE FACTS," said the full-page advertisements. The crux of the argument was that Wall Street centralization was a threat to the public weal, destroying prosperity, causing the withdrawal of jobs, and draining tax dollars from the region. "If the present tendency continues, it will make clerks of all of us in the Middle West," Eaton declared in a statement.[11]

The argument caught fire. In no time at all a crusade was mounted against the evils of Eastern trusts. The *Youngstown Vindicator* asked in a front-page editorial whether its readers wanted to live in "Youngstown, leader in the steel industry of America, or Youngstown, another pay roll town?" The proxy committee kept the merger on the front page for two weeks straight, forging a bond of pride among the citizenry. In any event, Eaton's role as David to the steel company's Goliath was immensely entertaining. "The excitement was of two kinds," explained an observer. "In Youngstown literally everybody was excited. In Cleveland, although there was plenty in the newspapers, excitement became acutely personal in the upper ranks of society. Within a few weeks not only Cleveland's leading men, but also their wives, sweethearts, and daughters were to have discussed the subject so exhaustively that any mention of it was to become almost nauseating."[12]

If Schwab and Grace knew the extent of the prairie rebellion that raged in the Middle West, they did their best to banish the thought from their minds. Their confidence was all the more lethal because it was born of habitual success. Armed with a public opinion poll by Ivy Lee that downplayed the strength of the opposition, Schwab and Grace arrived in Youngstown like conquering heroes. On March 26 the lavish private car *Bethlehem* rolled into the city, and Schwab and Grace received editors, not reporters, aboard the car. Before lunch they visited the Sheet & Tube general offices, where they permitted themselves to be photographed arm-in-arm with Jim Campbell. "So dour was the appearance of Mr. Campbell that many an anti-merger zealot pasted the photograph on office walls with the caption popularized by a local newspaperman: 'Two city slickers and the country boob.'"[13]

At a luncheon before the chamber of commerce, Schwab dispensed with modesty. He told the audience that his company had always purchased "opportunities" and that under his policies they always had blossomed into moneymaking ventures. In spite of its size, Youngstown was too regional to survive in the coming decade, where two or three steelmakers would conduct business on a nationwide basis. "We like Youngstown Sheet & Tube," Schwab said, "and we know we can be just as useful to it as it will be to us." Then shifting focus, he launched into a eulogy of Mr. Campbell, ending with a plea scripted by publicist Lee calling on everybody to stand by "Jim Campbell, right or wrong!" Grace spoke after Schwab. Steady as a metronome he ticked off statistics of the benefits of the merger—precise and calibrated to preclude any argument.[14]

Finally Jim Campbell rose. The strain of the past few weeks was apparent. In a voice soon choking with emotion he told the audience of the terrific struggle he had waged before he could favor a sale to Bethlehem. Youngstown had been his private and personal concern for so long, it was hard to give away. Tears streamed down his cheeks as he voiced his approval of the merger, saying it was the best for the company, although his mournful manner suggested the opposite.[15]

A rebuttal was issued the next day by Cyrus Eaton. Speaking before the same chamber of commerce audience, he repeated the charge that a look at the balance sheets of the two companies would show that Sheet & Tube was in considerably stronger shape than Bethlehem. If the latter's depreciation accounts were figured on the same basis as Sheet & Tube's (a calculation which the Price Waterhouse report to stockholders refrained from doing), Bethlehem would have to earn 1½ times as much as it had in the last five years to pay dividends that would match the value of Youngstown payouts. The basis of the exchange of stock—1⅓ Bethlehem shares for every share of Tube & Sheet—was unfair to the shareholders. Eaton pledged that if the proposed merger were withdrawn, and he implied it would be, he would be perfectly content to let the company continue as an independent concern and that he had no plans of a merger with Republic, even though a combination with Inland Steel might still be desirable.[16]

Grace refused to respond to Eaton. The board of directors had acted on the plan, and it now lay before the stockholders to make the final decision, he said evenly. "We don't blow hot one minute and cold the next. Any report that we intend to withdraw our offer is merely idle chatter. If you want to quote me on it, classify the report as silly or absurd, or if you can think of a stronger word, or words, use it. I don't see how a compromise could be reached. The stockholders either decide to merge with us, or decide they don't want to."[17]

Emotions were running high on April 11 when stockholders filed into Stambaugh Auditorium in Youngstown at 3:00 P.M. Only an hour before, Judge C. S. Turnbaugh had dissolved the final injunction against the vote owing to a dispute over a block of 91,000 shares. It was the seventh time in three days that the stockholders had assembled. Each of the previous meetings had been halted by legal wrangling between the Eaton and Schwab forces. A tense calm prevailed as the proxies were collected by election inspectors. At 5:00 P.M. the votes had been counted and the results were about to be announced when a reporter named Leslie Gould entered the auditorium. Grabbing the mike at the podium he announced that while he had been on the phone with his editors in New York, he had been dragged out of the booth by Sheet & Tube safety police. He protested his treatment in vigorous terms. A search of the telephones was conducted and it was discovered that all of them had been disconnected. When a few stockholders tried to leave the auditorium they found their way blocked by the private police. Ferdinand Eberstadt, Eaton's top aide, went to the front doors, found them locked, and returned to the stage, demanding, "Mr. Campbell, what crime have I committed to be kept a prisoner in this building?"[18]

Pandemonium broke out. When order was restored, Campbell disclaimed any knowledge of an order that had been given to the safety police to lock the doors and disconnect the phones; later it was stated that a company lawyer, fearing yet another injunction by the Eaton group, had made the request on his own authority. The vote then was announced. The merger had secured 857,821 votes, 58,000 more than required under Ohio law.[19]

*

Bethlehem's victory was a mixed blessing given the current state of business. After a spurt in January and February 1930, steel production had receded to 72 percent of capacity when the Youngstown stockholders approved the merger on April 11, and had pulled back to 69 percent a month later when members of the American Iron and Steel Institute assembled for their annual spring meeting in New York. There was undeniable pessimism at the gathering at the elegant Commodore Hotel. The confidence that industrial America was impervious to the jitters on Wall Street was evaporating as orders dried up and unemployment cycled upward in important manufacturing sectors. Having hitched himself to the front of the prosperity bandwagon for so long, however, Schwab was loath to let go. Since the crash, he had been conspicuous among business leaders who reassured the public that any slowdown would be over in 30 or 60 or maybe 90 days. Again he recited his optimism. "The record of 1930 will compare most favorably, from a business standpoint, with recent normal good business years," he divined to the 800 steel executives. "As a matter of fact, business is a lot healthier today than it was six or nine months ago, because of the inevitable housecleaning which has taken place."[20]

In the hallways of the Commodore the talk was of the Youngstown merger and the continued efforts by Eaton to spoil it. The irony of the situation had not escaped the members of the institute: For years Schwab had been cast in bronze in the popular press; now in a matter of three months a flashy interloper had painted the Steel King in the unflattering hues of a self-dealing, arrogant monopolist. The ruckus in Youngstown had spilled over into national politics. Democrats clamored for an investigation of the deal by the Department of Justice, and some Corn Belt Republicans concurred. "Why, the negotiations now going on in Youngstown are so brazen and so daring as to go back to the old days of the Standard Oil activities in ruthlessness and disregard of the public interest and the stockholders' interest," said New York Congressman Fiorello H. La Guardia in a House speech. There were even complaints about steel prices and allegations of price fixing. An uncomfortably bright light was turned on the trade.[21]

So there was great interest when U.S. Steel President James Farrell, sharing the podium with Schwab, warned that "the bloom was off the rose": In recessionary times mergers were unwise. It was high time for "everyone to give undiluted attention to his own business," Farrell concluded pointedly. In reply, Schwab gave a story about the plowman and the bottle fly. "My plowman was standing by the plow at Loretto and I was standing by the plowman, and a great bottle fly came along and settled on the flank of the horse. He took his hat and was just about to kill it, when I said, 'Don't kill that fly. That is what makes the horse go.'" New combinations had the same effect: "It is natural to the course of economic progress that larger enterprises should develop from smaller ones," Schwab said virtuously. "They are the bottle flies of the industry. They must continue."[22]

Two months later, Eugene Grace was in the courtroom of Judge David G. Jenkins at the common pleas court in Youngstown. The billion-dollar merger had reverted back to the courts; having lost the proxy vote, Eaton had filed suit seeking to block the merger on the grounds that Youngstown stockholders had been denied access to pertinent facts regarding Bethlehem's finances and use of corporate earnings. One such fact was Schwab's bonus plan for Grace and other general officers. Eaton argued that his imprudence in the payout of lavish bonuses limited the future dividends of Youngstown investors and amounted to "fraud and concealment" of earnings from stockholders.[23]

Doubtless it gave the financier great pleasure to explore that which corporate executives are most reticent to discuss. On Thursday, July 17, Grace spent the day on the witness stand, mute and with folded hands, as the counsels squabbled over the propriety of revealing his salary. Eaton's chief counsel, Luther Day, contended that the salary question was highly germane to the suit; Newton D. Baker, lead counsel for Bethlehem (and secretary of war under Wilson), argued that the company was under no obligation to disclose any of its internal matters of administration.[24]

Permitted to ask one question, Day asked Grace to state his annual salary as president of Bethlehem Steel. "Twelve thousand dollars," Grace replied. The amount was so unbelievably low that Judge Jenkins overruled Baker's objections. Day was allowed to probe further. Like any good lawyer who knew the answer to his question before he asked it, Day wondered about the executive bonus plan. He asked Grace how his bonus was calculated.

"The factor used to determine my bonus is one and one-half percent," Grace replied in his direct if not overly helpful manner.

"One and one-half percent of what?" Day wanted to know.

"I don't know," answered Grace.

"What was your entire compensation for 1929?"

"Objection," interrupted Baker.[25]

Judge Jenkins sustained the objection and adjourned the trial for the day. Among the newsmen in the packed courtroom the consensus was that 1.5 percent must refer to that portion of Bethlehem's net earnings. That would place Grace's bonus at $740,000 in 1929, a figure that *The New York Times* said "seemed incredible." If true, it would make Grace one of the country's best-paid executives.[26]

On Monday, Judge Jenkins overruled Baker and ordered Grace to disclose his bonus compensation. Grace testified that it was $1,623,753, or more than double the suspected $740,000. The 1.5-percent figure he had used was technically correct, but it was the minimum of the sliding scale. In fact, he had been paid an average of 3.319 percent of corporate earnings before depreciation and interest charges. During the previous five years, Grace had received $4,729,716 in bonuses in addition to his annual salary of $12,000. Furthermore, in spite of declining earnings in 1930, his bonus compensation for the first half of 1930 was slightly more than $700,000, or $3,850 per day.[27]

Through the 1920s, the Federal Trade Commission had slogged through hearings that suggested widespread price collusion in the steel business. Nobody had paid any attention. Now the salary of a single steel executive was exposed, and the nation was agog. "Mr. Grace, it develops, receives as much as Douglas Fairbanks and Charlie Chaplin," the *Times* noted with surprise. *Fortune* rushed into print comparisons with other magnates. It reported that Walter Teague earned $125,000 in 1923 running Standard Oil of New Jersey, a corporation nearly twice the size of Bethlehem; Walter S. Gifford, president of Bell Telephone, drew $250,000; and the late Elbert Gary never earned more than $500,000 a year as chairman of U.S. Steel. What was faulted by many commentators was the cloak of secrecy over Bethlehem's bonuses. *The Wall Street Journal* remarked that if such high bonuses were to be paid, the stockholders at least ought to know of it. But a search of the annual reports of the company failed to reveal any reference to the bonuses since the report for 1917, even though the company went to great lengths in the same reports to describe an employee pension program that cost a fraction of executive compensation.[28]*

Eaton then unleashed his legal bloodhounds on Charlie Schwab. They wanted to know how he administered his own pay. In a written statement Schwab answered that he received $150,000 a year as chairman of the board. However, what Schwab admitted and what subse-

*Grace's 1929 bonus would be equal to $10.7 million in 1987 coinage (the Consumer Price Index multiplier between these years is 6.6). Other Bethlehem officers were scarcely shortchanged, based on court testimony: C. Austin Buck, vice president of raw materials, got $378,664 in bonuses over his $10,000 straight salary, while senior vice president H. E. Lewis got $375,784. In all, Charlie's 14 officers-directors shared $3,425,306 in bonuses in 1929, an average of $244,665 per man. This resulted in a 114-to-1 pay differential between upper management and the mill worker, whose 1929 wages averaged $2,150.

quently came out in the controversy were somewhat different. Early in 1930 Schwab had received a $100,000 raise, making his annual pay $250,000. He said the increase was voted on by his officers-directors "without my presence or knowledge." Again this transaction was not disclosed to the stockholders.[29]

Lurid as the salary disclosures were, other material in the trial proved equally startling. For example, it was revealed that Grace had authorized an $800,000 loan to Henry G. Dalton, a Sheet & Tube vice president, in order to speculate in stock in the company. The Dalton loan had been recalled by Grace after he learned that the U.S. Justice Department was considering an investigation of the payment. All told, management and its allies had spent about $9 million to buy stock after March 7 to defeat Eaton's challenge and in anticipation of rising stock valuations. One businessman testified that W. J. Morris, financial vice president of Sheet & Tube, had paid him $152,500 for his 1,000 shares of stock, or $12,500 more than the quoted price on the New York Exchange. In another incident, the trustees of the Stambaugh estate, which held 12,000 shares of Tube, traveled to South Bethlehem to discuss the renewal of a lease at the Stambaugh building. After a three-year lease was signed by Tube management, the trustees proxied their votes for the merger.[30]

Significantly, it was brought out in cross-examination that the Stambaugh trustees had also tried to peddle their stock to Eaton. Vote buying was not isolated to one side. Under questions "hurled" at Eaton by Newton Baker it was revealed that the opposition had spent about $11 million to purchase stock to defeat the plan. The ex-divinity student from Pugwash, Nova Scotia, was shown to have used funds invested in Otis & Company ruthlessly in pursuit of his objective of blocking the merger.[31]

When the trial finally ground to a halt in October, four acrimonious months after it began, it was considered the most expensive legal battle in American history. It was said to have cost the principals $2.25 million in fees, and it claimed one fatality, a Sheet & Tube lawyer who committed suicide in his private office minutes before the trial was to open on its third day. He was reported to have been suffering from melancholia.[32]

On December 29, 1930, Judge Jenkins issued his verdict. He ruled that the pro-merger directors of Youngstown had seriously erred by failing to inform themselves fully of the ramifications of the purchase and to exercise their responsibility to protect shareholders. Judge Jenkins affirmed Eaton's contention that the proposed exchange of stock was not fair compensation to Sheet & Tube stockholders. He took considerable umbrage at Price Waterhouse for delivering a misleading report to stockholders and in general agreed with the testimony at the trial that the deal had been structured to benefit insiders in management. Affirming the legitimacy of the lawsuit, the judge set aside the merger.[33]

Even though Bethlehem announced that it would appeal the decision, it had suffered heavy damage from the trial and judgment. At the very least, many more months would be taken up in appellate review, and there was always the possibility of further damaging testimony. Almost immediately the verdict on Wall Street was that the litigation marked a watershed in the Schwab-Grace phenomenon: while they had survived the suit's revelations intact, they no longer commanded the same formidable presence in steel. The aura of the all-powerful Bethlehem Steel Corporation had been dimmed. Even an industry booster like *Iron Age,* which rarely veered from the party line that steel management was great, said the Youngstown dispute raised serious questions about steel conglomerations and had sparked a lamentable crisis of confidence among stockholders in the ability of management to protect their interests. "There has been not an increasing but rather a decreasing sense of directoral responsibility," the journal warned, saying that Judge Jenkins's opinion constituted "a notable addition to the bulky 1930 catalog of lessons" regarding business excesses.[34]

The blow to Bethlehem's prestige was compounded a few days later when a group of stockholders filed suit against Schwab, charging that the whole bonus program was an illegal and fraudulent misuse of company earnings. It asked for the return of $36 million in bonuses distributed since 1911. In addition to Schwab, 12 officers of the corporation were named as defendants in the suit filed in the Chancery Court of New Jersey, where the company was incorporated. A number of Wall Street notables were accused of failing to stop the bonus payments as members of the corporation's board of directors. They included Percy Rockefeller and William E. Corey, who had been given seats on the Bethlehem board following the 1923 merger with Midvale. Judge John H. Backes of the chancery court signed orders requiring the defendants to show cause why the payment of bonuses should not be enjoined. Backes also restrained additional bonus payments pending determination of the injunction motion.[35]

Schwab was shocked into action. In a statement to reporters, he waved aside accusations of fraud and financial favoritism. "The value of the system to Bethlehem is clearly shown by the fact that the cost of its executive management is less than that of any other important steel company of which we have definite knowledge, and its value is also reflected in lower manufacturing costs." Because he had not personally participated in the bonus program, he said he was "entirely free from all possible prejudice" in fixing the relative percentages of the bonuses for the officers involved.[36]

"It fell flat," said Ivy Lee of Schwab's rebuttal. Lee argued that the company's defensive stance must end. Reviewing the situation with Grace, the publicist suggested the preparation of material to "educate" the public as to how the bonuses were of benefit to stockholders and had not been done in secrecy. "I feel that although the public is somewhat agitated over the size of the bonus, there is even more agitation over the theory that it was a secret matter and something that was put over on the stockholders," he said in a memo to Grace.[37]

When the suit was filed, there was much speculation about whether it was part of Eaton's assault on Bethlehem management. In fact, it turned out to be an independent effort by four stockholders convinced that they were not given adequate dividend compensation by Schwab and his excessively remunerated subordinates. Their suit galvanized other shareholders, and the Protective Committee for Stockholders of Bethlehem Steel Corporation was formed. From a few dissidents an organized movement came together. One of the Protective Committee's first actions was an effort to notify all stockholders of the existence of the organization and to solicit funds for its operation. When Bethlehem management refused to make a list of its stockholders available, the group went to court and won an order compelling the company to supply the names.[38]

The chairman was worried. Judge Backes had issued a second order which restrained him and the other defendants from disposing of their personal stock holdings in Bethlehem. The judge agreed that the securities should be available to meet any claims should the suit be successful. Schwab was concerned he might be subject to personal liability as author of the bonus plan. The whole legal tangle was getting uncomfortably close to his own personal fortune. In a bid to put the dispute behind him, Schwab agreed to Ivy Lee's urgings and ordered a special booklet on the bonuses to be composed. It took the form of an 18-page letter to stockholders written by Lee and Grace.

Schwab signed it. The booklet had two purposes. Seventeen pages were devoted to a defense of the bonuses, while in the final paragraph Schwab stated, "I intend to submit to you for your approval at your annual meeting to be held on April 14, 1931, my administration of the bonus system of the Corporation since 1917. In view of the results accomplished, and of my strong conviction as to the value of the system, I earnestly recommend and request that you give such approval, and, to that end, that you promptly sign and return the proxy which is being mailed to you, if you cannot attend the meeting."[39]

Members of the Protective Committee were furious. To them Schwab had betrayed the sanctity of fiduciary trust, the very bedrock of capitalism, and now was trying to win immunity from their suit by a trumped-up vote of the stockholders. The dissidents first hightailed it to the chancery court to file new motions, then paced through Newark to attend the stockholders' annual meeting.

*

At the annual meeting Charlie answered the rebel shareholders with deep emotion and high theatricality. The Protective Committee, he said, had "taken ten years off my life. Boys, I want you to be a happy family," he implored. Tearfully he spoke of his patriotism as a dollar-a-year man and, in a dramatic coda, grasped the podium and declared, "God bless and prosper you all!"

But the dissidents were a hard-boiled group. "I have been a small stockholder of Bethlehem for several years," stockholder Elbert Miller said from the floor, "and in that time I have attended a majority of the stockholders' meetings, but I never heard of the bonus plan until I read of it in the newspaper accounts of the Youngstown litigation. Why after 14 years are we now asked to ratify it?"

"I can't answer that, it seems to be a legal question," Schwab answered lamely.

Another dissident pointed out that Bethlehem had paid out little of its recent earnings in dividends, saying it needed these monies for long-term growth. In the seven years from 1923 to 1929 $25 million had been distributed to stockholders. In those same years bonuses for top management were $13.5 million. "It is wrong in principle, it is outrageous in practice, it is contrary to justice to take this money out of these stockholders and put it into the laps of a few men," he stated.

Finally the chairman had had enough. Losing his patience, he blurted out, "I had the feeling that this damn company belonged to me, you know, and I went ahead and did the best I could."[40]

Didn't the Protective Committee realize that the million-dollar paychecks to Grace were the very fulcrum of personal motivation which had lifted Bethlehem to the highest ranks of industrial America? Couldn't it be understood that when the company was small, incentives had been the best way to achieve rapid growth? In the end the chairman's tactic of conceding nothing, changing nothing, boomeranged. With large blocks of stock held by institutional investors, the bonus program was approved by a wide majority. But anticipating this development, the Protective Committee had asked the court for an injunction to prevent the vote from releasing Schwab from liability in the suit. Judge Backes agreed with the stockholders, issuing a restraining order that permitted the proxy vote to be taken, but not to override the suit.[41]

The ruling compelled Schwab and Grace to come to terms with the minority stockholders. A settlement was announced on July 2, 1931. It involved five major changes. Henceforth bonuses for President Grace and other general officers were to be calculated after, not before, depreciation, depletion, and obsolescence. This provision narrowed the pool of available corporate funds from which bonuses could be drawn by an average of $15 million based on prior yearly earnings—in 1929 it would have reduced the bonus total from $3.4 million to $2.2 million. In anticipation of reduced bonuses, officers were to be placed on "fixed normal salaries" to be determined by a committee of directors. No bonuses were to be paid when dividends were passed on common stock; the total amount of bonuses was to be disclosed in the annual stockholders' report; and the division of the bonuses no longer was left to Schwab alone to administer. Instead he was required to submit a report to the directors, who were to pass on the proposed pay.[42]

The settlement was billed as a triumph for Samuel M. Untermyer, the "trust buster" attorney who had been retained by the Protective Committee to sort out the mess. (Untermyer's father, a wealthy New York merchant, owned a significant block of Bethlehem stock.) As for Schwab, while rid of the chancery suit, he was not immune from the poisonous darts of an increasingly assertive press. *The Nation* had gained renewed influence in the wake of the crash, and it hammered away at Schwab, reciting the disapproving opinion of "Bonus Charlie" from such mainstream publications as the *Wall Street Journal*. "Mr. Schwab may weep all the tears he will at the questions his stockholders ask him; he will never successfully defend such practices," said *The Nation* harshly.[43]

The New Republic, under editor Edmund Wilson, drew similar lessons. The "grim battle" over Bethlehem pay had exposed the hypocrisy of the materialist creed of the 1920s, the magazine said. It went on: "The proposition, seriously entertained during the recent prosperity orgy, that American capitalists, out of sheer humanity, were about to share their profits with the people, has been given a great setback. But the most devastating argument of all against such a proposition is the picturesque figure of Charlie Schwab himself. Helplessly he breeds Reds faster than they can be tracked down by the American Legion."[44]

*

What gave the bonus agitation an extra impetus was the state of the economy. At a most inconvenient moment, the engine of capitalism had gone into reverse. The recession which Bethlehem's chairman hadn't seen coming and didn't think important had gained enormous momentum in the second half of 1930 as consumption contracted and banks called in loans from overextended buyers. At the end of the year production of steel products had dropped off to 40 million tons from 55 million tons in 1929, a 27-percent loss. What's more, at the rate at which steel prices were dropping, *Iron Age* worried whether profitability might not be wiped out entirely. "An uninterrupted fall in prices [has] made it clear that the causes of depression were more far reaching than had been realized at the beginning of the year," the magazine noted in its year-end review.[45]

It was not until early 1931 that the full grip of the Depression was felt in this country. The withdrawal of U.S. dollars following the 1929 crash had caused serious financial problems in Europe. On May 11 a run on the Kreditanstalt Bank of Austria produced a chain of circumstances which subsequently forced the Weimar Republic to default on its reparation payments. The panic spread from the Reichsbank to the Bank of England. An estimated $150 million in gold was withdrawn from England, much of it sent to Paris. This in turn led to the recall of European credit from the United States.[46]

As Europeans demanded their gold, reversing the stringency of late 1929, American banks called in their loans and effectively paralyzed credit. The first major sector to buckle under the pressure was automobiles. In July 1931, motor car output fell 79 percent from July 1929. In all, production slid to 2.4 million units for the year, compared to

5.6 million vehicles two years earlier. The shrinkage of auto production struck the steel industry badly: two-thirds of Bethlehem's automobile sheet and engine market vanished by the end of the year.[47]

Then sales to railroads collapsed. "Records going back to 1915 disclose only two years in which equipment purchases even approached the low point in 1931," was the dismal news reported in *Iron Age*. As late as the fall of 1930, the rail mill at Sparrows Point was working off a backlog of orders, while the adjacent 110-inch plate mill was engaged in rolling steel frames and undercarriages for a fleet of multiple-unit electric commuter cars ordered by the Reading Railroad. In June 1931, the order book in both departments was blank and the rail mill was shut down.[48]

The same stunning reversal swept Bethlehem's sales in structural steel as the dollar volume of all types of building construction plummeted by 32 percent. The sound of riveting all but ceased in midtown Manhattan. As for light-rolled products, the country seemed to have lost its taste for Dole pineapples and other canned convenience foods. The pineapple pack for 1931 was 4.8 million cases, a record high, but the pack went largely unsold. As the company closed out a disastrous summer, with heavy monthly losses, there was no need for more tin plate from Bethlehem Steel.[49]

Drastic retrenchment was ordered by Grace and Schwab on July 15, two weeks after they settled the bonus controversy. But the brakes of financial control had not been applied in time. In August the businessmen found themselves with the lowest operating ratio in company history—32 percent—and despite their acumen with cost sheets, the production pipeline was beginning to leak red ink. Grace admitted to newspapermen in New York that he was at a loss on how to stem the flow. "Let us adjourn this meeting and go uptown to the Cathedral and pray. It may do some good." Schwab placed his faith in September, a month which usually marked a seasonal upswing in orders. The industry was "turning the corner" and "getting its second wind," he said. Instead, that month had the lowest operations to date; steel output dropped 10 percent from August's record low.[50]

The extent of the debacle was becoming apparent. On September 30, 1931, for the first time since 1909, the company posted a quarterly deficit after dividend requirements. In October it joined U.S. Steel in cutting hourly pay 10 percent across the board. With stunning rapidity the advances of the '20s were coming undone. Since the beginning of 1930, the company had spent $47,158,004 of its cash hoard on additions to capacity, mostly at Sparrows Point and Lackawanna. It had used up $20 million to buy the plants of the Pacific Coast Steel Company. Millions more were dispensed to buy Sheet & Tube stock; legal fees alone claimed $1 million, according to court filings.[51]

Each and every dollar of these expenditures came to haunt management. With capacity at unexampled lows, the expansions were wasted. Because the expenditures were based on the assumption of growth, the facilities did not incorporate any major cost-savings devices—in fact, the jumbo-size No. 3 open hearth shop built at Sparrows Point in 1930 proved too expensive to operate and would be idled through most of the Depression. The weak markets exposed the shortsightedness of Bethlehem's choice of capital investment in other ways. While management had obsessed over individual costs and ways to trim them incrementally, they had brushed aside new technologies that could have improved performance when output sagged. As *Fortune* pointed out, steel traveled as much as two miles at the big steel mills: "This enormous mechanism must go into production no matter how small a daily total of orders. It is like picking up a sledge hammer to drive home a tack."[52]

Overexercise of the very qualities upon which Schwab's business dominion had been founded was to sap his organization of its ability to adjust to shrinking markets. The economies of scale, which permitted Bethlehem to reap strong profits in times of heavy demand, disappeared when production dipped below 50 percent of capacity. Vertical integration became more a liability than an asset, as it required the company to expend millions just to cover fixed costs and overhead. In steel, the two disciples of stupendous size, Bethlehem and U.S. Steel, would suffer most grievously over the next few years, while smaller producers—National Steel, Inland, and Armco—would gain market shares at their expense.

In ironic counterpoint to Bethlehem's collision with the Depression was the progress of its appeal of the ruling by Judge Jenkins barring the Youngstown merger. In August 1931 the Ohio Court of Appeals began reviewing the case. But Bethlehem's common stock, which had sold for $102 a share in March 1930, had plummeted to $27.50, and Youngstown's common, once $140, was averaging $22. The company announced in October 1931 that, due to "changed conditions," it was dropping its appeal of Judge Jenkins's ruling and scrapping its merger with Youngstown.[53] Bethlehem could hardly afford to buy 3 million tons of capacity when its own mills were running at a loss.

PART III

THE ROOSEVELT REVOLUTION 1931-1941

CHAPTER 11

DEPRESSION SNAPSHOTS

In 1931, the steel companies of America had the physical resources to produce 71 million tons of steel. This was enough to make 103,200,000 automobiles, or to lay 403,410 miles of 100-pound track, or to put up 1,128 Empire State Buildings. There was an enormous reservoir of steel capacity—fully 55 percent more than during the Great War—but before it could be tapped, the industry had to have customers. Demand had dropped to a shocking low, with the result that instead of 71 million tons the industry had turned out only 29 million tons. Among the other unhappy developments in 1931: Ingot capacity had declined to 38 percent below the levels of 1930, pig-iron production registered the third lowest tonnage since 1900, and scrap fell to the lowest price on record.[1]

The steel industry, one of the vital organs of the country, was seen now to have become atrophied. After two years of sickening decline it was clogged with dead tissue, its pulse fading, and no one seemed able to do anything about it. It was beginning to be acknowledged that America could have quite satisfactorily gotten along with about half the mill capacity erected since the war; that indeed the overexercise of the economy's great productive resources, together with a sharp maldistribution of wealth, were root causes of the disaster. Dr. Lewis H. Haney, a consultant for *Iron Age*, wrote grimly that surplus capacity was "built up during the period of inflation when money was abnormally easy. As the prices of products decline, as standards of living contract, and as export trade falls to the vanishing point, this over-capacity and over-capitalization become more and more burdensome. One of the great problems which lies ahead is the problem of liquidating the over-expansion of invested capital. This will doubtless involve much reduction in equities."[2]

On the water-fringed plains of Baltimore Harbor one could see a whole industrial landscape laid bare: between Sparrows Point and Baltimore City almost everyone was dependent in one way or another on steel. When steel was disabled, grocers on Dundalk Avenue sold less food; when paychecks of 18,000 steelworkers dried up, fewer families could meet their monthly house mortgages and greater numbers crowded into apartments. Around the bungalow belt of Edgewater and Saint Helena For Rent signs were hung, while among the rowhouses of East Baltimore repossession agents stalked relentlessly with their bank papers and moving trucks.[3]

Sparrows Point's distress was symptomatic of the greater malaise. As a supplier of a basic commodity on two coasts, and owned by one of the nation's biggest corporations, the plant suffered grievously from financial constrictions across the continent, from the shutdown of wire spring works in Massachusetts to the slowdown at the canning plant of Castle & Cooke in South San Francisco to the drop to one shift a day at the Chevy assembly line on Broening Highway in Baltimore.[4] And what was happening on the peninsula reverberated backward thousands of miles to the captive ore mines, where depressed ore consumption hardly justified the cost of transporting the mineral riches of Cuba and Chile to Sparrows Point.

The movement of Cuban ore was the smallest in 37 years, except for 1921. The 12-month total was 250,236 tons, compared to 751,878 tons in 1929. The drop in Chilean ore shipments was even more conspicuous. The 1931 total was 817,556 tons, against 1,997,858 tons mined in 1929. In all, shipments had declined 52 percent from the average of 1,690,000 tons between 1925 and 1929.[5] Dr. José I. Rivero, editor of *Diario de la Marina,* one of the leading newspapers in Havana, dramatized the predicament of wage earners around the world in an imaginary dialogue between a father and son sitting in a bare room before an empty hearth:

> *"Why don't you light a fire, father?"*
> *"Because we have no coal, my son."*
> *"Why have we no coal, father?"*
> *"Because I have no work, my son."*
> *"Why have you no work, father?"*
> *"Because there's too much coal, my son."*[6]

*

As America fell deeper into the Depression, there was no safety net. A mere 151,000 of 32 million U.S. workers were protected by some form of unemployment insurance in 1932. There was no protection in steel. The industry had waved aside proposals dating back to President

Harding's 1921 Unemployment Conference calling for a system of setting aside a portion of wages and income for payment in periods of slack business and layoffs. Rejecting jobless compensation as a "dole" that pauperized workers, Bethlehem had developed a policy of employment distribution or "spread-the-work." Hours of labor were divided up among the mill force based on incoming orders. Soon nearly every employee not laid off (about 12,000 men) was working part-time on a part-time pay schedule. As orders slumped, so did the hours of work.[7]

"Things got so tight you didn't make enough money to pay for what you needed," remembers Charlie Parrish. "You lost all you done saved." From a high-water mark of 70 hours per week in 1929, Charlie Parrish watched his hours recede to five hours, three days a week, in December 1931 under the company's spread-the-work schedules. With his former wage of 50 cents an hour sliced back to 45 cents an hour, he was looking at $13.50 in gross wages every two weeks. That was barely enough to pay for his food, and now he had a pregnant wife to worry about. Visions of Goochland County reared up before him as he rode on the trolley line to the blast furnaces from Turner's Station, a hamlet across Bear Creek.

Was he lucky or was he trapped? That's what he was trying to figure out over the winter of '32. The work world he had known had gone topsy-turvy. Big Jeff had been "busted back" from straw boss to a simple laborer—now he gasped for breath while scraping up the slag on the casting floor. It was pitiful. And George Martin, sovereign lord of the furnace mechanical gang, had been moved down from foreman, a post which was eliminated, to master mechanic.

In his off hours Charlie picked up some odd jobs. He did some roofing for people he knew and took up farming again, growing enough sweet potatoes, turnips, onions, green beans, and tomatoes to keep Alice, his wife, busy through the autumn. "The men here tried to go to other factories, thinkin' that somebody be hiring again, but there was nothing in it," says Charlie.[8]

Across from Blast Furnace Row, the open hearth department smothered in its cloak of smoke. However atrophied, the big furnaces were still the lifeblood of Sparrows Point. While the cranes and machines performed their tasks with dumb certitude, the shop men had to deal with the heartbreak of it all. A reduced volume of orders spelled ruin for the incentive-based wages of William Walk Womer. "No, it ain't that we were goin' hungry then," Ben, his son, remembers, "but the money wasn't there and my dad and the other men were getting pretty upset."

Wages for the skilled shop men had risen to $8 or more a day in the late '20s. Now with production low and getting lower, the pay system ran in the opposite direction. Because of the inherent delays in production, William Womer's tonnage incentives had vanished, and the great weight of unproductive machinery bore down on him and his family in spite of his own personal efforts.

When No. 1 shop went dead from a lack of orders, the men were forced to accept the company's "act of God" rate. Ben explains: "Whenever there was a delay in charging the furnace or unscheduled down time, that was an 'act of God,' and you didn't get paid. Many a time my dad and the other men came home without getting any pay. And to think that the good Lord had done that to 'em!"[9]

As the Depression intensified, petty stealing and bribery became accepted practices. "A couple of them foremen had all sorts of rackets goin'," recalls John Duerbeck, then an 18-year-old screwboy in the hot sheet mill. "They'd come around asking you to contribute to this or that, like a horse they were bettin' on, or some bull roast they had tickets to. Hell, on some crews the guys each put down five or six bucks on pay day to keep the foreman happy."

The stress on cutting costs fractured the established job hierarchy as foremen juggled the employment rolls, according to Duerbeck. "The older guys started losin' out in the hot mills 'cause the company thought the young guys could do the work a little faster." One day, for example, he saw his name on the schedule board in place of Tony, a rougher who had worked at the job for years. "I guess Tony was about 50 and I was still only 18 years old. I went to the foreman and I said, 'Can't you let Tony stay on 'cause he has a family and I'm single.' And the foreman told me, 'You can go right now, John, but Tony ain't gonna get that job. I'm running this mill, not you.' "

While John worked, Tony, dismissed from the mill, found a job in Baltimore. "He got $5 a week as a baker's helper. He left his family and moved into the back of the bakery 'cause that was the only way you could survive on that pay. Now later he got a job at the Point as a laborer in the machine shop. But he was a different man. Beaten. He pushed the broom and he got to drinking."[10]

One problem was that there was nowhere to go; nowhere to escape. Nathaniel Parks tried. Laid off as a sheet catcher, he borrowed money for the bus fare to Buffalo, where he had been told the Lackawanna plant was hiring. There weren't any jobs, and his skills weren't easily transferable to other factory work. For a couple of months Nathaniel drifted around the Middle West, joining the army of the unemployed, hitchhiking on the highways and knocking at the doors of sheet mills in eastern Ohio.

Finally pulling out of it, Nathaniel returned to his parents' brick house on J Street to find his hometown in worse shape than when he'd left. "All the black people at the Point were broke. The company let them keep their houses, but the women and children were going out for heating fuel. You picked up the coke and soft coal off the slag heaps. You could pick up a couple bags a day; sell what you didn't need at 15 cents a bag."[11]

In the winter of '32, Joe Lewis was at a soup kitchen. When mechanical doors were installed in No. 1 open hearth early in the Depression, Joe found his job as door puller eliminated. When other jobs failed to materialize, he agreed to be the cook for George Koenig, a ward boss in Baltimore who had opened up a soup kitchen on Eastern Avenue. It didn't take Joe long to learn that fishheads were the cheapest way to make boiling water "taste like something," and that paprika gave the soup some color when Mrs. Koenig failed to pick up enough second-day vegetables from vendors at Broadway Market.

All his life Joe had been taught never to accept handouts or welfare. He couldn't wait to leave school at age 15 and start earning money at Sparrows Point. He had always kept his nose clean; stuck with his family on Clinton Street; confessed to the Irish priests at St. Brigid's Church; never harbored a subversive thought. But in those bitter days he knew something was wrong. It wasn't anything he could articulate precisely; it was only that he was a keen observer. It pained him to observe men and women come through the doors of the storefront who had lost everything and to see teenagers whose sullen defiance he knew came from being scared to death. "I get a little emotional thinking about it," Joe says slowly. "'Cause, well, these were my people, Poles, Irishmen, Germans, and they came to Koenig's because they were hungry and cold."[12]

Such scenes were in pitiful abundance in the industrial districts of Baltimore, second largest port on the East Coast and home of dozens of once-thriving manufacturing plants. The sign of the times was evident from the demonstrations for jobless relief at city hall; from the columns listing liquidation sales in the local papers; from the Friday and Saturday "specials" advertised in the *Dundalk Community Press* for steelworkers and their families at the Sparrows Point company store. By January 1932 the store was selling off luxury items at a fraction of their list prices and offering cheap staples for customers on tight budgets.[13]

*

"Have you heard Dr. Charlie Schwab's depression remedy?" asked Al Richmond. "Sweat! That's what it says in the paper. Sweat! Put it in bottles. It's good for what ails you. My God, Charlie Schwab has cut his workers out at Sparrows Point down to one day's work a week. And he's telling 'em to sweat. With brains like his running the steel industry, don't you think the workers can do a better job of it?" Richmond, a section organizer for the Young Communist League, delivered this pitch to passing workers on the sidewalks of East Baltimore in the winter of 1931–32. There were about eight YCL members and about twice that number in the Communist Party in the city, but they had given themselves a mighty task—to organize Sparrows Point.[14]

No other important industry had so cowed organizers, radical or otherwise, for so long. The AFL's Amalgamated Association of Iron, Steel and Tin Workers had the distinction of a name that represented exactly what it wasn't. Defeated at Homestead in 1892 and trounced in the 1919 steel strike again, the Amalgamated was not recognized in a single important iron, steel, or tin plant. With less than 4,000 dues-paying members, or under 1 percent of workers in the industry, membership was concentrated in a small and dying field of metal-making—horseshoes. Quite literally, the Amalgamated was a horse-and-buggy union in an age of mass consolidations.[15]

A considerable part of the weakness of the union could be traced to its exclusive and conservative character. From its birth during the Civil War as the National Forge, the union consisted of a small group of craft workmen who were interested mostly in maintaining control over their skilled jobs and comparatively high wages. The union's ideology and structure were no match for the technological and financial forces that forged the 20th-century steel business. Starting with the united phalanx of steel rail manufacturers of the previous century, the industry had pushed aside unions in every phase of operations that came under its jurisdiction. With great reluctance, the Amalgamated had opened its doors to unskilled steelworkers in 1911; during the grassroots surge in unionism during 1918–19, the Amalgamated had been led kicking and screaming into organizing drives in the Middle West by leftist activists.[16]

Richmond was a member of TUUL (Trade Union Unity League) and its steelworker affiliate, the Steel and Metal Workers' Industrial Union. TUUL was organized by the U.S. Communist Party in 1926 out of the Trade Union Educational League which had existed since 1920. The union was as much in rebellion against the stand-pat leadership of the Amalgamated as with the steel barons; at a convention in 1930 at Cleveland the group dedicated itself to "industrial unionism" which would include all workers, skilled and unskilled, in a given industry. "The new center must become the organizer of the masses of unorganized workers, as well as direct the work of the minorities in the reformist unions against the A.F. of L. bureaucracy," was the proclamation of the Cleveland convention.[17]

TUUL had handed itself the job of drawing together a militant workers' movement from the unemployed and underemployed in the spirit of the old IWW. Their patron saint was Karl Marx, not the bourgeois Mike (Grandmother) Tighe, president of the Amalgamated.[18] Yet how did one surmount the barriers of bigness, anonymity, and fear at strongholds like Sparrows Point? By throwing oneself into the day-to-day concerns of the workingman; by taking positions and demanding reforms in the name of the people. The radicals took careful notice of Schwab, for his pronouncements and predictions made fine fodder for their propagandizing. In a basement office on Eutaw Street, TUUL threw paper missiles at Schwab's fortress on the tip of the harbor, publishing a newspaper aimed directly at employees at the Point. The paper published a nationwide platform for steel reform, a program which included a six-hour day, abolition of company towns and stores, establishment of safer working conditions, the end of Jim Crow conditions for Negro workmen, and immediate cash and food relief for the unemployed.[19]

Filling out the mimeographed pages of the newspaper were letters from Sparrows Point workmen describing horror stories at the works. A sampler:

> Over 3,000 tons of sheet iron has to be produced every week at the Sheet Mill Dept. Most of the workers sweat over fire and gases in the hot mill. Three workers were so fatigued they fell down and they were brought to the hospital. One of them in particular, a doubler on No. 2, had been unconscious for sev-

> eral hours before anybody noticed him. ("Metal Slave," May 1931)
>
> I am only 18 but I do not consider myself a youngster since I have been working at the Sparrows Point plant, for there they age one considerably, the worry of keeping a home on the poor wages paid and the rotten speed-up conditions. ("Tube Mill Worker," undated)[20]

The accuracy of these accounts cannot be verified. While obviously aimed at attracting converts, the letters rendered the physical details of the mill precisely, and the sense of exploitation and drudgery they conveyed did not differ greatly from the recollections of decidedly unradicalized workmen. The very publication of a counter-*Iron Age* on the grassroots level was a remarkable achievement, given the autocratic history of Sparrows Point.

"Bringing misery out of hiding" became the activists' cry. With the conviction of the true believer, Al Richmond would accompany Carl Bradley, Baltimore secretary of TUUL, to "enemy territory," tiptoeing around Sparrows Point to visit leftist workers and eluding the eagle eye of the company police. Richmond remembered, "You approached the massive array of furnaces and mills and you thought: What a fantastic locale for a feudal castle, water on three sides and a narrow strip of land on the fourth. Heightening the impression of a feudal stronghold was a protective semicircle of military forts. You passed Fort Holabird and en route you also spotted Fort Carroll, perched on an island in the Patapsco River, due west of Sparrows Point, commanding the water passage to it from the Port of Baltimore."

The Communists did not soapbox in Bethlehem's domain, either in Sparrows Point or in the surrounding villages. "It was forbidden there and you knew you did not have enough steelworker muscle to shatter the prohibition."[21] So Richmond and Bradley would return to Baltimore and hammer away on the street corners, lampooning, ridiculing, but mostly asking the question: Why?

*

A chief exhibit of the critics' brief against capitalism was the steel industry. In the first months of 1932 the trade had replaced railroads as the most troubled sector in the U.S. economy. Nine out of ten of the 500,000 mill employees were on "short time" or laid off; the industry collectively lost over $70 million in a matter of six months, with the heaviest losses registered by U.S. Steel and Bethlehem. American companies were hit harder than steelmakers in any other nation. While the Depression had reduced foreign steel output from 68 million to 39 million tons between 1929 and 1932, the U.S. dominance of world steel was cut much further. With the ability to produce nearly 50 percent of all steel in the world, American companies handled only 27 percent of world steel sales in 1932.[22]

In contrast to the woes of U.S. producers was the steady advance of the Soviet Union. Isolated from international finance, the Bolshevik regime had managed to pull the Russian economy out of the chaos that had followed the 1917 revolution and civil war. Beginning with a minuscule 114,260 metric tons of steel in 1919, Soviet production steadily increased through the '20s, then continued upward as the Depression thrashed Europe and America.[23]

Evidence of what some economists regarded as a miracle of centralized planning took place in 1932. For the first time in history Soviet mills outstripped German, British, and French mills in monthly production, and the U.S.S.R. became the second-largest producer after the United States. Russia's progress in relation to America and its once indomitable "New Capitalism" was shockingly apparent in a comparison of recent steel outputs between the two countries together with those of other leading industrial nations:

Nation	***Production***			
	1925	*1928*	*1931*	*1932*
U.S.	50,840,747	57,729,481	29,058,961	15,322,901
U.S.S.R.	2,064,518	4,578,670	5,952,756	6,393,700
Germany	13,422,758	16,003,105	9,140,384	6,360,745
France	8,208,700	10,471,920	8,622,266	6,217,360
Britain	8,271,648	9,542,064	5,826,912	5,892,768
Belgium	2,809,400	4,305,132	3,422,934	3,075,348
Japan	1,433,294	2,101,081	2,075,885	2,643,776[24]

For those who had witnessed the Russian experiment with a mixture of apprehension and fascination through the '20s, Communist collectivism and the Five Year Plan seemed to be paying good dividends to its people. At the Soviet steel center of Magnitogorsk in the Central Urals, unemployment was unheard-of, while farther east in the foothills of the Great Altai Mountains a $250-million steel complex was under construction at Kuznetsk. When finished, the mill was projected to produce 1.45 million tons of rails, plate, and structural shapes a year, making it the largest mill in Asia and nearly equal in potential to Sparrows Point.[25]

*

On February 18, 1932, Schwab tried to lighten America's load by hosting a media event—the celebration of his 70th birthday. Before popping cameras and a horde of reporters he feted himself in his offices at 25 Broadway. The room was banked with red roses. On his desk were congratulatory communications from President Hoover, General Pershing, Senator James J. Davis, and scores of clergymen, university presidents, and leading business figures. Also displayed was a quilt which his 91-year-old mother had patched as "something useful" for her boy.[26]

Schwab led the reporters into the corporate boardroom where he sat at his accustomed place and Eugene Grace took the seat to the right. What followed was a "chuckling, laughing interview," according to the *Times,* heavy in nostalgia and cheerful asides. But behind the party atmosphere the chairman had a message. The Depression had placed him in the position of managing a corporation which, despite its mounting losses, seemed to have everything to lose from a change in the status quo. So he lauded the Republican administration and paternalistically pledged his support in ameliorating social distress.

"Our system can and will be improved. Particularly in the direction of a better system of pensions, care for the sick and other things beneficial to the working man." Sidestepping a question on the desirability of unemployment insurance (which he opposed), he said the steel industry had done much to improve the condition of labor. And in a pointed aside, he called on Americans to discount the Russian example of economic planning, saying it would not work in the long run.

Then he told reporters what was the biggest surprise of his birthday—a special 16-cylinder automobile that was the gift of Grace and the other officers. He said he was looking forward to driving it up to his mansion on Riverside Drive that evening. "It's nice to have rich friends," quipped the tycoon.[27]

While the chairman made good copy, it was the duty of Grace to scrutinize the accounts and make the appropriate adjustments. An important source of economizing, of course, was labor. In the months following Schwab's birthday, there were rumors at Sparrows Point of another wage cut. The workers expected it, and yet when it came there was an unexpected twist. Advance rumors had it at 10 percent;

Grace's announcement said 15 percent. The previous October, workers had been subject to a reduction of base hourly rates from 45 cents to 40 cents. Now the base was pegged at 34 cents.

The 15-percent wage rollback in May 1932 by the fourth-largest industrial concern was a major blow to the president that Schwab supported. Herbert Hoover had staked his prestige on agreements he had made with Bethlehem and other industrial organizations in 1930 to hold the line on hourly wages, lest consumer purchasing power plummet. While weekly wages had fallen due to declining hours, the wage *rate* had held up reasonably well in the first part of the Depression, and this had lent credibility to the president's domestic policies.[28]

Herbert Hoover had come into office with a well-defined philosophy of political economy. He believed firmly that corporate capitalism would seek the welfare of all and just as firmly disbelieved in government in the economy, which he felt would lead to socialism and an oppressive bureaucracy. His career as a highly competent technocrat, first as an international mining engineer and then as secretary of commerce under Harding and Coolidge, had blinded him to the realities of the Depression. Convinced that any man could succeed if he applied himself, he ignored the conclusions of his own Committee on Social Trends, which pointed out the acute hardships arising from unemployment among a working class removed from the land and dependent on cash wages. He stood aloof from efforts by Governor Franklin Roosevelt of New York and others to provide aid for the stricken. When the Senate passed a resolution calling for a special committee to study jobless insurance, the president had selected two reactionary Republicans to serve as his appointees. One of them, Senator Felix Hebert of Rhode Island, junketed to Europe for the avowed purpose of studying the issue. He returned to Hoover's summer camp at Rapidan, Virginia to announce, "I am firmly convinced as a result of my study of unemployment insurance in Europe that the institution of such a system in the United States would be the first step toward a national dole." Democratic Senator Robert Wagner of New York countered angrily that unemployment insurance was no more of a "dole" than life insurance.[29]

When it no longer could be denied that the economy was crumbling, Hoover reluctantly placed before Congress a plan of action. Passed on January 22, 1932, the Reconstruction Finance Corporation was established to provide emergency financing for railroads, insurance companies, banks, building and loan associations, and other major institutions. The RFC was based on the premise that if the government aided businesses at the top, prosperity would flow downward, relieving distress at the bottom. The philosophy of "trickle down" was applied only after the president had failed dismally to prevail on bankers to loosen credit in order to stimulate the economy. The bankers insisted that they could not act without aid from the government; even Hoover's secretary of the treasury, Andrew Mellon, refused to lend money through his Pittsburgh bank without some tax or credit benefit from the government.[30]

The RFC failed to get the economy out of its tailspin. In the absence of forceful leadership in Washington, most businessmen viewed the RFC not as a means to create more jobs, but as a stopgap measure to preserve their own enterprises and keep them from total collapse. The railroads, which had promised to expand construction with RFC funds, used the loans instead to pay off immediate debts and undertake piecemeal construction. The railroads were not necessarily to blame: red tape and delays by RFC officials contributed to the ineffectuality of the program. Work loans for the repair of freight cars, after requiring months of review, aggregated only $7.2 million countrywide and scarcely relieved unemployment. "Tonnages are not large and when placed are widely distributed. Some of the work is to be spread over a long period of time. The awards therefore have given little impetus to the [steel] mills," noted *Iron Age.* In all, as the RFC grudgingly allocated money for the repair of 40,000 freight cars, up to 772,565 cars and 11,660 locomotives were idle.[31]

The failure of the Reconstruction Finance Corporation caused another chain reaction. Almost immediately the steelmakers complained that the loans did not provide them with orders of sufficient volume to operate their mills at a profit. Bethlehem, for instance, took exception to the biggest RFC railroad loan of all, $27.5 million to the Pennsylvania Railroad for work on New York-to-Washington electrification. The loan permitted the release of steel material that had been rolled previously, which continued the drought at Sparrows Point and South Bethlehem. Protests were lodged over RFC loans to the New York Central and to Denver & Rio Grande Western, because they involved minuscule purchases of rails and structural steel. Likewise, the activities of the agency gave little impetus to the construction industry, whose contracts had decreased to below $2 billion in 1932, compared with $4 billion in 1931 and $5.87 billion in 1930.[32]

*

If railroads and banks weren't going to use the government loans to buy steel, then the steelmakers weren't going to follow the government's line on wages blindly. Schwab, who was nowhere to be found when the working men he professed to love were rewarded with a wage cut, popped up at the American Iron and Steel Institute to deliver a keynote address that suggested that further wage "deflation" might be in order if demand did not improve. He also insisted that the federal government must "protect" steel in the international marketplace, even though the Smoot-Hawley tariff signed by Hoover in 1930 had already increased steel import duties significantly.

"Our markets today are being upset by the dumping of foreign steel at sacrifice prices on American shores," Schwab complained. "The volume, to be sure, is a small proportion of our business [1.4 percent in 1931], but the activities of the marginal producer inevitably affect the total structure. This is a case where we must look to Washington for aid. The foreign producer who is getting only $10 or $12 per ton for his product is selling for less than he receives for his product at home, and is, therefore, within the dumping act."[33]

It was a proclamation of naked self-interest, a demand that the government raise tariffs on imported steel, bolstering industry revenues, in return for the industry's support of the Reconstruction Finance Corporation. "We have done our part," Schwab concluded. "We have put our house in order. The Federal Reserve is doing its part. But, above all, the Federal Government, which is the heart of our national structure, must balance its budget and restore confidence there." In apt double-speak, the speech was titled, "Steel Men Face the Future with Faith and Courage."[34]

What made Schwab's special pleading all the more repugnant to his detractors was the disclosure in financial material that he and his fellow officers had made no effort to follow the wage discipline that they had been imposing on their employees. Following settlement with the Stockholders' Protective Committee and institution of the modified bonus program, Grace had been placed on an annual salary of $180,000—the highest fixed salary in the steel business. In spite of the two wage reductions and heavy quarterly losses, Grace's salary scale was not altered. Schwab, too, made no effort to moderate his pay, continuing to receive his full $250,000 in annual compensation.[35] At a time when judiciousness and social responsibility were desperately needed in the business community, Schwab and his "boys" still grabbed for the extra buck.

Something had to give, and it gave in the political arena. Hoover became a discredited, even despised, figure. Returning to his home state of Iowa on a campaign trip, he was met by angry farmers holding banners that said, "In Hoover we trusted, now we're busted." Around the country people expressed their disgust with a vocabulary based on the president's name. Newspapers wrapped around unemployed workers for warmth were nicknamed "Hoover blankets"; a man without any money in his pockets was waving his "Hoover flag"; the shantytowns of the jobless were "Hoovervilles."

Franklin Delano Roosevelt helped himself to the votes lost by the repudiated president. On Election Day 1932 Roosevelt received 22.8 million votes against Hoover's paltry 15.7 million, capturing all but six states and winning the Electoral College by 472 to 59. Just as 1928 was unique in establishing a new high in popularity for the Republican Party, 1932 brought the greatest Democratic victory in history. In conservative Baltimore and Baltimore County, Roosevelt swept every working-class precinct, including the districts around Sparrows Point that had handed Hoover a solid margin of votes four years earlier.[36]

The defeat of Republicanism was a blow to the business community that was to last an eternity. In spite of solid backing by the Chamber of Commerce and other powerful business interests, the Dollar Decade of Republican rule had come to grief. Through the fat years, Bethlehem Steel had been at the head of campaign contributions to Republican candidates. Its $10,000 "gift" to Hoover in 1928 was described as the largest contribution by a corporation in that race. (The reporting of contributions was extremely lax and there were persistent rumors that Schwab had funneled huge amounts of cash to other campaigns, notably Charles Evans Hughes's unsuccessful challenge to Wilson in 1916 and the Harding win in 1920.) In 1932 the Bethlehem Corporation followed through with a $4,450 contribution to Hoover, a sizable sum, especially considering the state of the economy.[37]

In the weeks following the election, the nation was deep in gloom. Hoover would be in the White House until March and the poverty of his lame-duck leadership was writ in the continuing decline of basic manufacturing. Steel led the way. "No parade of statistics is needed to emphasize that 1932 was the most disastrous year in the history of the American iron and steel industry," *Iron Age* reported in its January 5, 1933, issue. "We described 1931 as a 'calamitous' year; what, then, shall be said of 1932?"[38]

The final months of 1932 established a new negative statistic at Sparrows Point—a low of 3,500 daily jobs at a facility that had once employed 18,000. On some days the crowd of men who queued up at the employment office was greater than those who passed through the clockhouse. It was a desperate moment. Charlie Parrish was down to a biweekly paycheck of $11.40 (38 cents per hour times 30 hours per pay period). He was doing anything—even blacktopping the roof of his boss's house—to earn some cash. Joe Lewis was ladling soup at Koenig's kitchen, while Nathaniel Parks picked through the landfill to heat his parents' house and traded in company-store coupon books on the local black market. "Say a guy got broke and wanted some money. He sold me a $5 book for $3. That way you made a little."[39]

As winter approached the *Dundalk Community Press* warned that volunteerism had reached its limits and local relief agencies were at the breaking point. "Relief and Welfare Authorities have estimated conservatively that by March 1933 more than 100,000 citizens of Baltimore city and Baltimore county will be suffering from the actual necessities of life." A week later Post 88 of the American Legion, Sparrows Point, pleaded in a full-page advertisement:

–HELP–

HUNGER, WANT and DESTITUTION
Confront Many Baltimore County Homes

> In order to help the CHILDREN'S AID SOCIETY to cope with the magnitude of existing conditions, ASSISTANCE IS NEEDED FROM EVERYONE.
>
> CONTRIBUTIONS of food, clothing, surplus or unharvested crops, fuel, jobs, shelter, new or used shoes, et cetera are EARNESTLY SOLICITED.[40]

The "specials of the week" offered by the company store for the Christmas holiday season included Heinz Beans at 6 cents for a small can. Rice was selling at 6½ cents for a 15-ounce package; lard at 7½ cents a pound; sweet potatoes at 5 cents a quarter peck; ham shoulders at 9 cents a pound; and generic brand chocolates 55 cents for 2½ pounds. "PROSPERITY ALWAYS SUCCEEDS DEPRESSION," the *Community Press* said hopefully on the eve of Christmas, but without some sort of decisive steps by the government, the Great Depression seemed insolvable. In a matter of three years the plant on the Chesapeake had run the gamut from full employment to thousands on the bread line, from piping prosperity to crippling losses. Everyone wondered what Roosevelt might do to try to pull industry out of its tailspin.

CHAPTER 12
THE NEW DEAL

Clasping the lectern on the steps of the Capitol on Saturday, March 4, 1933, Roosevelt addressed the crowd. "This is a day of national consecration," he began, "and I am certain that my fellow-Americans expect that on my induction into the Presidency I will address them with a candor and a decision which the present situation of our nation impels. This is pre-eminently the time to speak the truth, the whole truth, frankly and boldly. Nor need we shrink from honestly facing conditions in our country today."[1]

One hundred thousand people listened to the address on the grounds of the Capitol, and millions more tuned in on a coast-to-coast linkup of radio stations. The economic situation was deteriorating. Considerably more Americans had lost their jobs than the Hoover administration had cared to admit. Altogether between 13 and 17 million persons, or more than one-quarter of the working population, were jobless, and many millions more were on short time. Panic-stricken depositors had forced bank holidays or strict withdrawal restrictions in 27 states, including Maryland, and a run in the previous two days threatened to topple the biggest banks in New York and Chicago. Since 1929, 11,000 banks—44 percent of the total—had failed, wiping out $2 billion in customer deposits.[2]

To speedily restore a sense of calm and balance was paramount. Only that morning in his room at the Mayflower Hotel, Roosevelt had penned the phrase, "The only thing we have to fear is fear itself," and now he used those words to rally the nation. "Our distress comes from no failure of substance," he pointed out. "Plenty is at our doorstep, but a generous use of it languishes in the very sight of the supply.

Primarily this is because the rulers of the exchange of mankind's goods have failed through their own stubbornness and their own incompetence, have admitted their failure and abdicated." In a passage that seemed aimed directly at Charlie Schwab and other captains of industry, Roosevelt asserted: "Stripped of the lure of profit by which to induce our people to follow their false leadership, they have resorted to exhortations, pleading tearfully for restored confidence. They know only the rules of a generation of self seekers. They have no vision, and when there is no vision the people perish. The money changers have fled from their high seats in the temple of our civilization. We may now restore the temple to the ancient truths."[3]

The government would not stand back idly and watch the Depression worsen—he pledged immediate, constructive action to relieve suffering and start the wheels of industry turning again. If necessary, he would ask Congress for "broad executive power to wage a war against the emergency as great as the power that would be given to me if we were in fact invaded by a foreign foe." Within hours Roosevelt had assembled the leaders of Congress in anticipation of a special session for passage of emergency legislation. On the same evening he summoned his Cabinet together in the second floor library of the White House. Because of the banking crisis, the appointments had been confirmed by the Senate. The Cabinet members were to be sworn in immediately by Associate Justice Benjamin Cardozo. The officers lined up on the basis of the date when their department was founded."

Starting with Secretary of State Cordell Hull, nine men gave their oaths; then Frances Perkins stepped forward to accept the post of the tenth and youngest department, Labor. A social reformer, New York State commissioner of labor under Governor Roosevelt, and the first woman ever appointed to the Cabinet—her appointment had stirred the greatest comment when it had been announced a month earlier. Newspaper pundits had deemed Perkins a handicap because of her sex, and William E. Green, president of the AFL, said that organized labor would never become reconciled to her appointment.[5]

But Roosevelt was undeterred. He had persuaded Perkins to take the job after she had first said no, and he had listened carefully when she discussed what his "New Deal" would have to do for working people to be successful. According to Perkins, it would have to establish minimum wages to put a brake on additional wage cutting. It would have to provide direct aid for the jobless. It would have to abolish the child labor that lingered in some trades, and ameliorate the tremendous peaks and valleys of employment which she believed kept the laboring classes financially and psychologically dependent on their employers.[6]

Of top priority, she told the president, was redressing the imbalance of power that had grown between workmen and employers in large industries, especially steel. American democracy had always suffered from the uneven linkages between political and economic life, with Thomas Jefferson first recognizing that economic security was essential if each individual was to participate in the democratic process. According to Perkins, the low wages in steel and its absolute bar against unions endangered the nation's stability and democratic heritage. New policies had to be forged "based on the human and economic needs of the nation as a whole."[7] The new president pledged his support.

*

When Roosevelt tapped Miss Perkins for Labor, she had already devoted more than two decades of her life to the issues she would deal with in Washington. Miss Perkins had gained a worldwide reputation in her fields of factory inspection, labor law, and workmen's compensation. Among her previous posts were: executive secretary of the New York Consumer's League (1910–12); director of investigations, New York Factory Investigating Commission (1912–13); executive secretary of the Committee on Safety (1912–17) formed after the Triangle Shirtwaist fire killed 146 female operatives in New York; member of the New York State Industrial Commission (1919–21); chairman of the Industrial Commission (1926–29); and head of the New York State Department of Labor (1929 until her appointment to the U.S. Labor Department in 1933).[8]

Her storehouse of experiences would prove invaluable as she undertook some of the most delicate and volatile assignments of the New Deal. Except for Roosevelt himself, who, as promised, deferred to Perkins's judgment in most instances, the labor secretary became a key player in the political clashes and industrial upheavals of the tempestuous '30s. Indeed, it wouldn't take long for the male precincts of South Bethlehem and Sparrows Point to come face to face with this deep-voiced dynamo who was five feet, five inches tall (in medium high-heeled shoes), weighed 150 pounds, dictated too fast for most stenographers, and favored black felt tricorn hats.[9]

"Whether a high-caste New England conscience turns right, as in Henry Cabot Lodge, or left, as in Miss Perkins, it tends, with banked fires and a due civility, to go the whole way," wrote Russell Lord in his 1933 profile "Madame Secretary" in *The New Yorker.* Perkins had shown strong feelings for the downtrodden during her formative years. In her senior year at Mount Holyoke, she was influenced by a course in American history by Annah May Soule. Soule had the idea that her students should go into the local factories and make a survey of working conditions. In Dr. Soule's class, Perkins had an opportunity to use her scientific training to test hypotheses and draw conclusions from firsthand observation. The impact of a Holyoke education on Perkins was suggested years later when the college's alumnae quarterly asked her if she had any advice for current students. She had: "The undergraduate mind should concentrate on the scientific courses, which temper the human spirit, harden and refine it, make of it a tool with which one may tackle any kind of material."[10]

Following her graduation, she accepted a job at Ferry Hall, a girls' school at affluent Lake Forest outside of Chicago. She taught physics and biology and managed a dormitory. In her three years there she became deeply engrossed with religion and the Chicago settlement movement. In 1905 she joined the Episcopal church (the church would provide great solace and strength to her during the Cabinet years) and joined Jane Addams's Hull House. She made the rounds with a district nurse on tough Halsted and Maxwell streets, her Boston accent (labor was *laboh* and masses *mawses*) mixing with the Italian, Greek, and Hebrew spoken. She visited families in distress and families in alcoholic decay. She was stunned to find that in Chicago, where 25,000 women and girls worked in the clothing industry, most received only $2.50 to $3 a week, and some were cheated out of that by fly-by-night sweatshop operators.[11]

In 1907 a social agency called her to Philadelphia to follow up on immigrant girls there. Founded by Quakers, the group was concerned with dishonest employment agencies and indecent lodgings set up to entrap women, including Negro girls from the South. Perkins was paid $40 a month. Like many reformers she was preoccupied with the link between capitalism and the deplorable conditions of the slums. Why did the present economic system produce so much poverty and family disruption? Was it inevitable? Could it be avoided? Was it just? In this connection, *The Pittsburgh Survey* figured prominently in her thinking.

In 1909 the six-volume *Survey* began its serialization in *Charities and the Commons* magazine. The *Survey* took the first comprehensive look at a steel community. Through detailed interviews and house-to-house surveys it revealed that hours of labor were uniformly longer in steel than in less concentrated industries; that labor was more intense and pay for most jobs exceedingly low; that men became exhausted from the routine and that the burden on family life was correspondingly high; that fatigue increased the likelihood of illness and premature aging. John A. Fitch titled one article, "Old Age at Forty," and that became the rallying call of the reformers.[12]

The study did not have any immediate effect on Congress or the mainstream press; yet for a progressive like Perkins the study was a signal event, adding weight and meaning to her own observations and hypotheses. It not only focused her attention on the social costs of a runaway factory system, but convinced her that the faith of many reformers in free competition was misplaced. Taking night courses with Simon N. Patten, an economist at the University of Pennsylvania, Perkins became convinced that society had entered a period of "surplus civilization," where attempts to re-create Adam Smith's free market of small proprietors (for example, Wilson's "New Freedom" and trust-busting efforts) were futile. Large industrial organizations were here to stay. They had to be accepted and were useful, with the understanding that it was preeminently the role of democratic government to engage in the redirection of the economy to more stable and humane uses of its productive resources. Hers was a mixture of the central planning favored by socialists with a rock-ribbed faith in private property.[13]

In 1910, Perkins was offered the position of executive secretary at Florence Kelley's New York Consumers' League. It was as lobbyist for the league, fighting in Albany for a 54-hour workweek for female factory workers, that she was introduced to three legislators who would figure greatly in her professional life—Al Smith, Robert Wagner, and Franklin Roosevelt. All three were said to be experiencing the stirrings of progressivism. Perkins took to Smith and Wagner immediately. Of the young senator from Dutchess County she wasn't so sure. "I knew innumerable young men who had been educated in private schools and had gone to Harvard. He did not seem different except that he had political rather than professional or scholarly interests."[14] With his habit of throwing his head up while peering through his pince-nez, Senator Roosevelt gave the appearance of looking down his nose at people. He seemed dilettantish and smug. Perkins was, of course, to reassess her opinion of FDR; she claimed his narrow escape from death and long, painful convalescence following an attack of infantile paralysis in 1921 created a different man, with a humility of spirit and a deeper philosophy. "He was not born great, but he became great," she said.[15]

In Albany Miss Perkins was instrumental in gaining passage of the 54-hour law over upstate Republican hostility. She won over cynical Tammany leaders such as "Big Tim" Sullivan—whom less adroit reformers had alienated or shunned—through her determination and personable approach. "Me girl, I seen you around here and I know you worked hard on this and I know you done your duty," Big Tim growled to Miss Perkins by way of indicating his support for the 54-hour bill. She became associated with Smith and Wagner through her involvement in the Factory Investigating Commission. Once she took the two men to a candy factory where women worked under hideous conditions. Globs of hot chocolate and sugar filling hung in their hair. Fire exits on the upstairs floors were blocked by heaps of mops and debris. From such investigations, she wrote, the future governor and future U.S. senator of New York "got a firsthand look at industrial and labor conditions, and from that look they never recovered."[16]

*

"Increased purchasing power" was the idea that animated the Roosevelt advisors in the first months of the New Deal. They viewed the core of the crisis as many years of artificially depressed wages which together with technological unemployment had dried up consumer spending. Under this "underconsumptionist" theory the economy had been able to keep pace with its output only through use of installment credit and the inflated valuations of the stockmarket. When these stimulants were used up in 1929, as they were bound to be, the economy shrunk like a punctured balloon because wages had not kept pace with profits and workers had not obtained their fair share of the gains in productivity since the war. In Perkins's mind the only way the economy could revive was through greater purchasing power brought about by higher wages.[17]

"Prosperity was a two-faced goddess," argued Perkins, "and no one would be prosperous unless the workers had a modest prosperity. Their wages must enable them to be a continuous market for consumer goods, and in turn the makers and distributors of consumer goods would be a continuous market for steel and durable goods." This, of course, was at odds with the Hoover philosophy that prices and wages had to retreat in recession to levels that cleansed the system of its inflated values. As a consequence, a "floor" on wages and a "ceiling" on hours were critical to the administration's recovery program. Again from Perkins: "The fixing of a bottom in the falling of wages perhaps would be a stabilizing element for business. After a minimum wage provision with machinery for its operation was attached, the bill would have, of course, the effect not only of spreading the work, but also of increasing the total money going into pay envelopes of the country."[18]

It was also agreed that unregulated competition was a root cause of the Depression. Employers who had conducted their businesses fairly had been penalized, even driven out of business, by price-gouging and other "senseless" competition. The ideal of a competitive economy was seen as a dangerous illusion. "The cat is out of the bag. There is no invisible hand. There never was," wrote Rexford Tugwell, who continued, "We must now supply a real and visible hand to do the task which that mythical, nonexistent, invisible agency was supposed to perform, but never did." Finally, there was a growing consensus that the right of workmen to organize into unions and bargain collectively with their employers was essential to redress the imbalance of power between large corporations and unorganized wage earners. "If destructive competition [was] to be eliminated, something like uniform labor standards and working conditions must be maintained throughout the industry," argued Perkins. Her opinion was shared by Robert Wagner, chairman of the Senate Labor Committee. He vigorously pushed for the inclusion of collective bargaining in the recovery package. He was supported by Roosevelt "brain trusters" Raymond C. Moley, Felix Frankfurter, and General Hugh S. Johnson, who were drafting corollary legislation.[19]

From the push and shove of the "Hundred Days," the special congressional session called by Roosevelt, a remarkable number of measures was passed to meet the economic emergency. A general relief program was begun; the Reconstruction Finance Corporation was revamped and directed to loosen up its loan policies; the Home Owners' Loan Corporation was established to save small homeowners from foreclosure; farming relief and reform was launched with the Agricultural Adjustment Administration; under the Civilian Conservation Act, several hundred thousand unemployed young men were placed in camps across the country to engage in reforestation, flood control, and road building in public parks and reservations; federal insurance of bank deposits (first to $2,500, later increased to $5,000) was provided by the Glass-Steagall Banking Act; a Truth-in-Securities Act regulated the sales of stocks on U.S. and overseas markets; and the Tennessee Valley Authority began a program of unified development and conservation of the Tennessee Valley through flood control and production of power.[20]

But perhaps the boldest experiment of all was the National Industrial Recovery Act, the administration's response to the prostration of business. It called for a sweeping reorganization of business with the three-pronged objective of increasing employment, raising the wages of the lowest-paid workers, and allowing industries to eliminate unfair competition and restore profits. When Roosevelt signed the law, he declared its goal as "the assurance of a reasonable profit to industry and living wages for labor, with the elimination of the piratical methods and practices which have not only harassed honest business but also contributed to the ills of labor." Achievement of such goals was sought through the adoption of fair-competition agreements, called "codes," by which the companies of an industry agreed to conduct their business. The codes were to be framed by the industries themselves, largely through each sector's established industry association, but would require approval by the national recovery administrator before they became operative, and had to conform to uniform regulations. Through the codes, industry could stabilize, even raise, prices. And to prime the pump, Title II of the act carried an appropriation of $3.3 billion to finance a public works program.[21]

In return for the administration's suspension of antitrust laws, the codes were to provide for labor reforms and correction of longstanding abuses. They had to prescribe maximum hours and set minimum pay rates, prohibit child labor and sweatshops, and outlaw other practices harmful or unfair to employees. Furthermore, labor's right to organize and bargain collectively was to be guaranteed. The collective-bargaining provision was by far the most controversial aspect of the law as it took shape in Congress, for it was the first explicitly pro-bargaining declaration since Wilson's short-lived War Labor Board. In the interim, a worker's right to organize and strike had been severely circumscribed by state legislative acts and by injunctions against picketing handed down by conservative federal courts. The language concerning organized labor was stated in Section 7(a) of the act, and soon "Section

7(a)" became a household word in business and governmental circles. It stated:

> Every code of fair competition, agreement, and license approved, prescribed, or issued under this title shall contain the following conditions: (1) That employees shall have the right to organize and bargain collectively through representatives of their own choosing, and shall be free from the interference, restraint, or coercion of employers of labor, or their agents, in the designation of such representatives or in self-organization or in the other concerted activities for the purpose of collective bargaining or other mutual aid or protection; (2) that no employee and no one seeking employment shall be required as a condition of employment to join any company union or to refrain from joining, organizing, or assisting a labor organization of his own choosing."[22]

To fight the economic war, Roosevelt set up the recovery administration like a military operation. General Hugh S. Johnson was appointed commandant of the NRA, responsible for establishing the codes of Title I, while Harold Ickes was placed as chief of Title II Public Works Administration. The NRA got a motto ("We Do Our Part"), a symbol (the Blue Eagle), and a presidential mandate: Roosevelt wanted codes for steel and four other basic industries—autos, oil, coal, and textiles—completed within 60 days. The rush was on. Hastily setting up quarters in the Department of Commerce building, General Johnson issued a special bulletin to the American Iron and Steel Institute to draw up a basic code for the nation's steel mills by mid-July.[23]

After some discussion it was agreed that the Department of Labor would represent the steelworkers in the preparation of the code, with Miss Perkins delegated to present the steelworkers' case before the NRA. Perkins said she knew the White House was inviting political controversy, and she told the president so, pointing out that it was unorthodox for a government official to submit a case for a partisan position. "No more unorthodox than the NRA itself," Roosevelt replied. He said the codes had to be adopted and that recovery could not proceed unless the public was convinced that the interests of labor had been considered. "I think the Secretary of Labor ought to be the Secretary *for* labor," he added. "Go ahead. Do the best you can."[24]

*

On July 15 the Iron and Steel Institute presented the proposed code to General Johnson. It called for an immediate 15-percent increase in wages (raising base wages at Sparrows Point back to 40 cents an hour), but it also qualified the language of Section 7(a) in such a way as to sanction company unions. It stated that bargaining with employees through in-house employee representation plans was the preferred method of settling the joint problems of labor and employer and that all disputes should be settled finally by the decision of the highest officer of the company involved.[25] If accepted, this wording would have undercut the law's guarantee of labor's right to organize and bargain freely without employer interference. But if it were rejected in a way that caused anger in the industry, the whole recovery program might be thrown in jeopardy.

Three days later the administration made its response. Secretary Perkins placed a call to Eugene Grace and asked him to invite her, as part of her preparation for the code hearings, to inspect one of his mills. There was a long pause and then acquiescence. The plant she wanted to see was Sparrows Point, where, she wrote, "Bethlehem has its newest and, theoretically, most efficient plant." She put a similar request to Myron Taylor, chairman of U.S. Steel, and got invited to visit the Homestead works. The idea was to serve notice that "there would be no complacent agreement between the government and the employers." So the determined labor secretary (Cabinet salary, $12,000) set out to visit an enterprise whose profits had reaped its president the most fabulous bonus in the country ($1,623,753) four years earlier.

On Saturday morning, July 29, 1933, Baltimore newspapermen gathered at Camden Station to interview Secretary Perkins as she arrived on the B&O sleeper. Her previous day at Homestead had been eventful. After touring the mills she had been permitted to hold a meeting at the town hall. At the end of the meeting, as she was saying good-bye to the burgess of Homestead, there was a disturbance downstairs. A newspaperman whispered that many men were in the lower hall and on the sidewalk because the burgess had not permitted them to come in. She asked the burgess if she might not have the hall for a few minutes more to hear the men. Refusing, the burgess said, "You've had enough. These men are not any good. They're undesirable Reds." After other attempts by Perkins to reconvene the meeting in a local park were rebuffed by the burgess, she and the undesirables marched to the federal post office building to talk. The incident drew unfavorable comment about the intimidation and repressive policies found in towns where steel companies ruled.[27]

What problems did Perkins anticipate in "disciplining" the steel industry? one Baltimore reporter wanted to know.

"But the government is not disciplining industry," she replied. "Industry is disciplining itself."

"But the experiment is an unprecedented one, isn't it?" she was asked.

"The times are unprecedented," she answered.

"It is the end of laissez-faire, isn't it?"

"I don't know that it is," she said.

"A change of the modus vivendi?"

"They are what Al Smith calls $2 words," said Miss Perkins, who went on to express her opinions "in words which were not $2 words," said the *Evening Sun*.

"Discipline of industry by industry itself is the Democratic method. This is a measure for creating employment and purchasing power within this country. The old frontier times are ended. When people began to live in urban centers they began to discipline themselves. They decided not to throw the garbage anywhere and to submit to certain disciplines for the general good."

"There is a lot of evil in human nature," said a newsman.

"There is a lot of good in human nature, too," she answered.[28]

The Perkins tour of Sparrows Point was unprecedented. Never before had a high government official viewed the plant from the perspective of the employees. Stopping at the tin-plate inspection line, she inquired about working conditions. What were the hours? What were the wages? How much employment was there in a week and how much in a month? The tremendous heat present on the inspection line in midsummer was readily apparent; what steps, she wanted to know, had the company taken to improve the conditions? Were there many accidents, and did the company have safety devices to keep them down?

She put the same questions to workmen at the sheet-rolling mills. Red-faced men lumbered up to her and talked about their jobs. It was hard and hot, they said. After a shift they could pour water out of their shoes. She queried them about their wages. Several said they had made less than $600 in 1932. Asked what they considered a living wage for a man with a family, the answer was, "Lady, that would be between $30 and $35 a week."[29]

On the final leg of the tour she asked her hosts, general manager Stewart Cort and M. J. Larkin, special assistant to President Grace, for a room where she might talk to some of the men privately. She talked to about 20 workers from the pipe mill. "Do you want a union?" she asked bluntly. "Can you speak freely? Do you feel like speaking freely to me?"

There was a pause before one of the men spoke. "Yeah, I think it's all right. The superintendent here is a decent man. If he said this is private he ain't got no spies in here. I'll tell you frankly how I feel. We work here without a union. We got on all right until this unemployment came. We had good jobs. There wasn't much dissatisfaction about the wages or the hours. They were all right."

But he went on, "But, of course, there are lots of picayune troubles, fights all the time. I think we ought to have a union, but the company doesn't want a union, and so none of us are going to kill ourselves to have one."

Although she felt there was a general sentiment in favor of an independent union, there was also much confusion over what a trade union was and fear of how the company might react. Several men said they thought that the company's employee representation plan was a union, and a few said they didn't need a union. Others said a "real" union would be an improvement. With a union they would have a sense of independence. With a union they wouldn't have to take everything from the boss.[30]

The steel code hearings began two days later in Washington. Speaking on behalf of steel employees, Perkins faulted the industry for failing "to rise to the opportunity" to rule out poor industrial practices. The hours proposed under the industry code still permitted lengthy shifts during months of heavy demand to be alternated with very little work during other months. This was wholly unsatisfactory. While applauding the 15-percent rise in base wages, she said the proposed hourly base rates were still far too low to support a workman and his family. In her tour of steel country, she said, charges were made by workers of industrial espionage, and the incident at Homestead seemed to bear this out. Such matters would be closely monitored by the Labor Department, she warned.

Then she summed up the administration's position that steel wages had failed to keep up with industry profits. The ghost of the '20s was thoroughly thrashed. "Bureau of Census figures indicate that wages of blast furnace workers as compared with the value added by manufacture declined from 43 percent of the value in 1914 to 26 percent in 1929. In steel works and rolling mills the ratio of wages to value added by manufacture was 57 percent in 1914 and again in 1923, but in 1929 it had dropped to 47 percent. Four-fifths of this decline took place in the two-year period 1927–29. During these two years the value added increased by over $370 million, while wages rose only $88 million."[31]

"The old deal and the old way," she continued, was to use the fruits of rising productivity to invest in new plant and equipment and for the payment of outlandish bonuses for corporate officers. The dark-suited audience stirred uneasily. "Now there is an excess capacity," and measures were needed to adjust producing capacity to meet consumer demands. If steel was to be excused from antitrust laws and given other protection, then the industry was beholden to act in a publicly beneficial manner. Traditionally dividends were continued to stockholders in poor years out of surplus reserves. "Would it not be equally wise and just to make some of these reserves available for employees who must be laid off from time to time to stimulate purchasing power?" she wondered.[32]

Her resolve caught the industry off-guard. A spokesman for the Steel Institute, ex-Commerce Secretary Robert P. Lamont, read the proposed code into the record. As he reached the provision substituted for Section 7(a), General Johnson interrupted. The open-shop declaration was "inappropriate and of questionable legality," he said. NRA general counsel Donald Richberg seconded Johnson. Only Congress could alter the language of Section 7(a), he said. Lamont asked for a brief recess. In the anteroom he huddled with the board of directors of the Steel Institute, represented by Eugene Grace and William Irvin and Myron Taylor of U.S. Steel. He then returned to the platform. The directors had voted to withdraw the contested language. Did that signify that steel had accepted the labor provisions contained in Section 7(a)? General Johnson asked Lamont. Lamont nodded yes. Perkins, who had been prepared to lead the attack on the company union provision, took to the floor and shook hands with Lamont, congratulating the industry for its "patriotic and far-sighted policy."[33]

*

But the steel managers were scarcely reconciled to Section 7(a), and if Miss Perkins needed any reminder of their resistance, an incident two weeks later would crystallize the point. On August 15, 1933, a meeting was held in her office regarding what had become a routine matter in NRA codes: the adoption of a statement from employers, workers' representatives, and the government that the code met with their approval and that they would abide by its terms. The heads of the important steel firms—Eugene Grace, William Irvin, Tom Girdler, Ernest Weir, L. E. Block, and Hugh Morrow—came to her office. She also invited to the meeting William Green, president of the AFL, who had been appointed labor advisor for the NRA. When the steelmen caught sight of Green, just at the moment that Perkins called the meeting to order, William H. Moore, counsel to U.S. Steel, tried to draw her aside. A few moments later the steelmen rose in unison and filed out of the office.

An astonished Perkins and NRA counsel Richberg followed them out of the room. What was wrong? they asked Eugene Grace. "If we sit down with Mr. Green," he answered solemnly, "and if we sign a code that he signs, it will be assumed that we are dealing with organized labor. As you know, we have an almost sacred policy that we will never recognize or deal with organized labor." Perkins handed Grace a copy of the remarks that Green had prepared for the meeting. He was to make a laudatory statement regarding the NRA and to give his full approval of the proposed steel code.

Still that would not do. Grace said that if it became known in steel towns that they had spoken to the president of the AFL, it would imply union recognition. The back and forth went on for almost an hour. Green finally got fed up and left in a huff.

"I felt as though I had entertained 11-year-old boys at their first party rather than men to whom the most important industry in the U.S. had been committed," commented Perkins. She was struck by the social maladroitness of the executives—even surface politeness could not be extended because of extreme fearfulness and a dogged commitment to a "sacred policy." She saw in the incident the instinctive resistance to labor rights by some businessmen, even those who were about to sign a code with labor guarantees in it.[34]

While Perkins attempted to put a politic face on the incident, the AFL's Green was not so demure. He issued a press statement that denounced the action as a "challenge" to the government's authority. "In my opinion this raises the question of whether the government will surrender to private industry in its efforts to restore purchasing power through the NRA," he said. Roosevelt hit the ceiling when he heard of "the presidents' walkout," as the episode was dubbed by the newspapers. Roosevelt did not like to see industry acting in open disrespect of his labor secretary as well as complicating the difficulties in reconciling the various parties to the code.

The White House sent out word to Charles Schwab and Myron Taylor within hours: The president wanted to see them. The following afternoon the steel chairmen were invited into Roosevelt's inner office. Misreading the president's intentions, Schwab began by saying that he found it difficult to accept parts of the code because of his obligations to his stockholders. Roosevelt then asked pleasantly whether he had been looking after his stockholders when he paid the million-dollar bonuses to Grace and allowed his miners to live in coke ovens.[35]

When Schwab entered the White House in 1926 for an intimate chat with Calvin Coolidge, he had been hailed by Mr. Coolidge as a striking example of "how industry and ability win in this country." When the door to Roosevelt's inner office swung open, neither Schwab nor Taylor offered a word in protest of Section 7(a) to White House reporters. To a visitor later that day, Roosevelt remarked, "I scared them the way they never have been frightened before and I told Schwab he better not pay any more million-dollar bonuses."[36]

Three days later the industry signed the steel code with minimal fuss. In addition to the inclusion of Section 7(a) the final code called for a reduction of hours and an increase in wages. A basic eight-hour day, and 40-hour week were established. No employee was permitted to work over 48 hours in any one week, in order to secure maximum employment for laid-off workers. The provisions involving wages reaffirmed the 15-percent hike provided by most steelmakers in July. In no case were hourly wages to drop below 35 cents for the Eastern district encompassing Sparrows Point, Bethlehem, and Steelton, 37 cents in Johnstown, 38 cents in Buffalo, and 40 cents in the Pittsburgh-Youngstown and Chicago districts. In the South, however, wages as low as 25 cents an hour were permitted. At the code hearings Perkins had opposed regional pay differentials, saying the Southern rate reflected racial discrimination, but in the end politics won out—Southern Democrats warned that the states would secede before paying wages equal to the North. "The code as it was finally adopted was not a perfect code," she was to concede. "It left much to be desired, but at least it opened the door to a continuing improvement in the lives, the work and the wages of the people who work in steel."[37]

*

The code took effect on August 29, 1933. Prices were placed under a "basing point" system that fixed the charges of basic and semifinished products nationwide. Sparrows Point was one of the chief basing points. All four categories of pig iron—No. 2 foundry, malleable, basic, and Bessemer—were set at $17.50 a ton, $1 above the pre-code figure and nearly $4 above the Depression low of $13.56 in December 1932. From Sparrows Point and the three other Eastern basing points (Bethlehem, Birdsboro, and Swedesboro, Pennsylvania) the delivered price of pig iron was set at $18.76 at Philadelphia and slightly higher at other Eastern cities. There was a $10-a-ton fine for undercutting the published price.[38]

The same basing point price of $17.50 was established in the Middle West and New England regions with the exception of Duluth, Minnesota, and Everett, Massachusetts, where the price was $18. In the Alabama district, the controversial "Southern differential" remained, with $13.50 for pig at Birmingham. However, the price of Birmingham pig delivered to Northern ports was higher than the delivered price of local mills (e.g., $18.88 at Philadelphia and $19.13 at New York). This protected the Northern mills from price competition and, like the Rail Pool of yore, discouraged companies from wandering from their regional trade zones. Interterritorial "raiding" was one of the many business practices that had become legion since 1930. By the Steel Institute's own admission, purchasing agents had been bribed, invoices falsified, and bait-and-switch tactics used in the quest to sell off bloated inventories and stanch the flow of red ink.[39]

The swift cartelization of the industry under the Blue Eagle was remarkable. Reporters who entered the Steel Institute's code offices located at the Empire State Building in Manhattan were treated to the sight of junior officers from rival companies grouped behind glass-paneled partitions in daily consultation over everything from patent rights on steel treatments, to raw material and traffic operations. Twice a month the six steel presidents (of "walkout" fame) gathered to vote on policy matters and approve schedules of prices of hundreds of iron and steel products. They were joined by other steel figures, including George M. Laughlin of Jones & Laughlin, plus a squad of lawyers. During the code deliberations, Schwab had been listed as chairman of the board of the institute; in September 1933, following the resignation of Robert Lamont as institute president, Schwab assumed that job.[40]

In this cozy arrangement the only real check on steel prices was by the government itself. Two officials represented the government: Kenneth Simpson represented the president directly, and Donald Richberg represented General Johnson and the NRA. Even in the first days of the code, the administration and the steelmen did not coexist easily, and within a month's time a public dispute broke from behind the closed doors of the institute. At issue was the price of the old bedrock of steel sales, rails. Under the $3.3-billion kitty for Title II public works, the government was prepared to advance loans to U.S. railroads for 840,000 tons of rails and 235,000 tons of fastenings in order to rebuild maintenance-deferred rights of way and enhance employment. However, the New Deal's railroad coordinator was Joseph B. Eastman. Long the "dissenting member" of the Interstate Commerce Commission, with an encyclopedic knowledge of the railroad business and a conviction that government operation would be best, he said he wanted competitive prices from the steelmakers. The price for rails, he instructed, should not exceed $35 a ton, and he added at a White House conference attended by Grace, Taylor, and other steel notables that if they thought the price was too low, he would be glad to audit their books.[41]

Four companies were left in the rail business: U.S. Steel, Bethlehem, Inland, and Colorado Fuel & Iron. Through some coincidence of mathematics they all submitted bids of $37.75. "Collusion," Eastman charged. Not so, said the steelmen, who complained to a commercial newspaper that Eastman's price was not only unreasonable but "in poor taste." Eastman, a tall, stoop-shouldered New Englander, swore the steel barons would get their price over his dead body. On October 30, 1933, Roosevelt called the belligerents to the White House. Eastman arrived "clenched like a fist around a Puritan Bible," wrote *Fortune* magazine. No less committed to their own cause were Grace and Myron Taylor, accompanied by their high-priced lawyers, Paul Cravath and Nathan L. Miller, Republican ex-governor of New York.[42]

"What miracles the President wrought in that room no man can explain," said *Fortune* in frank admiration. "People say he greeted Cravath like an old neighbor who'd moved to other parts. People say he made everyone feel somehow, very much against their wishes, perfectly at ease. They say he told a long and rather involved joke about two Scotsmen which came to the point with the words: 'Well, split fifty-fifty.' They say that the steelmen went into a huddle and Mr. Eastman didn't murmur a thing about collusion and that when they emerged it was with some such price as $36.50. And that the President reminded them of the fifty-fifty joke and that in high good humor he reached for a pencil and did some strict ciphering as strict ciphering should be done. The President figured that one-half of $2.75 is $1.37½."[43]

The $36.37½ figure stuck. Between November 7 and 14 the Public Works Administration allotted some $51 million for the purchase of rails and rail fastenings by railroads, together with $84 million to the Pennsylvania Railroad for completion of the stalled Northeast electrification program and purchase of new passenger cars and locomotives for the service. As part of the Pennsylvania allotment, the railroad agreed to use its own funds to purchase 100,000 tons of new rails. The release of PWA money was a godsend to the battered cost sheets of Sparrows Point. The plant picked up the bulk of orders placed by the Pennsylvania and other regional carriers. It further profited from a pickup in steel sheet and plate orders placed by car-building plants for the Pennsylvania's all-steel passenger fleet and for electric locomotives. Combined with a moderate upturn in tin-plate orders, Sparrows Point's production drew up to 55 percent of capacity at the end of 1933, a significant turnaround compared to the dark days of the Roosevelt inaugural when it operated at a microscopic 18 percent. Capacity at the Steelton and Bethlehem plants had been even lower—12 percent. They too benefited, though not as greatly, from the placement of PWA orders for capital construction.[44]

In the first full quarter of the code's operation (October 1 to December 31, 1933) the steelmaker reported a quarterly net profit of $629,671. While decidedly modest, it represented a milestone nevertheless. After nine consecutive quarters of losses totaling over $30 million, Bethlehem was back in the black. The company, to be sure, was not healthy, but it had stopped the heavy hemorrhaging. Thanks to the higher prices and production levels under the Blue Eagle, it had passed through the worst storm in its history.[45]

*

For workmen the effects of the Recovery Act were mixed. Yes, wages were up (a 10-percent raise in March 1934 brought hourly rates back up to 1929 levels), and things weren't quite so bleak for people like Joe

Lewis, who had regained a job at the mill, and for Charlie Parrish, whose pay of $50 every other week kept enough food on the table for his family without resort to home-grown vegetables or surplus flour. Things, though, were still barebones, the hours low and day-by-day layoffs frequent.[46]

As the months went by it became clear that Roosevelt and his aides were not miracle workers, and that their blueprint for a surplus civilization remained far off. The economy was soft and steel orders erratic in 1934. At times Sparrows Point ran at a decent 60-percent clip, then just as suddenly dipped back to the discouraging 35–45-percent range. Daily employment never exceeded 12,000, two-thirds of the 1928–29 total. A delegation of steelworker wives and children called on Secretary Perkins in Washington and told of the company's practice of docking wages from returning workmen who owed debts to the company store. There was little the government could do if the debts were legitimately incurred, Miss Perkins reported.[47]

Under pressure from South Bethlehem the promise of free and unfettered union activity under the NRA was as elusive as ever. Under the pretext of celebrating the 15th anniversary of the establishment of the employee representation plan, Grace announced his unique interpretation of Section 7(a): that the ERP was the only legitimate organ of "bargaining" between the company and its employees. According to his statement, mailed to workmen, "No outside agency could possibly take the place of our employees' representation plan without destroying that all-essential direct contact and relationship." At the same time the local workforce was given a Hobson's choice of either voting in the company-sponsored elections or having their names submitted to the general office as absentee voters. The company then announced that 97 percent of the labor force favored the company ERP, a vote margin that suggested the kind of collective coercion practiced in the Soviet Union.[48]

Under repeated complaints that steel management was blocking legitimate organizing efforts, President Roosevelt created the Steel Labor Relations Board in June 1934. The board was designated to make rulings on all controversies arising from the application of Section 7(a). With no enforcement powers, though, the board was hogtied by legal objections from U.S. Steel and Bethlehem almost immediately. For example, to supervise a plant election to permit workers to choose whether they wanted to be represented by a nationally affiliated union, the board needed to compile an accurate list of workmen by reviewing the plant payroll records. This, it became increasingly apparent, the steel industry was loath to allow, thereby giving the board little choice but to go to court and seek an injunction to compel the company to grant the request. A court review would delay action by many months. Then at least a year would pass before the matter could be appealed to the Supreme Court, a delay that administration figures were afraid would undercut the momentum of the recovery program.[49]

By the fall of 1934 Secretary Perkins had to face a regrettable fact; by her benchmark of industrial justice, the company unions installed in every important steel plant were a mockery of Section 7(a); by the test of increased purchasing power of labor through better wages, the unions were worthless. The representatives did not have legally binding contracts with the companies; they did not negotiate on wages or have strike powers; they didn't even bother to hold membership meetings to solicit the views of the workers they supposedly represented. More than a year of haggling brought to sharper focus the vexing problems of guaranteeing labor rights in the face of organized industry hostility.[50]

At the end of 1934 President Roosevelt held a White House conference on the impasse over Section 7(a) and the Steel Labor Relations Board. All the key players were present—Grace, Girdler, Taylor, and Nathan Miller representing the "Big Six"; Secretary Perkins; Donald Richberg of the NRA; William Green of the AFL and Mike Tighe of the Amalgamated; and Judge Walter P. Stacey of the Steel Labor Relations Board. Roosevelt was his charming self. What he wanted to do, he said, was to give the recovery program a fair trial. But save for keeping management and labor in the same room without any walkouts, his achievements were minimal. From the start the conference was hopelessly deadlocked. U.S. Steel refused to permit the labor relations board to conduct representation elections at two of its plants for fear that the Amalgamated might win and thus gain a toehold in the industry. In light of steel's position, Green and Tighe refused to go 50-50. They rejected the president's call for a six-month cooling-off period and insisted on an immediate vote at the plants in question. When the steel executives balked at the principle of majority bargaining favored by Perkins and Judge Stacey, Roosevelt ended the conference in failure.[51]

CHAPTER 13

THE COMING OF THE CIO

The Recovery Act, which had saved Bethlehem and other steelmakers from ruinous losses, could not itself withstand legal challenge. In May 1935 the Supreme Court struck down the NRA in *Schechter* v. *U.S.*, the famous "sick-chicken" case. The Schechter brothers were poultry jobbers in Brooklyn who had been convicted of violating the NRA poultry code by selling diseased chickens and disobeying wage and hours provisions. The brothers appealed the conviction on the grounds that the law was illegal. The Court agreed in a unanimous decision. It ruled that the sweeping code regulations were an unconstitutional delegation of legislative power to the executive branch and to private groups, "delegation running riot," according to Justice Cardozo. Furthermore, the Court held that the Schechters' business was not subject to the regulatory power of Congress since it was outside the "flow" of interstate commerce.[1]

If the federal government could not prevent the sale of diseased chickens, how could it hope to solve the problems of the Depression? The New Deal was thrown into turmoil. Roosevelt withdrew from Congress his request for a two-year extension of the NRA, which was due to expire in June 1935, but not before giving the Court a tongue-lashing for its "horse-and-buggy" interpretation of the Constitution.[2] Two weeks later the NRA and its manifold codes for steel, auto, oil, coal, and other sectors expired.

The administration's stab at central economic planning was dead, but its activist philosophy remained. In the wake of the Supreme Court action, Roosevelt, Perkins, and the Democratic majority in Congress placed their support behind a labor bill drafted by Senator Wagner. On July 5, 1935, the president signed the National Labor Relations Act (Wagner Act). It established a three-member labor board with enhanced powers to guarantee the right of self-organization by workers, including the power to investigate company policies and to outlaw a list of "unfair labor practices." In laying out a constitutional rationale, Wagner wrote into the legislation a declaration that unfair labor practices "obstructed interstate commerce" by leading or tending to lead to labor disputes; the act therefore was aimed at diminishing the causes of these disputes.[3] (The lesson of *Schechter* had been learned—that progressive legislation had to be cloaked in constitutional arguments to pass court review.)

With subpoena power, an investigative staff, and the teeth to enforce its administrative findings (which the earlier Steel Labor Relations Board had lacked), the NLRB pointed to a change in New Deal tactics. Prior to *Schechter* the president had supported efforts to bring labor and management together in his recovery program; hence the public code hearings and White House conferences. But faced with the contravention of Section 7(a) by steel and other sectors, and the failure of federal conciliation to break the impasse, many administration figures argued that the government would do better if it encouraged workers to build up more militant organizations. Together with Agriculture Secretary Henry Wallace, Miss Perkins had pressed for basic changes in the Recovery Administration to make it less dependent on trade associations and more responsive to other groups in the economy. So long as low wages persisted in major sectors without the hope of improvement, the benefits of increased purchasing power had no chance. The Roosevelt administration consciously sought to balance the scales between big capital and organized labor.[4]

But did the established instrument of labor, the American Federation of Labor, possess the brains and brawn to organize industrial workers? The smart money said no. Despite the political support of the White House, the AFL was commonly described as a hopeless aggregation of hidebound unions. "The Federation has been suffering from pernicious anaemia, sociological myopia, and hardening of the arteries for many years," wrote *Fortune* in an influential article. The magazine offered its analysis of the AFL and the psyche of the American workman:

> *Because* its members are skilled craftsmen, "the aristocrats of labor," who trade in vested rights (in their jobs), and to whom such vested rights are as sacred as they are to the most conservative banker, they are not going to endanger these rights for the sake of any proletarian uprising of labor.... *Which means* that the great mass of American labor, which the machine age is reducing more and more to an unskilled status, is locked

> outside the Federation's doors.... *Therefore,* with the A. F. of L. laying down on the job, the revolt of labor must be led by the left-wing groups (chiefly Communist and I.W.W.) if it is to be led at all, but there is no permanency in a left-wing U.S. labor movement.... *And it is all because* the American workman is capitalistic by nature. That is why the A. F. of L., despite its many shortcomings, is the only major labor group, and that is why there will be no uprising of labor.[5]

Presiding over the federation was William Green, Samuel Gompers's dull, bureaucratic successor. Eugene Grace and his confreres had shown they could muscle him right out of a Labor Department meeting; surely they could keep the AFL out of their mills indefinitely. Green lacked a popular following and, what's more, did not have the stomach to assert his authority over the real powers within the organization. They were the presidents of the national unions, bosses of the old city-machine mold, who fussed over craft jurisdictions and froze out unwanted activists. In steel Mike Tighe's Amalgamated Association had become mired in internal squabbles with left-wingers and ex-TUUL members. So inept was his organization that out of a reported 100,000 steelworkers who joined the AA in 1933, only 8,000 were still paying dues a year later. Thirteen lodges were expelled by Tighe for "radicalism" and holding unauthorized conferences. Hundreds of workers who had joined the union with high expectations tore up their cards in disgust. "The Amalgamated was very, very weak. Actually it did nothing for the men," says Charles Barranco, a Sparrows Point wire drawer who had joined the union in 1933.[6]

At the AFL's 1934 convention in San Francisco, a faction led by John L. Lewis, president of the United Mine Workers, demanded that the federation organize industrial workers. The bloc was jubilant when the convention passed a resolution calling on the executive council to issue charters for new industrial unions, including a campaign to revive membership in the iron and steel industry. Yet the hollowness of the victory became apparent when the executive council delayed the issuance of charters to industrial unions and then inserted clauses that protected the rights of craft unions to "raid" the industrial unions for members. At the 1935 convention in Atlantic City, Lewis continued to press vigorously for the creation of industry-wide unions to organize workers in autos, steel, rubber, and cement. He said the failure of the AFL to organize production workers under the opportunities of Section 7(a) exposed two truths: the unwillingness of the old guard to lead a populist movement of disaffected workers, and the futility of splitting up workers along craft lines. When Lewis demanded a vote on the question of industrial charters, the delegates defeated the measure 18,024 to 10,933.[7]

Three weeks later Lewis invited sympathetic union presidents to the Mine Workers headquarters in Washington. He laid out a plan for the formation of a new organization, the Committee for Industrial Organization, to do the job that the AFL had spurned. Among those at the meeting were Sidney Hillman, president of the Amalgamated Clothing Workers, David Dubinsky of the Ladies' Garment Workers, Charles Howard of the Typographical Union, and Max Zaritsky of the Hat, Cap and Millinery Workers. Everyone at the meeting expressed impatience to get the ball rolling. Lewis submitted his resignation as vice president of the AFL, and William Green, faced with the defection of 800,000 of 3.3 million members, condemned Lewis and his allies for "organizing an insurrection."[8]

*

Audacious was John Lewis's agenda. Almost before the organization got going, Lewis announced that the CIO would undertake to organize the biggest workforce in America, the 500,000 employees of steel. The failure of the AFL to organize the trade was, of course, one of Lewis's most potent arguments for his new organization. But there was also a history of animus between Lewis's Mine Workers and U.S. Steel and Bethlehem: Since the mid-'20s their captive mines in Appalachia had been highly adept at breaking the back of UMW solidarity.[9]

In February 1936 Lewis offered $500,000 to the AFL toward a fund of $1.5 million for the purpose of organizing steel. The string which he placed on his offer was that the federation abandon its craft orientation and agree to organize under industrial lines through capable leadership. The proposal was turned down by Green. Then Lewis made the same offer to Tighe. Nearing 80 and ill, the president of the Amalgamated procrastinated, seeking a rival offer from Green so he could wiggle out of Lewis's bid. The AFL offer never came, though, and Lewis soon made quick work of Tighe. Tracking down the Amalgamated president "somewhat as a panther might stalk a moth-eaten alley cat," Lewis forced him to sign an agreement that gave the CIO full privileges to conduct the steel campaign, thereby reducing the probability of a rival drive by the AFL. Except for the right to issue charters to new lodges, the Amalgamated Association ceased to exist and Tighe passed into retirement. All power of the purse resided in Lewis and his handpicked deputies.[10]

Lewis established the Steel Workers' Organizing Committee as the operational arm of the campaign. Its initials, SWOC, had a nice ring. Almost immediately SWOC developed a strong momentum. The union welcomed volunteers and they flocked to makeshift offices to lend their help. To chair the drive Lewis picked his longtime aide and the vice president of the Mine Workers, the highly-regarded Philip Murray. Clint S. Golden, formerly with the Amalgamated Clothing Workers, was appointed director of SWOC's Northeast region. The Middle West and Southern regions were headed by UMW veterans Van A. Bittner and William Mitch. Filling out the national office were general counsel Lee Pressman, a graduate of Harvard Law School who had worked briefly for the Agricultural Adjustment Administration, Vincent Sweeney, a former Scripps-Howard newsman, Harold J. Ruttenberg, a brainy University of Pittsburgh graduate who handled research, and David McDonald, another Mine Worker who was appointed secretary-treasurer. SWOC rented space on the 36th floor of the Grant Building in Pittsburgh. Doubtless it pleased Lewis that his organization was situated eight floors above the suite of Ernest T. Weir, chairman of National Steel.[11]

On the eve of the steel campaign in 1936, Lewis was 56 years old. The son of a Welsh coal miner, he was born on February 12, 1880, in Lucas, Iowa. In 1882 his father helped lead a strike at the local White Breast coal mine, for which he was blacklisted and the family forced to move from town. At an early age young Lewis roamed around the West before he married a schoolteacher who was the daughter of a doctor. Myrta Edith Bell polished up his appearance and got him to read more, although he was never particularly well versed in Shakespeare and Plato as universally reported in press accounts. In 1910 Lewis was elected president of a Mine Workers local in southern Illinois. A year later he was in Springfield as the UMW's state lobbyist. His brightness and ambition attracted the attention of union leadership and in 1916 he accepted President John White's offer to become chief union statistician, moving his family to Indianapolis. The post allowed him to become expert in the business of coal. In 1918 he was named vice president of the UMW and in 1921 was elected president, the post he had occupied ever since.[12]

Lewis possessed red bushy eyebrows. His hair rose abruptly from his forehead in waves that rolled majestically from a part slightly to the left of center. A massive, jutting jaw accentuated his pallid Welshman's skin and made his face seem constructed of layers of hard rock rather than mere mortal flesh. Nothing in his public persona discouraged the impression that he was the owner of a superhuman will. He did not speak in public. He thundered. And his scowl was legendary. "Madame Secretary, that scowl is worth a million dollars to John L. Lewis," he once thundered to Frances Perkins. And he was right. It convinced workers that he was indefatigable and undefeatable, and that he would see them through to the end.[13]

The campaign to organize steel was kicked off on Sunday, July 5, 1936. An emotional rally was held outside of Homestead in commemoration of the martyrs of the 1892 strike. The following evening Lewis went on the air. Over a nationwide hookup of NBC radio stations he had purchased 30 minutes' time to speak on "The Battle for Industrial Democracy." At 9:30 Eastern time his basso profundo split through the static of crystal sets in living rooms around the country as people tuned in to hear what he would say.[14]

"I salute the hosts of labor who listen. I greet my fellow Americans. My voice tonight will be the voice of millions of men and of women employed in America's industries, heretofore unorganized, economically exploited and inarticulate. I speak for the Committee for Industrial Organization, which has honored me with its chairmanship."[15]

From his days as a UMW organizer he had used words as weapons in his arsenal of labor proselytizing. Now he turned his native eloquence against the steel concerns, blasting them for their treatment of labor. Although vast personal fortunes had been derived from the business, "it has never throughout the past 35 years paid a bare subsistence wage, not to mention a living wage, to the great mass of its workers." The impression floated by its publicists that it paid high wages had no basis in fact, Lewis claimed. He cited a study by the National Industrial Conference Board which ranked steel labor last of 21 industries between 1923 and 1936 in terms of hourly pay and actual money wages. Measured by average hourly earnings, unionized bituminous coal miners were paid 19 percent more than steelworkers, anthracite miners 27 percent more, and construction workers 20 percent more.

"Greater payments have not been made to wage and salary workers because the large monopoly earnings realized have been used to pay dividends on fictitious capital stock, to add physical values in the way of plant extensions, and to multiply the machines that displace human labor. Under the wildest flight of imagination, what greater injury could be done to steel workers by labor unions or any other legitimate agency than is evidenced by this financial exploitation by private bankers and promoters!"

Regarding the company representation plans, his voice dripped with sarcasm. They were no more than "pious pretexts" for denying workers their right of self-organization. "Their constitutions and by-laws are drawn by lawyers for the company. No changes can be made without the company approval. The officials are selected under company supervision. No method of independent wage negotiation is provided. No wage contracts have in fact been made between the companies and their employees under the company union plan."

He pledged to exert every ounce of his strength to the cause of unfettered unionism. "Organized labor in America," he snorted, "accepts the challenge of the omnipresent overlords of steel to fight for the prize of economic freedom and industrial democracy. The issue involves the security of every man or woman who works for a living by hand or by brain. The issue cuts across every major economic, social and political problem now pressing with incalculable weight upon the 130 millions of people of this nation. It is an issue of whether the working population of this country shall have a voice in determining their destiny or whether they shall serve as indentured servants for a financial and economic dictatorship which would shamelessly exploit our natural resources and debase the soul and destroy the pride of a free people.

"On such an issue there can be no compromise for labor or for a thoughtful citizenship. I call upon the workers in the iron and steel industry who are listening to me tonight to throw off their shackles of servitude and join the union of their industry."[16]

The "new unionism" that John Lewis put on the map required new avenues of communication. Having launched the steel drive by means of radio, the most modern of public relations devices, his organization soon was utilizing every method available for contacting large numbers of people. Publicity was identified by Phil Murray as crucial to the three stages of a successful membership drive. The first step, he said, was setting up "the organizing machinery" of staff, office, and volunteers "according to local conditions—but always under the strict and close supervision of the regional directors and the national office." Second was developing a quick-response publicity apparatus for issuing leaflets, replying to newspaper publishers, holding rallies, and keeping civic authorities informed of the union position. It was only after laying down this groundwork that SWOC could undertake the third stage of "organizing, signing up members [and] establishing lodges" at the plant level, Murray said.[17]

Within a month the fledgling organization had its own publication, *Steel Labor.* Sent in bundles to steel towns and distributed free, the weekly paper was a crisp, professional tabloid of eight pages edited by Vincent Sweeney. Its front section contained news of the campaign. There was a wrap-up of events from Washington and Wall Street, with stress on information relevant to steel. An editorial section included syndicated columnists of New Deal persuasion, political cartoons, editorials, letters to the editor, endorsements, and words of inspiration from John Lewis and Phil Murray. While nothing in the paper could be construed as tarnishing the image of unionism, neither could Sweeney be accused of engaging in class warfare or lurid sloganeering. A good newsman of the objective school, he kept the paper focused on wages and hours with some leavening attacks on company police and industrial dictatorship.[18]

The paper amplified SWOC's demands for a $5 day (or 62½ cents an hour), a 40-hour week, and time and a half for overtime. The wages were 30 percent above the 47-cents-an-hour rate that prevailed in the Eastern and Middle West districts, and the overtime demand was a strategic bargaining chip handed to SWOC by U.S. Steel. On July 23, the steelmaker announced it would go on a 48-hour week because of an upswing in orders. The boardroom pronunciamento was a gaffe of the highest order, antagonizing the men in the mills who saw it rightly as a way to deny them overtime pay after years of shrunken paychecks. *Steel Labor* denounced the hours extension as evidence that the industry wanted to go back to "the medieval days" of the seven-day week; Frances Perkins said it sabotaged the government's program to spread employment and increase purchasing power; and Congress was inspired to pass the Walsh-Healey Act, which limited hours to 40 a week without overtime on federal contracts, including navy contracts that U.S. Steel was bidding on.[19]

The CIO campaign inspired other voices. For years activists in Baltimore had complained about the antilabor bias of the A. S. Abell Company, publisher of the *Sun* papers. In 1936, Charles Bernstein did something about it. A Socialist with ties to the Workmen's Circle, he formed the *Maryland Labor Herald,* a paper dedicated to "voicing the demands" of the working man. The weekly had an important impact on the dynamics of the grassroots scene. In the opening months of the steel campaign the energies of SWOC went mainly into driving a wedge into U.S. Steel in the Middle West. The idea was to confront and organize the nation's largest steelmaker, then move to the smaller companies. If *Steel Labor* followed the broad strategy on the national level, Bernstein's paper expressed the local yearnings of a coalition of unionists, civil rights figures, and leftists.[20]

The *Labor Herald* paid great attention to the nuts and bolts of neighborhood organizing by publishing testimonials from churches and fraternal groups as well as publicizing the speeches of union supporters. The CIO drive placed emphasis on the have-nots of the mill, the immigrant populations that lived in East Baltimore and Highland Heights. Articles in the paper reflected the tedious process of reaching out: the many small meetings called at Finnish Hall, a center of activity; a SWOC bull roast sponsored by the Polish National Alliance; leaflets handed out after Mass at St. Casimir's by union men and their wives. "In those days, my God I can still remember, we had a slogan: 'Make Baltimore a Union Town,'" says Mike Howard, a volunteer organizer who went door-to-door through East Baltimore in 1936. "We were lit up with this idea because we thought that it would end all the crap you had to take."[21]

Every fourth workman at Sparrows Point was black. Race tensions were high: the brutal lynchings of black men in 1931 and 1933 by mobs on Maryland's Eastern Shore were a reminder of the perils of trying to unite blacks and whites in a single union. Since his organizing days Lewis had recognized the need of opening unions to Negro workmen—his United Mine Workers had the largest number of black members of any union in the country, and this had helped stop the practice of importing blacks to break coal strikes. In Baltimore, SWOC had to be known as more than the "white man's union." Under the leadership of Robert Kimble, Israel Zimmerman, and G. Fred Rausch, the way was paved for trade-union rallies in the black community. The first major rally took place at the corner of Eden and Monument streets in the heart of the black ghetto in August 1937. Thurgood Marshall, a 28-year-old lawyer for the Baltimore NAACP, endorsed the union drive. Also speaking on behalf of SWOC were John P. Davis of the National Negro Congress as well as organizer Zimmerman. Fifty-two Sparrows Point employees were reported to have signed SWOC cards, and more rallies were scheduled.[22]

Was SWOC getting its message across? Had it gained the support of the average workman? The record was mixed. Six months after the campaign began, Robert Kimble, the CIO director for Maryland, reported that out of 15,500 workers at Sparrows Point, 4,000 had signed SWOC cards, and new men were joining daily. Although Kimble had undoubtedly indulged in some membership inflation—3,000 was a more accurate figure, according to insiders—the campaign had still made an impact. Only a year before, no more than 400 Sparrows Point workmen were members of the Amalgamated.[23]

Even a partisan like Michael Howard confessed that progress was slow and haltingly uneven. "You had to break down years of fear," he said. The plant was so gigantic and the company's power so ingrained in people's lives that it was hard to get a toehold in some departments. At the coke ovens, where men worked in isolated gangs, the organizers found it hard even to get names of people who might be approached away from work. Forging solidarity under these conditions was daunting work, said Howard. "You have to remember: The men were still renting their houses from Bethlehem Steel, wives were still going to the company store for food. Changing around a universe like this takes a little time!"[24]

In its campaign against Bethlehem SWOC developed some clever tactics. Rather than attack the employee representatives directly, an attempt was made to win them over to an "outside-union" point of view. Phil Murray cautioned that it was important not to alienate any of the elements from which the CIO hoped to draw its membership. In 1935 there had been some agitation by company unions in the U.S. Steel chain, with the Braddock company union demanding a 10-percent wage increase and a worker representative on the board of directors. This policy cut both ways: if it promised to attract some of the "big men" in the mills, it also tended to fix the status quo which people like Howard fought against. Although Murray scarcely held the company unions in esteem, he saw the short-term advantage of trying to capture their leadership. He dubbed the process "biting at the heels of management."[25]

*

On March 2, 1937, the SWOC drive got an electrifying jolt when U.S. Steel announced that its chief operating unit, Carnegie-Illinois Steel, had signed a contract with the union. Wages were to be advanced to 62½ cents an hour, meeting the union demand for a $5 day, and a

40-hour week was established with time and a half for overtime. The settlement had been hammered out between Lewis and U.S. Steel Chairman Myron Taylor in secret talks that began in January 1937. A crucial compromise between the men involved the form by which the company would recognize SWOC. Taylor agreed to drop the company's policy of "open shop" in return for Lewis's agreement to waive SWOC's demand of exclusive bargaining rights, or "closed shop."

Under the so-called Taylor Formula, employees of the company could continue to be members of the company union, while SWOC members would be recognized as members of the CIO without "coercion or intimidation in any form." Most commentators agreed with Secretary Perkins, who called the settlement with the world's largest corporation (daily employment of 225,000) "the most important victory" of the decade for collective bargaining and unionism. Reports of the death of organized labor had been premature.[26]

A number of factors seemed to have influenced Taylor's change of heart. Foremost was economics: U.S. Steel had been hard hit by the Depression, and Taylor was unwilling to sacrifice the company's first increase in steel shipments since 1929 to take a selective SWOC strike while his competitors stayed open and ran at capacity. The chairman was deeply influenced by the course of union organization in the auto industry. In late 1936 the United Automobile Workers had targeted General Motors in Flint, Michigan, for a strike. After sustaining heavy losses for 44 days, GM had capitulated on February 11, 1937, and agreed to recognize the UAW. Taylor could not fail to appreciate the strength of the union among the workmen or the financial beating taken by the carmaker.[27]

Another factor that undoubtedly influenced Taylor was the presence of Walter Runciman, president of the British Board of Trade, in the United States during the time of the talks. In the face of Hitler's threat, Britain was engaged in a $7-billion rearmament program and Lord Runciman was in discussion with American manufacturers in regard to the disposal of contracts. It was widely believed that he made it known to Taylor and the U.S. Steel board that a guarantee of uninterrupted production would be required for the company to receive its share of military orders.[28]

Changes in the political climate at home also encouraged Taylor to rethink his opposition to SWOC. Roosevelt had won reelection over Alf Landon by the highest popular vote in history in November 1936. The landslide placed New Dealers in Congress and marked a shift in the political balance of many state legislatures. In Pennsylvania, where important steel mills were located, labor had gained considerable leverage. A Mine Workers officer was the lieutenant governor, and several candidates backed by Lewis had won races in the western counties. A bill providing for stiff penalties against employers found to be thwarting worker self-organization and bargaining was under debate in Harrisburg and was gaining support. In Washington, contracts for armor plate for the navy were stalled because the steel companies had not met the terms of Walsh-Healey. Secretary of Commerce Daniel Roper suggested to newsmen that the government might open a munitions plant if the 40-hour week were not adopted by private manufacturers. Still another worry came from the Labor Department. Asked by Taylor to issue an informal opinion regarding the company ERP, Secretary Perkins said that the ERP could not execute a labor agreement with the company because the employees had never given it authorization to represent them. Despite the advisory nature of the opinion, the decision was widely interpreted as narrowing the company's grounds for arguing that its ERP was an independent union under the provisions of the Wagner Act.[29]

That the cumulative impact of these considerations led to U.S. Steel's decision to recognize SWOC was closely tied to the background and perceptions of Taylor. Lewis's counterpart did not fit the standard mold of the American steel executive. Taylor had not become involved in the steel business until he was 49 years old. Educated as a lawyer and known for his interest in international art (he owned a Medici castle near Florence), Taylor had spent his early career in textiles, where he concentrated in the buying, financing, and selling of concerns. He had been invited onto the steelmaker's board of directors by J. P. Morgan, Jr., in 1923. Following the death of Elbert Gary, Taylor was made chairman of the powerful finance committee, and in 1932 was made chairman of the board. It was no secret that he was given the job by the House of Morgan to restructure the sprawling concern.

In 1934, Taylor reached outside management ranks to select a new vice chairman, Edward Stettinius. The new vice chairman was young (33 years old), of elite pedigree (his father had been a Morgan banking partner), and an auto man by training (formerly vice president of labor and public relations at GM). In 1933 Stettinius had begun a rapid rise to prominence in public life, accepting an invitation by General Johnson to serve as general liaison between the NRA and the Industrial

Advisory Board. Even after his appointment as U.S. Steel vice chairman he continued to commute to Washington as special advisor to the Industrial Board.[30]

Stettinius was especially cognizant of the changes under which government, industry, and labor were passing, not to speak of the political clout exhibited by Roosevelt in 1936. His career pointed to the spirit of accommodation that was gaining sway in certain high circles of business, especially in banking, where the defeat of unionism was not the major priority. The younger Morgan and the venerable Thomas Lamont, both directors of U.S. Steel, were not committed to defeating Lewis if it earned them only the wrath of Washington and endless disputations.

The role of the Roosevelt administration in facilitating the SWOC contract was only hinted at when the settlement was announced. In her reminiscences, Miss Perkins related her behind-the-scenes role in nudging U.S. Steel to the negotiation table.* Not long after the 1936 election, Perkins called Stettinius and arranged to meet with him at her New York club, the Cosmopolitan. "I wanted to go over some of these troubles in an entirely informal and human way and just canvass the situation," she said. She engaged him on a subject that had concerned him as vice president at GM.

"We talked about the general philosophy and the general wisdom of men who don't read very much, but who think and talk. We decided that it was the fashion to join [unions] and that the strength of the organization of the steel company, the all-powerful character of U.S. Steel, had much to do with it. They had seen U.S. Steel wipe out what were strong steel companies and take them over. There naturally came to be a dread in the workman's mind that he too could be blotted out. We talked about the people who had been discriminated against. I told him about the people who had come to see me at Homestead, who were crazy men in many ways, but their great grievance was that they were blacklisted by the U.S. Steel Company. They couldn't get a job anywhere in any steel mill or in any steel town even selling groceries. The U.S. Steel Company wouldn't let them."

"You're right about that," she quoted Stettinius as answering.

*Perkins's reminiscences, taped between 1951 and 1955 by Dean Albertson, were under seal at Columbia University until ten years after her death. She died in 1965 and the transcript was made available to researchers in the mid-'70s. The material includes many behind-the-scenes incidents not previously known, such as the one which follows.

"There has been very high-handed action in the past. I won't attempt to deny it. I know that's been the case. We won't even go into the reasons why it happened, but the owners were frightened too." The executive's conciliatory attitude persuaded Perkins of the wisdom of probing further. "We talked a long time. I finally said, 'What I would like to ask you is this. I will never repeat what your answer is, any more than you will repeat what I have said to you today. Do you think, now that you are president of U.S. Steel, is it even thinkable to you, that U.S. Steel Company should ever deal with and recognize a union in its own plants?'"

"Oh, I see no reason why it shouldn't," Stettinius said matter-of-factly.

Stunned by his response and all that it implied, Perkins ended the conversation by expressing her hope that an honorable settlement could be arranged which would preserve company earnings and labor peace. Not long afterward, she arranged to see Myron Taylor at the Cosmopolitan Club. She conveyed the same thoughts. Taylor thanked her for her concern and reiterated his interest in acting in a business-like manner.[31]

*

When U.S. Steel signed with SWOC, the major independent producers treated the news with shock and disdain. When Allegheny Steel, Crucible, Sharon, Wheeling, and Jones & Laughlin followed U.S. Steel's example, the ranks of the antiunion independents were thinned, but their determination to stop the Lewis drive was unfazed. Bethlehem, Republic, National, Youngstown Sheet, and Inland Steel joined in a coalition that became known as "Little Steel" as distinguished from "Big Steel" (U.S. Steel). Little Steel was hardly little in manpower or capital resources. In 1937, Bethlehem's average daily workforce had bounced back to 80,312. Republic followed with 46,000. Next in line were Youngstown with 23,000, National Steel with 14,000, Inland with 11,000 and Armco, who cooperated with the bloc, with 12,000. Together they controlled 38 percent of U.S. steelmaking capacity and had capital assets of $1.9 billion.[32]

Yet for all its tremendous size, Little Steel had less in common with modern enterprises such as U.S. Steel or AT&T than with small business. Little Steel had not expanded, so much as it had grown larger along lines laid down years before. Inbreeding at Bethlehem and the other companies had become so pronounced that by 1937 few of Little

Steel's top executives had any experience in business outside of steel, not to mention service in government, with the exception of Schwab's eight-month stint as shipbuilding czar in 1918. To a man Little Steel's chiefs were seat-of-the-pants operational types who had come up through the ranks along two career routes: as bright, efficient "doers" who had won the seal of approval of the company patriarch, or as descendants of the founding families. In *Succeeding* Charlie Schwab had pronounced his dislike of managers with college degrees or "big ideas" in his company. "Most talk about 'super-geniuses' in business is nonsense," he pronounced. "I have found that when 'stars' drop out, their departments seldom suffer." His company was run out of a small Pennsylvania city by a board of insiders who looked out over the mill smoke and were forbidden to sit on even the local bank board lest it distract them from company affairs.[33]

The same sort of managerial insularity prevailed at the other Little Steel companies. One or two men ruled supreme. Outside of Eugene Grace, Tom Girdler of Republic was the grittiest of the lot. Owl-faced and irascible, he reveled in his nickname, "Boot Straps." Born in rural Indiana, Girdler had gone to Lehigh University, where he was two years behind Eugene Grace, another country boy with burning ambitions. After graduating from Lehigh with a degree in mechanical engineering, Girdler knocked around the steel business before becoming assistant general superintendent at Aliquippa, a company town founded by "the Family"—the Jones and Laughlin clans who had founded Jones & Laughlin Steel. Denied the president's slot by George Laughlin, he accepted Cyrus Eaton's offer to run Republic. And he ran Republic with despotic thoroughness following the collapse of Eaton's financial empire in 1933.[34]

Girdler's reaction to U.S. Steel's defection was swift. He and Grace announced that Little Steel would grant the same $5 day and 40-hour week. By matching the Steel-SWOC settlement, they hoped to reduce agitation for union recognition by their own workmen. They had another reason for digging into their pockets. They had been advised by their lawyers that the wage settlement might satisfy the duty-to-bargain provision of the Wagner Act, which required employers to bargain in good faith with employees, but left open the question as to whether they had to sign a written agreement with labor representatives.

On March 18 John Mates and Israel Zimmerman of the Baltimore office of SWOC sat down with Francis G. Wrightson, assistant to general manager Cort, at Sparrows Point. They requested the opening of negotiations and proposed using the Carnegie-Illinois contract as a prototype for talks. While there was no substantive disagreement over the terms of the contract, Wrightson refused to sign it, saying his authority was limited to refusing to sign. From South Bethlehem, Grace issued a statement saying that the company had already engaged in collective bargaining—with its employee representatives. (This in spite of the Perkins advisory that ERPs were not legally entitled to execute a contract under the Wagner Act.) The pattern was identical at the other Bethlehem mills. The company was prepared to meet with the union to "bargain" about matters already settled, but refused to sign any agreement with SWOC, thereby continuing to deny recognition to the union.[35] Posters were put up at Sparrows Point which read:

> **To our employees:**
>
> In their effort to get you to join their union, the C.I.O. organizers are saying—
>
> —that you must join their union in order to hold your job;
>
> —that there is a rush to join the union, therefore, you had better sign up before it is too late;
>
> —that responsible Government officials want you to join their union; and
>
> —that your employees' representation plan does not provide a legal or effective method of collective bargaining with the management.
>
> Do not be deceived by such false statements; there is no truth in them.

The manifesto went on to say that wages as well as the good conditions they worked under were the exclusive result of the effectiveness of the employee representation plan. The company would steadfastly adhere to its method of employee relations in the interests of employees, their families, stockholders, and the public. "Outsiders have not been necessary in the past," the message warned. "Nothing has happened to make them necessary now."[36]

*

It would be revealed later by the Senate Committee on Education and Labor (LaFollette Committee) how Bethlehem was planning to keep outsiders from its mills. Company police at Sparrows Point and other plants were supplied in early 1937 with several boxcars worth of guns and ammunition. The munitions included army-type machine guns, submachine guns, army rifles, shotguns of the regular, repeating, and sawed-off variety, as well as pistols and revolvers of all makes and calibers. Bethlehem also purchased more tear gas equipment than any law enforcement agency in the 1936–37 period.[37]

The company hired Pinkerton's National Detective Agency to spy on SWOC organizers. Between July 1, 1936, and March 31, 1937, plainclothes Pinkertons were dispatched to Baltimore and Sparrows Point, Johnstown, Bethlehem, and Lackawanna from the Philadelphia office. Documents subpoenaed later by the government stripped the veneer of respectability from such operations. The Pinkertons were not assigned to guard property or supply other legitimate security services; on the contrary, the detectives shadowed SWOC organizers and compiled dossiers on employees suspected of union sympathies, often purchasing their information. To conceal management's involvement in the espionage scheme, over $16,000 in bills to Bethlehem from Pinkerton were disguised through a third party. Typical of the cryptic entries discovered by government investigators was this notation regarding spy operations in Baltimore dated July 10, 1936:

> Our client is interested in general conditions in Baltimore; i.e., activities of Communists, other radicals and outside disturbers who may come to Baltimore to annoy and disturb their loyal employees.
>
> Client authorizes us to keep them advised along this line.[38]

Boot Straps armed his mills to the teeth. Over $75,000 was spent by Republic Steel on munitions, according to the findings of the LaFollette Committee. By the end of May 1937 the cache at the Republic police departments included 552 pistols, 64 rifles, 245 shotguns, 143 gas guns, 58 gas billies, 2,707 gas grenades, 178 billies, 232 nightsticks. Furthermore, Girdler armed himself with propaganda. He ordered 43,800 copies of Joseph P. Kamp's *Join the CIO and Help Build a Soviet America.* Thus equipped, he put into effect his strategy to stop the union.

In April rumors were circulated in Republic plants where the union was strongest that the company would shut the plants if employees persisted in joining the union. The union continued to gain members.

In May Girdler followed through on his threat. The Massillon mill in Ohio was nearly shut down, and heavy layoffs started at the Canton plant. By May 25 there were 9,000 men out of work in these towns, locked out by the company.

On the afternoon of the 26th, Phil Murray held a "war council" in Youngstown. Representatives from the plants of Inland, Youngstown, and Republic were present. They reported on local conditions. Union strength was reported as high as 95 percent in some mills, and 60 percent or below in other facilities. But the membership drive had been stalled by the company lockout. Because unified action was deemed indispensable, Murray thought it advisable to try to close down Little Steel. Union organizers believed that SWOC could bottle up the Chicago district and many plants in Ohio. There was considerably less confidence in the union's strength in the East. After consulting with Lewis, Murray issued a strike call, effective at 11:00 P.M. on May 26, against Republic, Inland, and Youngstown. A decision regarding Bethlehem was deferred until June 11.[39]

Americans who looked at a newspaper four days later were shocked by what had happened in Chicago. On a field near the South Chicago plant of Republic Steel, police had engaged a group of strikers and opened fire. Ten marchers had been killed and 75 others shot or beaten. At first the incident was portrayed as a riot. The *New York Times* reported that a mob of strikers armed with clubs, bricks, steel bolts, and other missiles had attacked Chicago police who had tried to stop them. The marchers were "a trained military unit of a revolutionary body" determined to take the property by force, according to the *Chicago Tribune.* But eyewitness accounts and the film of a Paramount News cameraman who was present at the encounter soon contradicted the official story. The film showed policemen firing into the crowd and then charging the retreating demonstrators unprovoked, swinging their billies and hurling tear-gas grenades. Isolated demonstrators trapped by the police were clubbed mercilessly. One unarmed man was filmed being beaten by officers until he lost consciousness.[40]

There was little evidence that the marchers had hurled anything more dangerous at the police than verbal abuse and a few rocks. Of the ten marchers killed, none had frontal wounds. Seven received the fatal shot in the back; three in the side. "The nature of the marchers' wounds indicated that they were shot in flight," said the LaFollette committee, which conducted hearings on the incident. Police assertions that they had been assaulted by a revolutionary army (it was "between us and the Communists," said the Chicago police captain) were not borne out by the oral or photographic evidence. The LaFollette committee concluded that the police action was "characterized by the most callous indifference to human life and suffering."[41]

The outbreak of violence in Chicago stirred deep passions. "The Memorial Day Massacre," *Steel Labor* called it, and the union paper put the blame on Republic's chairman for conspiring and inciting the Chicago police. Girdler's propagandist, Joseph Kamp, wrote that the strikers were bent on supplanting government by mob rule, but had been repulsed by the "good, old-fashioned type of American government established by our forefathers."

On June 9 violence erupted in Youngstown when Republic police broke up a SWOC picket line. Two days later the Youngstown City Council authorized the expansion of the city police force. A total of 144 special officers were sworn in, most of whom were employees of Republic and Sheet & Tube. There was bloodshed at the Republic plant at Canton where a "John Q. Public League" was established to encourage workmen to go back to work under armed escort. SWOC headquarters in Massillon was later shelled by a force of Republic and municipal police. Three union members were killed. A striker was fatally injured in Beaver Falls, Pennsylvania, by a tear-gas projectile.[42]

On June 11 the strike spread to Johnstown. Eight thousand Bethlehem workmen walked out, forcing a partial shutdown. The strike was called to put pressure on Bethlehem management. "Grace is the power behind Girdler," John Lewis said, announcing the action to reporters. Grace and Schwab were encouraging Girdler to continue his resistance, and "should be called for a public accounting," said Lewis. "The warfare should be stopped by the simple act of signing an agreement."[43]

But the Bethlehem organization was no more disposed to settle than its allies. Returning from Germany, where he had taken the rest cure at Nauheim, Schwab called the Johnstown strike "a phase" and said it was hurting the country's chances of recovery. Saying the industry had gone through worse labor agitation at Homestead in 1892, he vowed, "We'll right ourselves in the end." A Citizens' Committee was established in Johnstown under the auspices of Sidney D. Evans, special representative for Bethlehem, in order to raise funds for a "back-to-work" movement. Chief among the company's backers was Daniel J. Shields, Mayor of Johnstown. Shields had already had a remarkable career. He had spent two years in federal prison on bootlegging convictions in the '20s and in 1937 had liens and judgments against him in the Cambria County Court totalling $46,000.[44]

[44]On June 16 Mayor Shields made a radio address in which he declared that Johnstown was under a state of siege. Falsely he accused union members and nameless outsiders and anarchists of beating up local citizens and threatening women and children. "From this moment on a 'Back to Work Movement' will gain momentum," he said. "You wives, mothers and sisters of our steel workers, fear not. Idle threats are being made, by the cowards, but let me assure you that I, as your mayor, am in a position to crush the lawless, the Communist, the anarchist and preserve and protect the homes of you good citizens."[45]

On the next day the mayor received $10,000 in cash from Bethlehem Steel, handed over by Evans and deposited in a safe-deposit box rented for the mayor by Citizens' Committee chairman Francis Martin. On the next day, Fulton I. Connor, a city councilman, telephoned Martin and said that Shields needed $15,000 more. Martin then called Evans, and Evans delivered the cash, wrapped in a brown paper package, to Morrison J. Lewis, assistant cashier of the National Bank, who carried the money to the mayor. All told, Bethlehem funneled $36,450.25 to Shields, the final cash payment being handed to the mayor at City Hall by mill plant manager C. R. Elliott.[46]

In return Shields kept up a constant stream of invective. On June 18 he wired President Roosevelt that he was convinced that the union was a Russian organization. He also had information that it planned "dynamite explosions" of bridges, "kidnappings," and the destruction of "me and my family." A day later Grace issued a public statement denouncing state and federal officials for not using their police power to protect company property against the agitators, and said he was reserving the right to sue the state of Pennsylvania for any damage caused by strikers at the plant. The Citizens' Committee hired Thornley and Jones, Inc., of New York, and Ketchum, MacLeod & Grove, of Pittsburgh, to conduct a national newspaper campaign and radio spots against the strikers.[47]

Trying to defuse the situation, the president appointed a special Steel Mediation Board. It consisted of Republican stalwart Charles P. Taft II, Lloyd K. Garrison of the University of Wisconsin Law School, and Edward F. McGrady, assistant secretary of labor. The board called an emergency session on June 19 in Cleveland. After conferring with the Little Steel presidents and with Lewis and Murray, the board issued a recommendation that the union call off the strike. It asked the parties to allow the National Labor Relations Board to conduct elections at the plants. If the union won a majority of votes, the Carnegie-Illinois agreement recognizing SWOC would be signed; if not, the agreement was "to be torn up" and Little Steel would continue nonunion. Lewis and Murray accepted the terms. The steel companies rejected them.[48]

*

Never before had a labor dispute been so maddening, Miss Perkins related in her Columbia reminiscences. During labor disputes in other industries she had been able to keep informal channels open to management, but in the Little Steel strike Perkins was locked out. Management had hunkered down. Grace and Girdler would not talk to Washington nor to Perkins's intermediaries, and through the sheer force of their personalities they carried along the other independents as well as, for the moment, the Iron and Steel Institute.[49] On June 17 Roosevelt had telephoned Girdler. Speaking "in perfect confidence as a gentleman," Roosevelt had explained his plan to create the Taft Mediation Board and asked for Girdler's cooperation. Girdler's response was made in a ranting statement before a Senate committee: "I won't sign a contract with an irresponsible, racketeering, violent, Communistic organization like the CIO and, until the law requires me to do so, I am not going to do it." A reporter asked him if he would agree to Roosevelt as final arbitrator. Girdler thundered back. "No!"[50]

The steel stalemate had become a terrible setback for the administration. At a June 29 news conference Roosevelt said he had just spoken with Taft. "The majority of people are saying just one thing," he said: "'A plague on both your houses.'" He did not elaborate, but told reporters they could quote him directly, waiving aside the rule that his remarks were not to be printed verbatim. A White House official said the president's remark was an expression of exasperation with both sides. It was a signal that he was washing his hands of the matter.[51]

Bethlehem's tough stance fell like a sledge blow at Sparrows Point. Members of the SWOC local watched the strikers in Johnstown bend under the pressure of the national propaganda campaign of the Citizens' Committee and the "back-to-work" movement. Soon the strike would be broken and many union members fired. On Sunday, June 20, John Mates took to the podium at Finnish Hall in Baltimore to address an estimated crowd of 3,000 Sparrows Point workers. "Although there have been rumors of strikes here, we are of the opinion that negotiations with the local management can be continued on a sane, reasonable basis without the need for any loss of time on the part of the workers," he said. This was a face-saving way of conceding that the local did not have the clout to confront the number-2 steelmaker at its tidewater stronghold.[52]

But Mates and the other speakers vowed that the company would not get away with its strongarm tactics. Under a motion by William Hobbs, president of the local, the membership voted to protest Bethlehem's "mass intimidation" of workmen to the National Labor Relations Board. Hobbs said the men had to stand together and wait for a decision under the Wagner Act.

"It was a low, low time," recalls organizer Mike Howard. "As far as we were concerned, Bethlehem was operating above the law. We had all these supposed rights, but it looked like the government wasn't going to stand up to the pressure."[53]

CHAPTER 14

TECHNOLOGY'S WRENCH

A revitalized union movement was not the only pressure the old guard at Bethlehem faced. Important changes in the manufacture and markets of steel had caught Schwab and Grace embarrassingly unprepared. Their problems were twofold. Due to gross overexpansion in the 1929–31 period, Bethlehem was burdened with fixed costs that dragged down profits. The great size of the corporation prevented it from responding to changing conditions with the flexibility of smaller concerns. It carried many redundant operations that its competitors were not saddled with. Equally debilitating was the psychology of hugeness, which impelled management to hang on to all of its assets in an attempt to prevent the deterioration of the investment value of the corporation. Unable to summon up the courage to abandon antiquated departments—or to modernize them—the company depended on a market upturn that would enable even old equipment to be run at a profit. This strategy in turn made it harder to operate at a satisfactory level of capacity.

Compounding the weakness of an overextended structure was an overdependence on the capital-goods markets, a legacy of Bethlehem's origins in the older metal trades. Grace and Schwab had concentrated on the manufacture of heavy-rolled steels until well into the Depression, when it became apparent that there was reduced demand for these products. In 1933 over 55 percent of the corporation's finishing capacity was devoted to structural shapes, heavy plates, and rails, by far the highest ratio in the business. As early as 1927 these products had peaked and then began to recede in relation to such light-rolled products as tin plate, sheet, and automobile strip. The Depression accelerated this trend. A noteworthy change occurred in 1932 when railroads, still the largest single consumer of steel in the 1920s, fell into fourth place behind the auto industry, construction, and the metal container business.[1]

Although the company had won a number of important construction contracts since 1931 (management did not oppose the New Deal when it came to getting government business in navy shipbuilding, PWA rail-laying, and construction of the steel framework and towers of the Golden Gate Bridge, opened in 1937), business in heavy steels showed a relative decline through the decade. This was a major reason why Bethlehem's earnings-to-sales ratio was consistently below that of National Steel, Armco, and other companies specializing in light-rolled steels.[2]

When partial prosperity returned to the industry in 1936, the metal container business boomed like never before. The heavy demand for tin plate would have seemed to assure a bright future for Sparrows Point, the sole outlet for Bethlehem's substantial tin-plate sales. In reality, though, the mill was under serious competitive threat. Increasingly, canmakers wanted an improved type of tin plate for consumer products—a tin plate that Sparrows Point did not have. The mill was losing business to other steel companies that had adopted the new technology. It became apparent that unless Bethlehem built a modern mill at the plant it risked the defection of important customers in its Eastern and Pacific coast markets.

*

With its enormous resources, Bethlehem might have set the pace in the art of steelmaking. But except for the innovative Grey wide-flange building beam, introduced in 1909 and important in early skyscraper construction, and its lucrative monopoly on munitions patents before and during the war, the steel company that Charlie Schwab had built was noted neither for its product development nor for breakthroughs in basic steelmaking. Both in the formation of U.S. Steel and the expansion of Bethlehem, Schwab's design had been geographical in nature and his brilliance that of welding together mines and mills in a unified mass. Research and product development did not play a part in this strategy. "Normal brains" Schwab called the managers he had selected to get full production out of the departments over which they ruled. There was no room at the top of the organization for scientists or inventors; Bethlehem's research department, established in 1926, was very small compared to those of other corporations. President Grace shared his boss's disdain for inventor types. In his usual direct manner, he saw to it that the research department avoided all "exotic" experimentation in favor of solving practical matters of daily production. Making the supply of lubricating oil go twice as far as before was the kind of research that excited Grace.[3]

The comment attributed to Andrew Carnegie, "pioneering don't pay," summed up the type of technological decisions that Schwab and his assistants had made when installing equipment at the Sparrows Point tin mill. They had 48 "hot" mills for tin-plate manufacture and 18 standard sheet mills. The former rolled light sheets, as thin as 1/100th of an inch or less. The latter handled sheets over 1/16th of an inch in thickness. Although the mills were large by industry standards and driven by powerful electric motors, the basic technological process had undergone little change since Mayor John Hanbury had introduced the rolling technique in Wales in the 1700s, replacing trip hammering. Each mill looked like a jumbo-size wringer washer, with a pair of opposing rolls on a stand. The mill was under the responsibility of a master roller, who often trained a crew of assistants with specialized functions.[4]

The process was highly labor-intensive as well as physically demanding. Operations began with a semifinished "sheet bar" of low-carbon steel. The bar was 8 inches wide, 30 inches long, and varied in thickness from ⅜ to ⅝ inch. Heated to cherry-red rolling temperatures in a small adjacent furnace, the bar was placed by a "heater" on the rolling stand. A "rougher" grabbed the bar with his tongs and inserted it sideways between the rolls. As the bar emerged on the other end, it was grasped by a "catcher" with tongs. While he was lifting the bar to return it over the top roll, the rougher fed a second bar into the mill. Then he took the first bar from the top roll and gave it a second pass. As this took place a "screw boy" twisted a screw mechanism on the stand to bring the rolls closer together.[5]

These movements were repeated until the two bars were drawn out to about 20 inches in width. Then they were "matched" (one placed on top of the other) by a "doubler" and given another set of passes. After the sheets were widened to 30 or 40 inches they were pried apart by "sticker pullers" and doubled again (to four sheets) and placed in the furnace for reheating. The next rolling involved a pack of four sheets and they were passed through the rolls until they reached 30 or 40 inches. Once again they were doubled to eight and charged into the furnace for reheating. On the last set of rolls the pack was drawn down to the desired gauge.[6]

The sheets produced by the hot-pack method had rough, irregular edges. After a brief cooling period the pack was placed on the floor and the sheets pried apart by a "singles boy." They were trimmed on cutting machines by a "shearsman." Because a scaly rust oxide formed on steel worked in contact with air, the finished sheets were taken to the pickling unit for cleaning in sulfuric acid. In spite of the enormous manpower used by the department (2,200 men labored around the clock at Sparrows Point in eight-hour shifts), there was a good deal of wastage, and problems with quality. The surface of the steel was microscopically uneven due to the manipulation of the sheets through the rolls. Furthermore, the adjustments made to the space between the rolls, from which the gauge of the sheet was determined, was done according to the master roller's experience and visual judgment. Minute variations in gauge occurred with each batch of sheet produced. Needless to say, there could not possibly be the same precision in manual rolling as in other finishing operations where grooved rolls cut the steel under rigid mechanical controls.[7]

Efforts to roll sheets through a succession of rolls, as was done with rails and structural beams, had been tried as early as 1865. Until recently, they had all failed. In 1892 a continuous mill for rolling sheets had been constructed at Teplitz, Germany, but, plagued by breakages and nonuniform thickness, it had been abandoned. After 1902, U.S. Steel's tin-plate subsidiary, American Sheet and Tin Plate, had tried twice to debug a variant of the Teplitz mill. This venture had been discontinued in 1910.[8]

The attempts in Germany and the U.S. had been unsuccessful because of the erroneous belief that a sheet should pass through rolls that were perfectly flat. Discovery of the crucial principle of "progressive convexity" was the missing element. What this meant was that a slight concavity given to the rolls, which imparted a slight convexity to the sheet, caused the molecules in the steel to run straight, allowing the edges of the sheet to proceed at the same rate as the middle. This eliminated the old problem of breakage. As the sheet was rolled out, the convexity in the rolls had to be reversed and slowly straightened out so that the sheet emerged from the final roll flat and straight.

Figuring out a rolling procedure and then converting it to the real world of high-volume output was not an easy proposition. But after American Sheet's failure in 1910, the major steel producers didn't even try. The "impossibility" of producing wide sheets continuously through rolls became an article of faith in the trade. The man who was to prove established opinion wrong was John Butler Tytus. Tytus had received his early training outside of the industry and then found a hospitable environment for experimentation at a steel company with a fraction of the financial resources of Bethlehem. The son of a Middletown, Ohio, papermaker, he received a BS at Yale University in 1897, then went to work at his father's company. After seven years he tired of the paper business and went looking for something different. Transfixed by the slow, heavy work he witnessed on the sheet lines of the American Rolling Mill, he asked for a job. Charles Hook hired him as a "spare hand." Almost immediately Tytus began searching for ways to improve the process. He said to Hook, "Some day, Charlie, we'll be making sheets in long strips like they make paper." Hook encouraged him to pursue the project in his spare time.[9]

During World War I Tytus got the chance to do some practical experimentation. The company was concentrating on forgings and other war material and the sheet mills were nearly idle. He was permitted to use one of the mills as a laboratory. In 1921 he had progressed far enough to convince Hook and the Verity family to put out several million dollars for the construction of a continuous mill. Tytus selected 100 of the best millwrights and rollers he knew at Middletown and began construction of a new mill at Ashland, Kentucky. Everyone was pledged to secrecy. Over the next two years they labored with the heartaches and complexities of developing a dependable high-tonnage operation. Rolls broke, housings snapped, and several million more dollars were needed before the mill could be revealed to the world in 1924.[10]

The Ashland mill validated the commercial practicality of rolling wide steel, producing quality metal up to 36 inches in width and as thin as 0.065 inch. Yet it was scarcely continuous. Four stops had to be made among the 19 rolling stands to reheat the sheets. The mill also experienced difficulties in regulating the speed of the rolls. Despite these shortcomings, Tytus had demonstrated the importance of convex rolling and the coordination of rolling stands into a single machine unit. His invention has since been ranked by experts as among the greatest breakthroughs in basic steelmaking, rivaling in importance the integrated iron and steel works developed by the railmakers in the 19th century and the by-product coke oven pioneered by the Germans at the turn of the century.[11]

Important refinements to the Tytus process were made by engineers A. J. Townsend and H. M. Naugle of the Forged Steel Wheel Company of Butler, Pennsylvania. The Townsend-Naugle mill opened at Butler in 1926 was truly "continuous," permitting uninterrupted rolling on the finishing stands. Armco quickly purchased the company and thereby gained control of all the basic patents for what was known as the continuous wide hot-strip mill.

Notwithstanding the superiority of the new rolling technique, Armco found no takers when it offered to share the process with other steelmakers through licensing. Under the leadership of the Iron and Steel Institute, which in turn was dominated by U.S. Steel and Bethlehem, the important producers snubbed the Armco offer. Perhaps the most arrogant stand was taken by Schwab and Grace. In 1926, the very year when the Butler and Ashland mills were in successful operation, the Bethlehem board of directors authorized the expenditure of $5 million to build 24 new hot mills at Sparrows Point. "I will never be happy until you come and tell me it is the 100th tin plate mill started at Sparrows Point," Schwab told Frank Roberts, the general manager.[12]

Schwab wanted to freeze steel to a common technological denominator, but pioneering equipment could not be arrested permanently. In 1927, Ernest Weir broke ranks, and opened a new mill under the Armco licenses at his Weirton plant. Incorporated into Weir's plan was the Steckel mill, another new idea, which produced thinner sheets of a finer finish. Later that year U.S. Steel, the oft-criticized laggard in new technology, completed a semicontinuous mill for its tin department at Gary, Indiana. Instead of using small sheet bars, U.S. Steel fed 2,000-pound slabs into a train of rolls and reduced the slab down to 16 gauge. This was not thin enough for tin plating, so hot-pack rolling was continued, but the Gary mill pointed up the possibilities of charging heavier loads into the mill, increasing yield.[13]

The next milestone was reached in 1929, when small Wheeling Steel developed an improved type of cold-reduced tin plate. In cold reduction the strip, having been rolled down in the hot strip and cleaned in the pickling vats, was reduced another 80 percent to strengthen the sheet's molecular structure and impart a hard, shiny surface. The strip was passed unheated, or "cold," through a series of tandem rolling stands. Previous cold rolling had been done on two-high stands with two rolls exerting pressure on the sheet as it passed through. Wheeling Steel adapted the four-high concept of Townsend and Naugle to achieve a new dimension of compressive power. Two rolls were used as work rolls between which the sheet passed and the other two rolls were backups to support the work rolls and to keep them from springing loose under the pressure.

A salutary outgrowth of the Wheeling method was a continuation of the development of longer runs of strip. Whereas the original Tytus mill produced strips several hundred feet long, the newest mills could handle strips that rolled out in two-mile lengths. All of this required a great deal of space for runout tables and coiling equipment. It also increased friction in the stands which had to be controlled by an array of lubricant baths and circulating water, and monitored with improved gauges and other instrumentation. These problems, though, were overcome.[14]

Expensive, yes, and intricate, but the end consequence of this chain of mechanical improvements was a product that had a better finish and was easier to roll into can bodies or stamp into parts. If a goal of good management is to use available capital to save on the cost of production, even if only by a few cents, the continuous-strip system was a marvel. The quoted price of hot-rolled strip dropped 28 percent between 1924 and 1935, from $51.40 a ton to $37 a ton, a feat all the more remarkable since other products, notably the heavy-rolled items, were neither better nor consistently cheaper during this period.[15] The lower cost of tin plate greatly encouraged canmakers to look for new outlets for their product.

*

The canning process began with the tin plate that was made in the steel mills. Five companies dominated the nation's supply of tin plate. In order of size they were: U.S. Steel, with eight plants in the Chicago and Pittsburgh districts and one plant at Pittsburg, California; National Steel, with plants at Weirton and Clarksburg, West Virginia, and Steubenville, Ohio; Bethlehem, with all of its capacity at Sparrows Point; Republic Steel, with mills in the Warren-Youngstown area; and Wheeling, based in the Pittsburgh district. There were a number of smaller producers, among them McKeesport Tin Plate (a Pennsylvania company with interests in National Can Company), Crown Cork & Seal, Youngstown Sheet, and Inland Steel.[16]

The tin plate made in the steel works was shipped in base boxes of 112 flat sheets of various grades and weights. Two companies had come to dominate the metal packaging business, American Can and Continental Can. Together they purchased over half of the steel-mill output and made some 8 billion of the 12 billion cans manufactured in 1934. Both were amalgams of hundreds of once-independent companies. American Can had been formed out of 175 concerns by Daniel Gray ("Czar") Reid and William B. ("Tin Plate") Leeds at the turn of the century, and had grown ever since. In 1934, American owned 51 factories in the U.S. and Canada and made about 45 percent of all cans. Its rival, "Con Can," had started out as a merger of several East Coast companies, and had gone on an acquisitions binge under Chairman Carle Cotter Conway in the two years before the Depression. Nineteen independents were gobbled up, including Southern Can (of Baltimore), U.S. Can, New Orleans Can, Manhattan Can, and Sociedad Industrial de Havana, which produced 80 percent of the cans in Cuba. In 1934, Continental controlled 38 plants and held about 25 percent of the container market.[17]

Canmakers used highly refined machinery to turn tin plate into containers. The flat plate was cut into strips and fed into the body-maker, where it was wrapped around a mandrel to form open cylinders. The open cylinders, called can bodies, were soldered along their seams and handed over to a whirling machine whose discs fitted out the bottoms and tops of the can and then curled around the edges of the cylinder to form a rim. These transactions came at a rapid-fire pace. A single body-maker could knock out 300 can bodies a minute at the Continental plant in Baltimore.[18]

The can companies made cans, but they *never* filled them. This was the job of the food manufacturers whose names were affixed to the paper labels of the cans. The big names of the '30s were Campbell's Soup, the largest food packer on the globe, Van Camp, California Packing Corporation, Hormel, Libby, and Castle & Cooke, which had purchased controlling interest of ailing Dole Pineapples. Cans were sold mostly under long-term contracts, usually lasting three years or more, providing for the purchase of all or a stated portion of a customer's requirements at prices that were subject to adjustment with changes in certain material and other costs. No. 2 cans sold at $20 per thousand, with discounts for large orders. A few food companies went their own way and made their own containers, notably Heinz and Borden. Overall Americans were using 100 cans per person a year.[19]

As early as 1929, members of American Can's research department had begun experimenting with the packaging of "near beer." They had tried canning the beer in regular and enameled cans. Each had exhibited major flaws. The regular cans had buckled and ruptured during pasteurization, spewing out their contents, while the enameled cans underwent a meltdown, turning the brew milky white and undrinkable. Researchers had tried to circumvent the buckling problem by increasing the air pressure on the outside of the can during pasteurization in order to balance the pressure from the inside, but the procedure proved too expensive. So it was with great interest that they looked upon the improvements in tin-plate manufacture in the 1930s. The metal produced by the hot-strip method together with Wheeling's cold-reducing innovation was strong enough to stand the strain of brewing. American Can was so pleased with the properties of Wheeling's cold-reduced plate (brand name: Ductilite tin plate) that it booked the company's output for a full six months during the height of the Depression.[20]

Yet a major hurdle remained. Beer was strongly attracted to metal, and the absorption of iron molecules in the brew made it cloudy and foul-tasting. Metal turbidity was the name of the phenomenon. To prevent it, brewers had to swab down their tanks with a thick adhesive substance called brewers' pitch. Unfortunately, the pitch could not be made to adhere to the sides of a metal can. The researchers were stymied until they came upon a new substance invented by Union Carbide & Carbon Corporation. Vinylite was an artificial plastic copolymer that could stick to almost anything. Sprayed inside a tin can, then baked at 300° F, Vinylite made a topcoating that laboratory tests showed could withstand the punishment of pasteurization. (Most brewers pasteurized by holding the beer at 140° F for 20 minutes, although some used a speedier method of flash pasteurization. Since tin plate heated up quicker than glass bottles and also cooled off quicker, the brewing process had to be modified somewhat to accommodate cans.)[21]

After six years of research and development, American Can had managed to produce a lined container suitable for the beverage. Now the company wanted to make a big splash by wrapping its cans around a famous national brew. But the three big brewers, Anheuser-Busch, Pabst, and Schlitz, didn't think their customers would accept suds in a can. So American Can struck a deal with the small Gottfried Krueger brewery of Newark, New Jersey. American agreed to install the can-filling machines at Krueger free of charge and pay for all costs related to a start-up, provided that Krueger would put the cans into a selected market. If the can succeeded, Krueger would pay for the American Can equipment out of its earnings. If the can failed, the agreement would be null and void and the equipment removed. Richmond, Virginia, was selected as the test market. The idea was to see whether canned beer would catch on in a low beer-consuming district, where competition was also hot for what business there was. Within a week of its introduction, 47 percent of the local distributors were handling the canned beer.[22]

In the next few months Krueger in a can shook the industry. Sales in Richmond jumped 550 percent over pre-can distribution and the cans walloped bottled Budweiser and other national brands. Other small brewers begged American to can their beer, but the company held off. They wanted to snag the big fish. The first of the nationals to come around was Pabst. In July 1935 its Export brand was packaged in snappy silver-and-blue cans and tested out in Rockford, Illinois. It was an instant success. Shortly thereafter, Schlitz's familiar brown-and-yellow label was wrapped around a can. This was a Continental can, a cone-topped affair with a bottle-type cap. It was quite popular until American Can's resourceful research department invented the "quick and easy" flip top. A sharply pointed lever was affixed to the top of the can and the end seam used as a fulcrum for popping the lever open.[23]

By the end of 1935, 160 million cans of beer had been sold nationwide. Nearly all the major brewers were canning a portion of their stock and finding beer consumption rising in rough proportion. Bottlemakers denounced the cans, citing taste and the bugaboo of metal turbidity, but retailers loved them. Shipping a case of bottled longneck Kruegers from Newark to Richmond had cost 23 cents, including the return rail freight on the empties. Shipping a case of Krueger cans cost 8 cents, and there were no empties to return.

"The brewers didn't ask for it. The public never dreamed of it. The bottlemakers wouldn't believe it," *Fortune* noted gaily, but one year after its introduction, canned beer was generating approximately $90 million in revenues for the metal container business.[24]

The beer can clinched it: The reluctant giant of steel came around. In 1936, Grace and Schwab authorized the largest capital expenditures for Sparrows Point since the booming '20s. The purpose: to erect a continuous strip and tin-plate mill. The Bethlehem board of directors dipped into the company's depleted treasury and forked out $16,193,549 for a cold-reducing mill for the production of ductile tin plate, "the best in the business," Grace promised. Ground breaking for a continuous hot-strip, however, was delayed until a nasty squabble could be resolved. Bethlehem had refused to apply as a licensee of the Armco process, claiming that it had developed its own unique continuous roll system. Armco wasn't impressed, and sued for patent infringement. The suit was finally settled in July 1937, when Bethlehem agreed to pay all standard fees and royalties for the Armco patents, thus winning the distinction of becoming the last major U.S. steelmaker to be licensed under the process.[25]

The $20-million continuous-strip mill that was slated for the Point followed the Schwab prescription of size, a commitment to the working of massive tonnages. The idea was to construct a mill with enough capacity to replace all of the hot mills (which numbered 66). Total annual capacity was set at 780,000 tons. Seventy percent of the capacity, 540,000 tons, was to go for tin plating and the rest for sale as strip for the automobile industry and appliance makers. Using special synchronous motors designed by Westinghouse Electric the company hoped to turn out long runs of strip at an average rate of 2,000 feet a minute.[26]

*

Through 1937 and early 1938 the 2,200 men of the hot mills watched the continuous-strip take shape alongside their shed. While they sweated and swore as usual, the building next door filled up with equipment brought in by trains. As their mills clattered away to the beat of iron tongs against red-hot sheets, the silence of the new building was broken at intervals by high-pitched whines as the rolls were tested and the runout table and coiler were tinkered with. After many weeks of tests, the Westinghouse and Mesta Machine Company engineers departed. Then came the real shock. A group of men, hired from outside the plant, started operating the new mill, supplemented by several dozen "Looper" management trainees. Who were these outsiders and what were they doing? the sheet-mill men wondered.

One of the Looper trainees was John Morgan, a graduate of Harvard University with a BS in mechanical engineering. He had been asked to work at the new skin-pass unit in the mill. He remembers the "transition period" where the new equipment was debugged as the most exciting point in his career, which would span the next 35 years at Sparrows Point.*

"Men were hired from Wheeling, Weirton, and Gary to run the mills, and the management trainees, the Loopers, were on the lines to learn. We were paid a salary of only $99 a month. I started in the skin-pass unit with the baler. I was using a small crane and working with coils that were 12,000 to 14,000 pounds. You learn technology with your fingers. It doesn't come by listening to somebody talk. You learn it by doing it. I left that job rather quickly and became a speed operator in the skin pass, running steel at around 1,200 feet a minute. We had what they called 'stickers,' sheets that would weld together and then tear out and wreck the mill. They often would start tearing slowly, so you'd reduce the mill's speed, open up the reels, and beat the defective material with sledgehammers. Sometimes you were able to break it loose while keeping the mill rolling. But sometimes you had to take out the whole coil and scrap it. I was rolling in six months. Now I was not a good roller, but it was a question of opportunity and I took it. Truly to become a good roller takes 8 to 10 years. But everything was go-go because Bethlehem wanted to get the production out. The new tin plate was marketed under the name of Beth-Colite, and we got things to work smoothly. Later I was appointed special engineer for the whole tin-mill division."[27]

If the new technology advanced the careers of middle-level management, it wreaked havoc in the ranks of the blue collar. Tin-mill work was hot and backbreaking, but it was well compensated and provided a route of advancement for men whose working-class origins precluded positions in supervision. The mill hands terribly resented the

*Owing to a longstanding policy by management not to speak for attribution on "internal" matters, the engineer requested a pseudonym.

outsiders and the college kids like John Morgan who had moved up at their expense. Crew by crew, the hot-mill men saw their numbers recede as production shifted to the automated mill, which needed fewer than 350 men to operate around the clock, as opposed to the 2,200 employees at the hot mills.

"What shocked and upset everybody was that the skilled men and rollers, men who had years and years of service to the company, weren't even given a tryout on the new mill," recalls rougher John Duerbeck. "I mean in those days the roller was God, he ran the hot mill and supervision didn't mess with him. Our hot-mill superintendent, Mr. Wilson, was a decent man and he protested. But he couldn't do much except tell us he was sorry. He told us, 'This is the policy from Bethlehem, you rollers and the rest ain't gonna get the good jobs on the strip mill because the big bosses think you might slow down production.' Hell, the men didn't like that 'cause the company had always given us the impression they'd take care of us. You know, you stick with us and we'll help you out. Now this was how they helped us out. My roller was Mr. Pete Roach and they told him if he wanted to work for Bethlehem Steel his job was as a laborer at the plate mill at five bucks a day. And this guy had been making $25 to $30 a day! And my heater, Mervin Gaen, was made a laborer in the erector gang. The foremen were screwed too. The company knocked everyone down to the bottom of the pile. We were considered no more than new employees."[28]

Forced to look for a job in another department, Duerbeck says he was told that a "beautiful job" was opening up at the blast furnaces. "This foreman really painted it good, so I put in a transfer and went down there. I was in my best gray suit, white shirt and black and white shoes on my first day because I was led to believe it was a clerk's job. So the foreman there hands me overalls and a dunce hat [hard hat] and soon I'm working in the furnaces as a laborer for the millwright gang. So we tear apart a cyclone and I get half gassed. I was so sick I couldn't wash up after work. And this was to be my beautiful job!"

Some hot-mill men took their pension and others quit, but most of the younger employees figured they had no alternative but to accept a transfer to jobs that came open in other departments. The family traumas stemming from the changeover were so great in the steelworker communities that men and women, daughters and sons, recalled the strain and struggle vividly years later. The experience formed the framework for the workmen's response to future technological change and automation:

> ANTHONY MAGGITTI (catcher): "All my relations worked at the tin plant. I got a job in 1931. Got laid off in 1932. Went back in 1933. The wages was good, $64 [a week] if the tonnage was good. Then the new machines came and the foreman says to me, 'You're on the hot-strip mill now,' and I find I was brought down to the laborer's scale. My first child was just born and I was making like $5 a day. My wife was nervous and there was men walkin' around the mill in a daze. I remember I said to my roller who was moanin', 'You gotta make up your mind, John, on what you're gonna do. The money ain't never gonna come back.'"

> BILL FORBES (singles boy): "I started when I was 18. The singles boys would take tongs and separate the sheets that stuck during the hot-pack [process]. The mill was run on and off for months and months, you'd work a few days in a pay period. When our mill was finally closed in 1939 I was put on the shears in the new mill. They gave you no break-in time—it was, 'You do it or you get the hell out.' In 1940 I got in an accident, had the toes on my left foot sheared off. Here I looked down and there was a blob with my little toe jammed, almost welded, on top of the mess. I was off a year. When I got back I didn't really want an on-line job so they put me in the shipping department."

> CHARLES RIPKIN (jobber): "I worked on the jobbing-sheet mill, biggest hand mill of all. We drawed out sheets 60 inches wide and 220 inches long. They weighed 190 to 210 pounds each. The men today couldn't do that! I had calluses on my hands, you could put a cigarette butt to them and it wouldn't hurt. When the hot mills shut down, they made me a scarfer. We had torches and burned off the defects marked by the inspectors on the slabs coming over. I made the changeover, but I know men that didn't."

> SIRKKA LEE HOLM (daughter of shearsman): "My father came to the hot mills in 1933 from Warren, Ohio. There were 50 Finnish families who came and they all got work in the mill. There was a benign aspect to Bethlehem Steel then. You had relations with the foreman and they were mostly

okay. You didn't see any of the big bosses making the decisions. My father and the other men knew things had to change, that you can't fight the new machinery coming in —but the way Bethlehem did it, it was awful. I mean men thrown out from what they'd done for years without even so much as a thank you. You weren't lulled into thinking anymore that here was a benevolent employer."[29]

If management could demote high-paid men to laborers, what would stop them from demoting others if new machinery gave them a chance? And if the company could bring in outsiders whenever it was convenient, what kind of job security did the average employee have? The consternation over the continuous-strip mills was infectious: In the saloons of Dundalk and East Baltimore, technological unemployment was discussed and denounced, and workers in other departments became highly knowledgeable of the problem as the tin-mill refugees took jobs in their crews.

Amazingly, Grace and Schwab had nothing to say about the issue. Were they aware of the resentment building up at Sparrows Point? Or were they kept in the dark by the insulating layers of plant management? Or did they simply view the matter from the perspective of the cost sheets placed before them? Shrewdly sizing up the displacement issue, Phil Murray took the matter of technological unemployment before SWOC's Wage and Policy Committee. The union eventually drew up a program to help the estimated 85,000 employees displaced or downgraded nationwide by the wide-strip mill. The SWOC platform called on the companies to transfer men displaced from the hot mills to temporary jobs until permanent vacancies opened up, with a 3-percent increase in earnings for those whose wages had dropped by 10 percent or more. Further, for those employees who could not be absorbed in other jobs, the union asked for dismissal pay equal to 10 percent of earnings for a ten-year period.[30]

Bethlehem and the other companies did not respond to the program, and this handed a potent new weapon to SWOC. The local Sparrows Point drive, on hold since the 1937 confrontation, bounced back in the wake of the displacement issue. Edwin B. Abbott, local journal agent for SWOC, hammered home the point of the company's indifference to the suffering and hardship of "2,000 men on the bread line" and the worthlessness of the "sham company union" in protecting workers' rights. Every evening union members went out in a sound truck to rally support along the streets of Baltimore's neighborhoods. Draped across the truck was the banner, A WORKER WITHOUT A UNION IS LIKE A MAN WITHOUT A COUNTRY. According to Abbott, writing in *Steel Labor,* "In all of our lodge activities—both political and social—we always bear in mind that our primary interest is in forming a strong, powerful union among Bethlehem steelworkers here. We are driving forward, boosting our lodge and planning for the future."[31]

CHAPTER 15

END OF AN ERA

Unionism did not fade away as Schwab had predicted during the Little Steel strike; on the contrary, it steadily gained believers. The number of men who signed SWOC cards increased at Sparrows Point in the last two years of the decade. Yet the union was a suppressed movement. Refusing to recognize or bargain with SWOC, management propped up its employee representation plan with perks and propaganda. As a consequence, the focus of SWOC settled into a drive to dislodge the company unions from steel country, a battle that shifted the theater of action from the mill gates to the government hearing room. In a barrage of petitions to the National Labor Relations Board, Lewis and Murray charged that the Little Steel coalition was engaged in unfair labor practices.

Bethlehem's extreme animus made it a logical target. A petition drawn up by SWOC counsel Lee Pressman was presented in August 1937. The filing revived the mass-intimidation complaint of angry Sparrows Point workers in the Little Steel strike, alleging that Bethlehem had threatened to discharge, lay off, or demote workmen for joining or assisting SWOC; that workers had, in fact, been discharged at the plants because of their union affiliation; that the company maintained arms and utilized Pinkerton guards for the purpose of intimidating workers; and that the "back-to-work" movement at Johnstown constituted antiunion racketeering. The petition further charged that the company ERPs were designed to obstruct and hinder the liberties of employees to engage in union activities and grossly violated the "self-organization" clauses of the Wagner Act.[1]

The NLRB began proceedings on the complaint in September 1937. Between then and April 1938, an examiner held public hearings in Pennsylvania, New York, Maryland, and Washington, D.C., accumulating over 12,000 pages of testimony from interested parties, including Edwin Abbott and other unionists at Sparrows Point. "We don't want to run the mills," said Thomas Updike, a worker with 29 years of service at Johnstown. "We know that we need bosses, but we want honest, fair-minded bosses. We don't want more than is rightfully coming to us. We want the company to make a fair profit. We want the steel we help make in [our] plant to be the best steel in the world."[2]

The hearings were delayed in December 1937 when counsel Hoyt A. Moore, of Cravath & de Gersdorff, filed a petition on behalf of Bethlehem asking for dismissal of the case. Joining Moore were the counsels for the steel plant ERPs—a total of 17 company lawyers versus two lawyers representing SWOC and three representing the NLRB. The petition was denied on January 10, 1938.[3]

In April 1938 a dispute developed among the lawyers concerning the company's application for a subpoena *duces tecum*. Bethlehem wanted SWOC to produce all agreements which it had entered with U.S. Steel since 1937. The union objected, saying the information was proprietary and irrelevant to the case. The spat delayed the proceedings by three months, as trial examiner Frank Bloom grappled with the legal issues involved, first denying the Bethlehem motion, then holding oral arguments, then denying the application again. In July 1938 an agreement was worked out that provided for the release of certain documents to lawyer Moore for in-camera review. A month later Bethlehem filed objections about the conduct of Bloom and demanded dismissal of the case for a second time. In a statement released to the national press, Eugene Grace pledged to contest any government finding which interfered with the "freedom" of his employees and their ERPs.[4]

On November 7, 1938, Bloom delivered his report. He concluded that Bethlehem had engaged in the unfair labor practices alleged. The decision was met by a howl of protests from Grace. The company filed voluminous briefs with the NLRB and oral arguments were presented before the board. By this date (March 1939) the case had mushroomed into the most litigated in board history, covering more pages of testimony and more legal argumentation than any other proceeding the panel had heard. A decision was rendered on August 14, 1939, two years after the case was submitted.[5]

The ruling was an almost complete vindication of SWOC's position. Using testimony from the hearings as well as subpoenaed documents from the Pinkerton Agency and other sources, the NLRB ruled that management had engaged in a conscious and long-term effort to interfere with and intimidate employees interested in unionism. Carefully the board went through the history of the ERP and examined its features in keeping with the board's mandate of setting legal precedent in labor law. "Under all of these circumstances, it is idle to suggest that the Employees' Representatives served the will of the employees, or that the Plans represented that self-organization of employees which is contemplated by the Act. And it is fruitless to argue that, if the employees so desired, they could change the Plans." The record of illegal use of undercover agents, together with cash payments to Mayor Shields, exposed a management committed to maintaining a phony union through repression, fear, and bribery, the board said.[6]

Having found Bethlehem guilty of a pattern of unfair labor practices, the NLRB took the step that distinguished it from the Taft Mediation Board of 1937, the Steel Labor Relations Board of 1934–35, and the Wilson War Labor Board of 1918–19: It backed up its words. The panel issued cease-and-desist orders that prohibited Bethlehem from the practices found to be unfair and further ordered the termination, or "disestablishment," of the ERP as an obstacle to the rights and liberties of Bethlehem workers.[7]

Philip Murray was jubilant. The ruling signalled "the end of the company union system in American industry," the unionist proclaimed grandly. At South Bethlehem they were expecting the worst. An appeal was filed in U.S. Court of Appeals in Washington the following day by Cravath & de Gersdorff. The appeal questioned the constitutionality of the Wagner Act through a narrow reading of the interstate commerce clause. The company contended that even if the labor violations alleged by the panel were true, they had no effect on interstate commerce because each of the mills employed a workforce that was within state lines.[8]

The argument was reminiscent of the *Schechter* case except that the disputant was not a Brooklyn poultry jobber but the fourth-largest industrial corporation in America, whose steel products had been shipped in 1936 to all 48 states as well as 71 foreign nations and U.S. territories. The company was seeking to reverse the tide of decisions by the Supreme Court that had established the legality of the Wagner Act since 1937. The weight of recent precedent seemed manifestly clear. But because the rulings did not square with its labor policies, Bethlehem threw its legal troops into the breach.[9]

There was deep irony in the course of events. Since 1933 the officers and directors of the corporation had played an all-or-nothing game. Faced with an administration that did not perceive public goals and steelmaker goals as identical, the company had spurned attempts at compromise and made precious little effort to accommodate itself to federal policy. (The same management was not troubled by the idea of calling on government to bolster industry profits through higher tariffs on imported steel, increased navy and public works expenditures, and other special considerations.)[10] At minimum, Bethlehem's high-profile resistance to unions had provoked Congress to pass new labor laws, the Wagner and Walsh-Healy Acts among the most notable. It had also spurred remarkable changes in steel country. Organized labor had undergone an unexpected revival. From the "horse-and-buggy" days of Mike Tighe's Amalgamated there had arisen a militant CIO with sophisticated organizing machinery and considerable public appeal. Much as management might like to deny it, its bludgeon approach in the Little Steel strike of 1937 had enhanced the power base and heroic stature of John Lewis and Phil Murray in the eyes of many workmen, while the company's involvement in bribery and dirty tricks at Johnstown had backfired in the court of public opinion. The contrast in labor policies between Bethlehem and its longtime rival in size and precedence was striking. U.S. Steel had made peace with labor. It had signed a CIO contract (and renewed it in 1938) without a strike or violence. Its stock had risen on the New York Exchange and Myron Taylor was hailed as an "industrial statesman." Taylor's moderate course was continued by vice chairman Edward Stettinius, who succeeded Taylor as chairman of the corporation in 1938.[11]

*

Keeping the flames of the old way lit at Bethlehem was Charlie Schwab. Entering his 77th year in 1939, he epitomized an earlier epoch of self-made men, an industrial mogul in the grand style, although one under increasing attack. "It hurts me—it hurts me very much—to be branded as nothing but a greedy, selfish, self-seeking, mercenary, merciless fellow," Schwab told B. C. Forbes. Facts concerning the war profiteering of his company had come out in the hearings of the Senate Special Committee Investigating the Munitions Industry (the Nye committee). According to the records, Bethlehem had been paid over $100 million in profits for supplying weaponry to Britain and France before America's declaration of war. Brought before the committee, Eugene Grace had acknowledged that his million-dollar bonuses had started in this period. The applause of the public always had been important to Schwab; the wave of '30s pacifism had wounded him deeply. He did not try to defend himself in detail, except to say to the *Times:* "War munitions makers were most prized during the war, but now they are being criticized. It is ridiculous and most extreme to charge that war munitions manufacturers encouraged the conflict."[12]

The circumstances surrounding Bethlehem's submarine contract with Britain in defiance of U.S. neutrality laws did not come out specifically in the hearings. (It was a secret that Schwab would carry to his grave.) But it was revealed, to blaring headlines, that a consortium of manufacturers had paid William B. Shearer, a right-wing propagandist, to disrupt the Geneva Conference on Naval Disarmament in 1927. According to Shearer's testimony, he was hired by Bethlehem, Newport News, and New York Shipbuilding after he had talked to Schwab and expressed his concern that pacifists had gained the upper hand and threatened private U.S. shipbuilding interests with curbs on proposed navy cruisers and battleships. (In response to Shearer's damaging testimony, Schwab reported "no recollection" of having ever met Shearer, who was paid $25,000 by the shipbuilders to direct anti-disarmament activities at Geneva.) Was it feasible for major nations to disarm? a reporter wondered. "No, I don't think it is a possible thing to proscribe war," said the steelman.[13]

Returning home from his annual sabbatical in Europe in 1936, Schwab volunteered the information that Germany was in "fine condition" and that Hitler was "really popular" because he was credited with "bringing order out of chaos." His remarks caused a sensation in light of the Shearer testimony and disclosures involving Schwab's late publicist, Ivy Ledbetter Lee. In 1934 the House Un-American Activities Committee had reported that Lee was in the employ of the Nazi government and I. G. Farben, the Fascist dyemaker. The revelation had come less than two weeks after Hitler's "blood purges," where a thousand men had been killed for political purposes. "Poison Ivy," the notorious nickname given to him by Upton Sinclair, haunted the publicist as American newspapers accused him of being a paid defender of Nazi atrocities. In an odd twist, Lee refused to use the publicity techniques he had developed for his clients to his own advantage, instead retreating into a shell of silence. Not long after returning from Baden, Germany, he had slumped semiconscious at a board meeting of the Pennsylvania Railroad. He had suffered a cerebral hemorrhage. He died several weeks later. "He has made his millions the last 20 years and now the world knows how it was done," wrote the U.S. Ambassador to Germany who knew Lee.[14]

Without a skillful publicist at his side, Schwab found himself besieged with stockholder suits and unfavorable publicity over his promiscuous use of corporate cash and stock. First he was plunged into controversy by the disclosure of a management stock ownership plan. Just prior to the Youngstown merger announcement, the directors had approved a plan to sell 220,000 shares of Bethlehem stock to management. Ninety-two managers participated, including 12 directors. Eugene Grace took about 50,000 shares. They were sold to the managers at a cut-rate price of $91.60—nearly $20 below market prices. It looked like a sure way to make still more money outside of the lucrative bonus compensation plan. The managers who approved the sale were insiders who had information of the plan to absorb Youngstown. If the merger had succeeded, Bethlehem stock would have appreciated greatly in value. If the price for Bethlehem common stock had risen to the pre-crash height of $150, the insiders would have been able to sell their shares at a tremendous profit.[15]

What no one had anticipated was the Depression. After the price of Bethlehem common had dropped to under $25, the insider board of directors decided that it was the responsibility of the other shareholders to take the loss off their hands. They approved a plan to have the 220,000 shares canceled, the stock returned to the company, and the money paid into the program by the executives (including interest) returned to them in cash. A number of stockholder suits were filed to enjoin the action. While all of them were dismissed ultimately on technical grounds, one judge denounced the scheme as an unscrupulous, if legal, "heads I win, tails you lose" proposition. In 1938 Schwab won approval for cancellation of the management stock plan.[16]

But this did not silence the attacks of disgruntled stockholders. Leopold Coshland complained in the 1935 annual meeting that Bethlehem's top officers were excessively compensated. In a year when stockholders received no dividends and steelworker wages were as low as 50 cents an hour, Schwab drew more than $200,000 as chairman of the board and Grace got $180,000. "A few years ago," Coshland said, "Mr. Schwab said he was giving up many of his activities on account of his advanced age, but that Bethlehem would always remain nearest his heart. He was wrong in his anatomy. He had held it nearest his stomach, and we have been his meal ticket."[17]

Schwab kept the dissidents at bay through management's control of the proxy votes of large shareholders, many of whom were institutional investors. When Coshland and fellow dissident Lewis Gilbert offered a resolution which would limit the salaries of Schwab and his subordinates to 20 percent of the company's net earnings, or a total of $110,000 in 1935, the motion was rejected by a vote of 2,370,220 to 355 shares. Defeated but undaunted, the dissidents later called on Schwab to retire. According to Gilbert, the chairman had "outlived his usefulness" to the corporation. Upon hearing Gilbert's remarks, Eugene Grace leapt from his board seat and moved menacingly toward ringleaders Gilbert and Coshland. "Don't come around and feign friendship," he shouted. Fisticuffs were averted only by the quick action of several other directors. They physically restrained Grace until he cooled down, according to the *Times* account of the fireworks.[18]

Schwab took up his own defense during the 1938 stockholders' meeting. Listening to the dispute over his pay, his eyes glazed over with tears. Then he rose from his board seat, the old actor determined to stir the audience's sympathy one more time.

"I have devoted my life to the Bethlehem Steel Corporation. I intend to continue to devote what few years are left me to the company, if the directors are willing.... I do not intend to give way to the suggestion of serving without pay. For many years I worked and labored for this company. This company owes me a debt which it can never repay. This suggestion is the one sad thought in my 60 years of business experience. Will you, Mr. Gilbert, withdraw it?"

"As a personal favor?" asked Gilbert.

"Yes."

"As a personal favor," Gilbert said, "I withdraw the suggestion."[19]

There was compelling reason for Schwab's performance. Unknown to all but a few intimates, he was desperately in need of cash. Upkeep of his Riverside Drive and Loretto mansions was a tremendous drain on his resources. It was money he could ill afford to spend when the value of his Bethlehem stock had tumbled and losses from his various get-rich schemes of the '20s mounted. Four examples of his speculative reversals were disclosed later: $100,000 lost on the bankrupt Lehigh Valley Silk Mills, $50,000 on a chain of roadside hot dog stands, $250,000 in West Virginia Zinc Smelting Company, and about $750,000 in General Aero Corporation, a promotional venture to develop a transatlantic airport at Atlantic City, New Jersey.[20]

Adding to his cash woes was his compulsion to gamble. "I shall go broke for I love to gamble," he once confessed to E. Clarence Jones. During his winter sojourns abroad, he played maximums at the Monte Carlo casino for a week or two at a throw, reportedly gambling as much as 60,000 or 70,000 francs a day, while back in New York he gambled at the Whist Club, site of high-stakes bridge games among a select clientele. To fund his pleasures he began taking out loans. Working only as an insider could work, he secured money from the Chase Bank in New York, where he had served as a member of the board in the '20s. He also received $300,000 in loans from the little National Bank of Ebensburg, Pennsylvania, relying on his influence in Cambria County, where both his Loretto estate and the Johnstown steel works were located.[21]

But he did not stop with bank loans. Court records filed later in connection with his estate showed that he had even approached the priests of his hometown and received a $25,000 loan from St. Francis, a small Franciscan institution. Schwab was the most renowned alumnus of the school, having gone there to the equivalent of 11th grade before leaving Loretto in 1879. In 1935 he promised to bequeath the academy $2 million upon his death in return for the $25,000 loan. He further obtained $22,000 from the Carmelite nuns in town. Here too he had an inside connection—his youngest sister. Mary Jane Schwab had entered the order and lived as Sister Cecelia in the cloistered monastery of St. Therese of Lisieux which was built at Loretto with funds provided by Rana and Pauline Schwab, the tycoon's wife and mother.[22]

In 1936 Schwab's agents approached the La Guardia administration in New York, hoping to interest the city in buying his mansion for the official residence of the mayor. He offered to sell the property, building, and furnishings for about $4 million, a great reduction from the original cost, and promised to contribute to any fund that might be used to raise money to purchase the property. But Fiorello La Guardia took great pride in the fact that he lived in a modest walkup in Lower

New York. He ridiculed the idea of moving into the robber baron's seraglio, telling newsmen that the city much preferred having the tax revenue from the mansion. What La Guardia was not aware of was that Schwab had already stopped paying property taxes. Relying on a law that permitted a property owner to default for six years on New York real estate before the city could seize the property, Schwab had stopped his tax payments back in 1933.[23]

Although his health had deteriorated, Schwab was as restless as ever. He stayed only a few weeks a year at Riverside, leaving his wife to spend the winter months at the mansion. Grossly overweight, with her limbs swollen even further by inflammatory rheumatism, Rana lived out the last years of her life a lonely, waddling figure trapped in the biggest private residence in New York. According to writer Robert Hessen, "Every afternoon, weather permitting, Rana went for a drive, accompanied by [a niece or sister who] happened to be visiting her. Her chauffeur drove along the west side of Manhattan and then north for 20 or 30 miles. The route never changed, nor did her observations about the mansions and monuments along the way. She cherished this daily routine; it gave her an occasion to wear one of her fur stoles, and to see something of the outside world without being seen."[24]

In January 1939, a week after Rana died in her sleep, Schwab permanently shut the Loretto estate and released its employees. The Renaissance château, together with the golf course, farm, gardens, and Normandy Village, was placed on the market. Schwab wanted only $50,000 for the property, but even at that price no buyer could be found. Several months later, he gave up his Riverside Drive estate and moved into temporary quarters at the Ambassador Hotel on Park Avenue. On August 9, after touring extensively in Europe, he was stricken by a heart attack at the Savoy in London. A heart specialist, Dr. Edward Gordon, treated him. After two weeks of rest, he was deemed fit to travel across the Atlantic with the doctor in attendance. Joseph Kennedy, U.S. ambassador to Britain and onetime assistant manager of Schwab's Fore River shipyard, arranged for the steelman to be taken aboard the S.S. *Washington*. The liner would be the last important passenger ship to leave Europe before Hitler's invasion of Poland. Arriving in New York on August 31, Schwab was carried off the liner in a stretcher. He was taken to his apartment on Park Avenue where Dr. S. A. Brown, his friend and physician, treated him. Periodic bulletins issued from the apartment indicated that the magnate was gaining strength. On the evening of September 18, 1939, however, he died suddenly of coronary thrombosis.[25]

Over his lifetime Schwab had accumulated a fortune variously estimated at $50 to $200 million; in its obituary *The New York Times* lauded his life as "a romance" which epitomized the age of "American opportunity." It was only after his death that estate accountants discovered that his legacy also included insolvency. Listed debts, including property taxes owed to the City of New York, amounted to $2,262,280. Against these debts were assets of $353,810. The Rockefellers' Chase Bank was the single largest creditor. At Loretto there were obligations of $300,000 against assets of $51,094. Charlie and Rana left the National Bank of Ebensburg holding $180,000 in unsecured loans. Ultimately the bank wrote off $250,000 in losses.[26]

So where did his money go? It proved impossible for the accountants to figure out exactly what had happened. He had kept most of his accounts in his head. There were many transactions whose history of withdrawals and deposits could not be determined with any certainty. His use of different brokers to conceal what he was up to in the stock market further complicated matters. What might be said was that the millions he had earned since his days at Carnegie had simply been squandered, used up in buying sprees or obsessive gambling. By one account, the mogul had lost $40 million in the 1929 market debacle. *Time* magazine reported that he had speculated heavily in stocks in the market revival of 1936, and that this money had vanished when the market crashed in 1937.[27]

Left behind were broken promises as well as worthless stock and gaudy properties of minimal value. Most potentially damaging of the debts uncovered by accountants was the $47,000 owed to the priests and nuns of Loretto. Partly in anticipation of his promised $2 million bequest, St. Francis Academy was involved in an ambitious building program, including construction of a college science center already bearing the name Schwab Hall. To the shock and chagrin of the priests who were then struggling under a $250,000 building mortgage, Schwab's will did not specify a bequest to St. Francis or for any other charitable purpose.[28]

Scrambling to protect the reputation of the man who had plucked him from obscurity, Eugene Grace allocated corporation funds to pay Schwab's debts to St. Francis and the Carmelites. Lawyers from the company were assigned to help the executors of the will, Edward

Schwab, Willard Mitchell, and the Empire Trust Company, to dispose of his property. Eventually the 1,000-acre Loretto site was divided into several parcels and auctioned off. Following a scare when it was rumored that Father Divine, the Negro preacher from Harlem, was interested in Loretto, the main building and property were sold at public auction to the Friends of St. Francis, a group of private citizens organized to purchase the estate to enlarge the holdings of the school. The purchase price was $32,500. The mansion was renamed Mt. Assisi Monastery and made a residence hall for Franciscan priests. Today about ten priests live in the mansion, whose materialistic accoutrements and voluptuous statuary have been replaced by religious articles of devotion by men under vows of poverty, chastity, and obedience.[29]

The Riverside Drive mansion had no buyers and stood vacant until 1948, when Chase Bank sold the lot to apartment developers. A complex named Schwab Towers was built on the site. *Loretto II,* his $100,000-plus private railroad car, was sold to Abe Cramer, a Pennsylvania coal broker, for $5,000 in 1940. Vandals had already stolen the valuable tapestries and elaborate set of silver. Cramer then leased the car to Zack Terrell, the owner of the Cole Brothers Circus, for use by another rusting icon, ex-prizefighter Jack Dempsey. Dempsey had signed a contract to tour with the Cole Brothers so long as they gave him a private car. The Schwab car became Dempsey's home in 1941 as he barnstormed across rural America in the company of lions, elephants, and midgets.[30]

*

Months before the appeals court in Washington took up a review of the company's appeal, Phil Murray devised a strategy to impel Bethlehem to recognize the union. He was determined to use the delay occasioned by the appeal to put the machinery of organization in place at all of the producer's mills. He wanted to be in a position to exert clout from all points, having learned from Johnstown the disadvantages of getting sucked into a strike involving a single plant. The extremely high degree of concentration in the industry seemed to make such a centralized union apparatus almost inevitable. He wanted to preempt the company, back it into a corner from which it would have no choice but to concede the wisdom of a company-wide master contract.[31]

Murray was a student of the business side of steelmaking. Whatever he didn't know about the cost of raw materials, overhead, and labor, his research staff under Harold Ruttenberg could find out, and he prepared himself for the day when he could argue with employers on their own turf in behalf of improved wages and benefits. "He was a very practical organizer," Miss Perkins remembered. "When he had a job to do he figured it all out. He estimated human motives, estimated who would want to settle badly, why they would want to settle first, where his pressure could best be applied, and so forth."[32]

He had a free hand in SWOC affairs after John Lewis ceded operational control of the organization and returned to his Mine Workers. On November 22, 1940, Murray was elected by delegates to the presidency of the CIO, succeeding Lewis, whose power had been eroded by his grudge battles with Roosevelt in the 1940 election and his strident isolationism. Murray, who remained chairman of SWOC, positioned the steel union as strongly favoring national defense and a two-ocean navy, making SWOC an important ally of Roosevelt as he dealt with the political crises arising from Hitler and the European war.[33]

Murray established the Bethlehem Organizing Committee, and in October 1940 kicked off a drive to organize the company's plants. Van A. Bittner, an outstanding organizer from the Middle West district, was placed in charge of coordinating the campaign. Jack Lever was placed in charge of Bethlehem's Pennsylvania mills (at Bethlehem, Steelton, and Lebanon), and Nick Fontecchio was selected to head a ten-man organizing unit assigned to breaching the walls of Sparrows Point. A veteran of the Chicago labor scene, cigar-chomping Fontecchio had been at the scene when police opened fire on marchers in the notorious Memorial Day Massacre.

For the first time, the national office concentrated on tailoring a campaign to the unique aspects of the Bethlehem firm and addressing specifically the problem of organizing districts where union sentiment was weak and the open shop tradition ingrained. The ten organizers assigned to Sparrows Point had a realistic view of the obstacles that confronted them. Oscar Durandetto, one of those organizers, remembers: "Sparrows Point was different than the Middle West. In western Pennsylvania and Ohio there were roots of unionism through the coal miners connection. Many sons of miners around Pittsburgh worked in steel so they were very pro-union. The guys in Baltimore came out of another mold. There was race and there was the 'Southern way of life,' both obstacles, but then there was Bethlehem. The company treated its men bad, done underpaid them and thrown up this company union. Really the company was our best organizer. The rank-and-file sentiment was anger."[34]

SWOC's Wage and Policy Committee had drawn up a program to ease the plight of steelworkers laid off or downgraded by the continuous-strip mills, a vital local issue in Baltimore. Durandetto and the other organizers continued to hit on the company's "treachery" in rallies around the city. Quickly and shrewdly, they tapped into the vein of growing dissatisfaction with the wage incentive system. According to SWOC, the pay system was no more comprehensible, or fair, in 1940 than when it had been denounced by the Wilson War Labor Board during the First World War. There were tremendous discrepancies in take-home pay between men who did similar work in different departments. Even those who did well under the system did not understand it. Because incentives had a bigger place in the Bethlehem wage structure than at any other steelmaker, and because incentives were based on an entire department's productivity, SWOC argued persuasively that workmen were penalized for down time and other problems over which they had no control. Research director Ruttenberg figured that hourly employees at Bethlehem averaged 10 percent less than their Middle West counterparts. These figures were not union propaganda: A survey by the Bureau of Labor Statistics also found that owing to "incentive" deductions at Bethlehem, a worker could earn as little as 56 cents an hour as opposed to the Middle West's 62½ cents minimum.[35]

Race was a major sticking point for organizers at Sparrows Point. For years the black worker had been a pliable, submissive part of the job force. But the farm boys who had come North to the mill in the '20s were older now. Many of them had children who had been to school and who had gained a measure of self-confidence and strength in numbers. SWOC assigned organizers to bring the 7,500 blacks at the Point into the union. Oscar Durandetto was in charge of organizing blacks on the "steel side." "Phil Murray told me this one time: 'A good organizer will get somebody else to do his work for him,'" he says with a fond chuckle, "and that's just what I did. I went up the mill, stood down them company goons, and I talked to men outside the clockhouse, as was my legal right."[36]

One of the men Durandetto approached was Charlie Parrish. "First thing I'd seen with Oscar—he weren't afraid," remembers Charlie. The organizer didn't look around to see whether the company guards might be listening. He made it clear that he was convinced that he was as good as any of the company's men, and this had a liberating effect on Charlie. Second, Durandetto took a personal interest in him. Though they talked only briefly, Charlie says he was impressed by how much the man knew about what was going on in the department without acting superior in any way. He slipped a few pamphlets into Charlie's hands. *What SWOC Means to You* was the title of one of them. That evening Charlie and his wife Alice read it over:

> 1. A Ten-Cent-an-Hour Wage Increase. A union ALONE can assure workmen a fair share in the profits of industry.
> 2. Job Security. A union contract is the steel worker's guarantee that he will be promoted when eligible. It is his guarantee that his seniority rights will be protected.
> 3. Vacations-With-Pay. Under a union contract your vacation-with-pay is GUARANTEED and not subject to the whims of an individual.
> 4. Under a union contract machinery is set up to guarantee a peaceful and just settlement of all grievances with the final step the employment of an impartial umpire. This machinery has been operating successfully in 700 steel companies where the SWOC has signed contracts... contracts that the SWOC regards as sacred.

Other material denounced the "gunman rules under which you are working." Colored workmen, it continued, "must work for their own freedom, and the surest way to do this must be through the unity of all workers of all races."[37]

A mass meeting for black steelworkers was held at SWOC offices at 712 North Bond Street on February 20, 1941. Charlie was there. "I went to the meeting and I listened to what they said. And what really made me a union man was this: The big man don't care nothing about the little man. I learned about Phil Murray and about the contract. You were supposed to fight any injustice where the contract was broken. And I thought, I was gonna fight."

Discrimination had become intolerable. "My boss, he only got worse. He'd tell the white fellas, 'Look, I don't care what time you knock off. I don't want them niggers to knock off.' No break time, no pay for overtime 'cause he wasn't gonna approve it. It got so I trained the new men and the boss then put them on as helper A and me, I'd still be helper B."

Charlie signed the union card and agreed to become a volunteer organizer in his shop. It didn't take long to discover that other workmen harbored pro-union thoughts. They were tired of being pushed around. Tired of the ERP rep telling them he was working on their grievance and needed a little more time. He soon joined forces with Hural Thompson, another volunteer organizer. They made a good team. "See, we'd start talking union in the locker room 'cause it was Jim Crow and there'd be no bosses around, and the boys got to signing the cards. One would say, 'Yeah, I'll join.' Then the next say it. I had a couple of boys, try to pay me off so they could get out of joining, and I said that ain't right. There ain't gonna be any of that payoff business with the union. See, people had to get out of that frame of mind, that dependency thing. You got to stand up for what's right, I tell the fellows."[38]

*

Union activism centered in East Baltimore and the steel hamlets of Turner's Station, St. Helena, Essex, and Edgewater. These were the communities that had borne the brunt of layoffs at the start of the decade and felt the sting of job displacement at the end. Diverse ethnically and segregated racially, the neighborhoods had a shared sense that the industry's vaunted paternalism was not for them and never would be.

"The houses at Sparrows Point, that was boss's land and for the hourly families that wanted to be bosses," says Ellen Pinter, secretary for the local SWOC lodge before she married Frank Pinter, a crane operator whose savvy at organizing ethnic workers was legendary. "The whole setup was from another age. I mean, there was still the company store just like some West Virginia coal camp. We all said, the workman isn't going to sell his soul to the company store no more!"[39]

Organizers like Frank Pinter related the union to the local concerns and worries of average people about their lives. The fact that SWOC was winning a diverse group of workmen was indicated by the mix of people Pinter was able to meet on the common ground of unionism. Among them were Mike Howard, a member of the local Communist Party; Oscar Durandetto, a professional organizer and dedicated "anti-Commie"; Joe Lewis, a native Baltimorean who never harbored a subversive thought but nevertheless signed the SWOC card "because the workman by himself didn't have a chance"; and Charlie Barranco, a steelworker who proselytized for SWOC after Saturday-night boxing matches at his fight club.[40]

The SWOC drive was opposed by the residents of the Point who enjoyed the "company way of life." Bred on Bosses' Row, Elizabeth McShane, wife of assistant open-hearth superintendent Colegate McShane, believed in the rank and precedence in town, saying it fostered unity and made for a wholesome environment for children. On the eve of the union drive she recalled the town in terms of Sinclair Lewis's '20s chronicle of small-town life. "I had read *Main Street* and it took me back." Because the descriptions of social life were so uncannily precise, the book's satire did not bother her at all. In fact, it made her nostalgic. "I could pick out the people, the whole social setup. There was a certain menu for the bridge party and if you didn't have a certain type of thing on your menu, well, the luncheon was just unsuccessful."

According to Elizabeth's neighbor on Bosses' Row, Mary Barnhart, the company protected the townsfolk from outsiders. "I know there were people who tried to come in and tried to be agitators," she says. "If they caused a problem I wouldn't know. All I know was that my father always said that if it weren't for those bureaucrats down in Washington, things would go back to normal."[41]

But normalcy did not return, regardless of how much the residents desired it. Pressure on management was intensifying on a number of fronts. On April 7, 1941, the NLRB started proceedings against Bethlehem and President Grace for continuing to recognize the ERP at its Fore River shipyard. On the same day, a group of stockholders at the company's annual meeting introduced a resolution calling on the company to "obey all labor laws" and undertake collective bargaining in good faith. While the resolution was easily defeated, the meeting was disrupted by other embarrassing questions. Why was the company provoking strikes when the country needed all the steel it could get for national defense? Alexander Crosby, a stockholder, wanted to know.

"Why does Bethlehem Steel refuse to sign with any labor unions?" he added.

"We have not refused to sign. We have some. I haven't them with me," Grace answered incoherently. Boos and catcalls filled the room.[42]

On May 12, the company's appeal of the NLRB ruling was dismissed by the U.S. District Court. This set in motion a series of filings by SWOC to the labor board requesting elections on the question of union representation. In June a schedule of elections was worked out after tense negotiations between the board and Bethlehem. The talks were held under the gun of a new ruling by the National Defense Commission that government military contracts be withheld from companies that were not in compliance with federal laws, including the Wagner Act. The ruling forced management to choose between the possibility of losing defense work or going forward with plant elections, which were only a preliminary step in the process of reaching a collective-bargaining agreement. The company chose to accept NLRB elections, all the while reserving its right to resist future developments it didn't like.[43]

A month later came news of a break in the ranks of Little Steel. Faced with the same choice between labor compliance and loss of defense contracts, Youngstown Sheet, Inland, and even tough-talking Girdler at Republic opted to abide by the results of a labor board cross-check of SWOC membership cards against their plant payrolls. The tally revealed that more than 50 percent of the hourly workforce belonged to SWOC. By law the union was entitled to become the exclusive bargaining agent for the 72,000 employees of the three companies.[44]

Bethlehem was now the principal holdout in the industry. Bittner ordered brief, surgical walkouts to demonstrate the union's vigor. On the night of August 18, pickets were thrown up at the gates of the rod and pipe mills at the Point. An American flag was unfurled across the gate by the picketers. The flag did not touch the ground, but it was low enough that cars using the road would have to run over it. Car after car drove up to the flag, stopped, then backed away. Around 11:00 P.M. there was a brief scuffle with police, and eight arrests were made. The line stood. After a full shift of lost production the job action was called off by Fontecchio at 7:00 A.M. The point had been made "in the only language the company knew," says Durandetto.[45]

*

Night still hung heavy over the peninsula when Bennett Schauffler and 35 other agents from the National Labor Relations Board drove into Sparrows Point and took up their positions at the mill. The time was 4:00 A.M. The date, Thursday, September 25, 1941—"Vote Day." The government officials were joined by 49 observers for management, 49 observers for SWOC, and 49 observers for the "Independent

Steel Workers' Association," a nonaffiliated union which had obtained NLRB certification to be on the ballot.[46]

At 5:00 A.M. 10 polling places were opened at the plant. Over the next five hours, 11,700 steelworkers voted. Employees of the night shift, who were going off duty when the polls opened, voted first. The day shift employees followed, released in small groups to minimize interference with production. As a precaution against irregularities, the color and form of the ballots were kept secret until the polls were opened, and were specially punched when they were handed out.[47]

Anthony Maggitti was one of the tellers at the tin mill. The election was held in the shipping department. "The company was real quiet and there were men from the NLRB all around supervising and observing," recalls the former hot-mill catcher. With skids of finished tin plate glittering shoulder-high beside them, the workmen smoked cigarettes and filed toward the battery of curtained polling booths with almost solemn movements. Every couple of seconds one of the curtains jerked open and the next man was signaled to the stall by Maggitti after he cross-checked the brass badge with the eligibility list.

Across forests of track John Duerbeck raced down Blast Furnace Row. "Hey, make sure you guys get over and vote," he yelled to the work gang in the tool shanty. "Heck, we've already been there," exclaimed Charlie Parrish as everyone laughed. "Yeah, it was funny," recalls Duerbeck. "I was definitely excited on election day, runnin' all over to get out the vote, and damn if the fellows weren't in and out of the booths before I got over."

Near the open hearth, where 2,400 men came on and off duty every day, Mike Howard was smiling. Everyone was giving the thumbs-up. "The enthusiasm was palpable," he remembers. "You could feel this mass of men going forward and voting their true mind. This was democracy and it had taken some six, seven years to accomplish."

The story was the same at the wire and pipe mills. "We had a tight brotherhood," says Charlie Barranco. The union guys were wearing their SWOC paper hats, while the company's men were in a funk. "We had a superintendent named Bulldog Johnson. Boy did he hate the union. So he stayed in his office all day and locked the door."

"Mr. Roosevelt had told the working man it was okay to organize and he did," explains Frank Amann, a bricklayer and volunteer organizer. "I been around since the '20s when you couldn't say the word *union*. I knew damn few men that wanted to go back to them days."

"They had millions of dollars to fight us with, and lawyers," adds Joe Lewis. "By getting a union together, by paying dues, now we had our own lawyers."[48]

The polls closed at 10:00 A.M. They reopened at 1:00 P.M. to accommodate the men on the 3:00-to-11:00 shift. At 6:00 P.M. the ballots were taken to the basement of the medical dispensary in town. NLRB agents counted the votes in the presence of the observers. SWOC officials crowded into the basement room. There were only two management representatives, John L. Wynne, special assistant for President Grace, and Francis Wrightson, assistant to general manager Cort. Most of the superintendents waited out the results in their offices. At 8:00 P.M. Bennett Schauffler announced that SWOC had won a lopsided victory, receiving 10,813 votes, or 68.7 percent of the total, against 4,198 votes, or 26.7 percent, for the independent association, and 731 votes, or 4.6 percent, for "neither." Nearly 90 percent of the eligible employees had voted.[49]

District director Fontecchio termed the victory a milestone in Baltimore history. He singled out the black vote as crucial. According to his tabulations, 6,000 of 6,500 blacks had voted, and the overwhelming majority went for the union. In a statement issued to the *Afro-American*, Fontecchio said, "Because of the great number of colored workers at the plant, had their vote been against us we would have lost the election. I want it known that we appreciate their support and we're going to work for the common good of all along the common policy of the CIO which frowns on all race, creed or color discrimination."[50]

Five days later, Sparrows Point shipyard workers voted to join the Industrial Union of Marine and Shipbuilding Workers (CIO) by a decisive margin. And over the next four weeks the corporation's other open-shop outposts at Johnstown, Steelton, Lebanon, and South Bethlehem fell into the union camp. Again the vote margins were decisive —2 to 1 or better over the independent unions and 8 to 1 or better over the choice of "neither." As a result, when SWOC officials started to negotiate a master contract with Bethlehem in the fall of 1941, they boasted the most solid rank-and-file support in the industry.[51]

The CIO campaign had placed an unprecedented strain on democratic institutions and had tested the mettle of many individuals and organizations. Without the steady hand of Frances Perkins and other federal officials, chaos, even class warfare, might have erupted after the 1937 confrontation. The orderly vote at Sparrows Point was a triumph of moderation and law in an age of growing extremism and violence. But neither policymaker nor workman had much chance to catch his breath. Those autumn days of 1941 brought ominous radio bulletins of a world in crisis—Europe in Hitler's grip, Japan looming menacingly over Asia. War had returned among industrial nations, causing the demand for American steel to soar.

PART IV

THE FAT YEARS 1941-1957

CHAPTER 16

IN THE COMBAT ZONE

The same geographical advantages that propelled Sparrows Point to industrial maturity during the First World War made it the logical place to make steel during the Second World War. As far back as 1936, Sparrows Point had been selling pig iron and scrap under a contract with the emperor of Japan. Japan's growing military machine was dependent on foreign sources for raw materials, and its pig-iron production was woefully behind its capacity to make steel. The Sparrows Point shipments went to the state-owned Yawata works and to other plants on the main island. Largely owing to U.S. exports of raw materials, Japan had been able to produce over 5 million tons of steel in 1936, double its output of only five years earlier.[1]

In July 1937 the Japanese army invaded China. By the end of the next year, the Japanese controlled half of the country. In 1940 all shipments of war-essential goods to Japan were terminated by Washington fiat. By that date, however, orders from Britain had more than compensated for the loss of Japanese business. Since Hitler's invasion of Poland in 1939 the British War Office had ordered a reported $300 million in military supplies from Bethlehem. The company was making more shrapnel, big guns, and warships for Britain and the U.S. Navy that at any time in the last war. For the first time since 1929, the Chesapeake plant was running at full capacity for a sustained period.[2]

Nineteen forty-one was a year of even more urgent catch-ups for all U.S. steelmakers. The Lend-Lease program to supply material to the Allies, approved by Congress on March 11, was succeeded by President Roosevelt's declaration of a national state of emergency on May 27. Then came Germany's surprise attack on Russia in June. Dr. Gano Dunn, the White House advisor on steel, predicted a crippling shortage of the metal beginning in late 1941 and intensifying in 1942.[3] This inspired Washington to plan for an expansion of domestic steel capacity. Federal financing for 10 million tons of additional capacity was disclosed to Eugene Grace and Benjamin Fairless, president of U.S.

Steel, in a meeting attended by Jesse Jones, chieftain of the Reconstruction Finance Corporation (RFC), and others.[4]

Sparrows Point was targeted at once. Its amphibious position, one foot on land and the other in salt water, gave it prime importance in the distribution of steel to the Allies and U.S. armed forces. In October 1941—one month after the union was voted into the plant—Jesse Jones entered an agreement on behalf of the government to pay for 1 million tons of steelmaking capacity at Bethlehem. Over half of the money was earmarked for the expansion of Sparrows Point. Prominent among the list of improvements was construction of a seventh blast furnace, 61 coke ovens equipped to extract benzene and other war-essential coke chemicals, an electrolytic tinning line, and three new 200-ton open hearths. The tab: $20,206,572.82.[5]

President Grace drove a hard bargain in negotiations with Jones. The Bethlehem contract with the Defense Plant Corporation, a subsidiary of RFC, broke out of the pattern that other major companies had signed. It was a "scrambled facilities" contract where the government financed additions and modernization of existing facilities. Previous DPC financing in aviation, tanks, and munitions involved entirely new plants. This contract permitted Bethlehem to pay rent to the government based on the degree of use of the facilities both during and after the war—opening up the possibility that the government might wait 30 or 40 years before Bethlehem's rentals would repay its original investment. Jones was none too pleased about these terms, according to an inside account. "He might have cracked down on Bethlehem publicly, but to Jesse that would be coercion. 'After all,' he observes, 'this is still a democracy.' So he proceeded to bargain. On October 16, DPC announced that 'it would finance' Bethlehem." (As it turned out, Bethlehem purchased the DPC-financed capacity outright after the war, providing the government with a good return.)[6]

Jones felt obliged to inform the White House of Bethlehem's terms before committing himself. He was told to go ahead because of the dire need to get ship construction under way. Not only was the navy screaming for armor plate; a crash program had begun to treble the nation's production of cargo ships. Sinkings of Allied shipping by U-boat "wolf packs" had reached a deadly crescendo of 500,000 tons a month in 1941. In effect the Nazis were subtracting from the sea lanes more than the current replacement capacity of the Allies, threatening to sink Lend-Lease.[7]

In February 1941 Admiral Jerry Land, chairman of the U.S. Maritime Commission, awarded Bethlehem a contract to build a 13-way shipyard on the west side of Baltimore harbor at Fairfield. The complex was constructed in conjunction with a long-idled Pullman plant which was used as a fabrication plant. Fairfield was designated to manufacture the EC-2 Liberty Ships, or "Ugly Ducklings," a cargo tramper that, if technologically prosaic, could be mass-produced at a reasonable cost. The initial award called for 312 vessels. The ship plate and structural steel for keels and inner bottoms were to be supplied by Sparrows Point, with specialized parts subcontracted to other manufacturers.[8]

One month after Pearl Harbor, Bethlehem had become the nation's top war contractor. Total orders amounted to $1,327,500,000, or about $100 million more than the number-2 war contractor General Motors. The corporation was engaged in the manufacture of bomb casings, 16-inch AP (armor piercing) shells, gun forgings, airplane parts, and navy warships. "Currently Bethlehem is in effect the steel skeleton of Mr. Roosevelt's famous arsenal of democracy," said *Life* magazine.[9]

*

The American Krupps was back in the thick of things, but under different circumstances. Although some of the main protagonists were the same as in the First World War (Grace in South Bethlehem, Churchill in London), the trafficking of munitions now was regulated by public agencies. Lend-Lease bureaucrats rather than House of Morgan bankers directed the sale of war goods abroad; the "privatized" war of Charlie Schwab was replaced by an alphabet soup of federal agencies that shifted domestic industry onto a war footing. These agencies accomplished the phenomenal task of logistics and problem-solving with dispatch and a remarkable lack of governmental scandal or corruption.[10]

One agency that kept America's production humming was the U.S. Conciliation Service under Frances Perkins, the seasoned secretary of labor. The office was credited with keeping many steel-mill production schedules on course. On average the Conciliation Service received 20 urgent calls a day from management and labor in the Pittsburgh steel district alone. "Four U.S. Conciliators = One Battleship" was the standing rule of thumb of the agency. Donald Nelson's War Production Board attempted to short-circuit labor flare-ups by insisting on factory-level union-management committees to study production practices, discuss grievances, and generally let off steam. Admiral Land's Maritime Commission spurred shipbuilding with star-studded launchings and tons of ticker tape. Not particularly impressed with the "incentive" of high profits for industry when ordinary citizens were taken from their families and jobs to fight for nominal compensation, Land rode herd over the shipbuilders with workaholic vigor and issued mock complaints to the press. "I'm damn sick of talking to men who are willing to provide the ocean if we'll provide the brains, the money, and the material," he groused.[11]

Behind these players was the inspired leadership of the man in the White House. In the days following the shock of Pearl Harbor, Roosevelt forged a consensus between steel labor and capital. In White House meetings the president secured a "no-strike, no-lockout" pledge from Phil Murray of SWOC and the presidents of the major steel companies, including Grace and Girdler. Further, management and labor ceded to his proposal for a National War Labor Board to arbitrate wage and hours disputes in steel. This was significant since the union and the Little Steel operators had been unable to reach a contract following the 1941 NLRB elections. For some time Phil Murray had called for 12½ cents more an hour at Little Steel, a proposal which the producers had rejected. On February 6, 1942, Perkins certified that a dispute existed in the matter and sent the dispute to the new War Labor Board. The board then empaneled a fact-finding committee to investigate steel wages.

The committee certified that the real weekly earnings of steelworkers had declined by 13.3 percent since 1940. However, responding to the president's call for all segments of society to limit their incomes, the board declared that it was in the national interest for steelworkers to forgo a hefty pay increase, however justified it might be. Workers were awarded a 5½ cent raise (to maintain their earnings as of January 1, 1942), with no further claim to higher wages during the war. The so-called "Little Steel" formula became the cornerstone of the administration's wage stabilization program. Basically it froze wages during the war at 1941 levels. Although disappointed, Murray accepted the decision.[12]

This agreement was followed by other measures from Washington which limited the power of steel management to determine prices. There was great fear that steel shortages would induce panic buying, driving up prices as in the First World War. To relieve inflationary pressure on steel, the War Production Board ordered the termination of hundreds of steel-based consumer products for the duration of the war. Included on the prohibited list were steel pots and pans, dish racks, household spray containers, metal chairs and tables, clock cases, automobile accessories, signposts, fountain pens and pen holders, finger bowls, mailboxes, and cash registers.[13]

Several months later the administration established the Controlled Materials Plan. Along with companion laws, CMP funneled steel production into the zones of combat with a thoroughness that made its depression ancestor, the National Recovery Act, seem primitive. After some notable glitches and much confusion, the system began working quite well. Orders for the U.S. armed services, Lend-Lease, and domestic needs were handled through a central clearinghouse which appraised requirements and approved allocations. To assure cooperation by the steelmen, Roosevelt invited Eugene Grace, Benjamin Fairless, and other top executives to participate in the program. Grace accepted. He was considered a valuable addition to the materials board and made tangible contributions in expediting steel production in cooperation with the Bureau of Ordnance. As one observer wrote hopefully of the new-found management-government alliance: "That Roosevelt and Grace are never likely to collaborate does not mean that they cannot co-operate at long range, with mutual respect, to mutual advantage. A little contact with the President, such as he has been exposed to lately, may broaden Grace's point of view. Association with a man like Grace may help the President to understand those practicalities which are, after all, the things that make any freedom, let alone four of them, conceivable."[14]

Net corporate profits permitted under CMP were modest. It was noteworthy, considering Bethlehem's profiteering in the First War, that Bethlehem's after-tax net actually fell from 8.1 percent in 1940 to 1.7 percent in 1943 and 2.6 percent in 1945. Federal excess profit taxes claimed over $200 million of Bethlehem's earnings, accounting for much of the low net. Nevertheless, due to the unprecedented volume of business transacted and IRS tax credits on wartime capital spending, net was up five times above Bethlehem's dismal showing of the '30s—an average of $32 million a year versus $6.5 million a year in 1930–39. And for long-suffering shareholders the war years were a relief: Bethlehem's $6-a-share cash dividends for common stock during each of the war years were the best since 1929–30.[15]

*

Apart from regulated profits and civilian oversight, there was another difference between the industrial aspects of the two wars. This concerned technology. The revolution in armaments prior to the First World War had been a revolution of heavy equipment. The change that took place after 1939 was the dawn of an age where lightness and electronic sophistication increasingly replaced land- or sea-based bulk as the deciding factor in warfare. This was not a sudden transition, but rather a gradual shift. The rise of airborne assault was a case in point. Slowly but surely fighter jets and bombers replaced big navy guns as the major offensive weapon on the sea. Launched from "flat-tops," the single-engine Grumman Wildcat, Douglas Havoc, and Martin PBM became the primary striking weapons of the navy in the Pacific. This proved crucial in the Battle of Midway. Planes in turn altered navy strategy from a straight-ahead sea offense to a wheeling pattern where destroyers circled around the flat-tops, protecting them from attack as they launched warplanes against targets often 100 miles from the fleet.[16]

Greatly enhancing the aviation revolution was a branch of electronics. Radar (*ra*dio *d*etection *a*nd *r*anging) proved an invaluable warning system during the 1940 Battle of Britain, alerting the Royal Air Force of impending Nazi raids. Adapted by the British and U.S. navys, radar eventually rooted out the U-boat menace plaguing the North Atlantic sea lanes. It also became a key ingredient of the offense, permitting precision bombing at night and in foul weather.[17]

The impact of the new weaponry as it pertained to Bethlehem was subtle but highly significant. While steel remained the foundation *material* of war, the Bethlehem Steel Corporation ceased to be on the forefront of war technology. Before the First World War, the company had been a member in good standing of the Armor Plate Pool that had divided up European military sales and had patent control over a multitude of military products. But in the development of radar and the refinement of warplanes and missiles the company played no contributory role. Flight technology was dominated by many small firms that had sprung to life after the First World War—Glenn L. Martin, Boeing, Douglas, Vought-Sikorsky, Grumman, Northrup & Ryan, Lockheed, Fairchild, and Brewster—and reached maturity soaring on the wings of government orders. The growth of the radar industry also was primarily a function of the federal government, which supplied the critical mass of money and scientists to develop the technology. The light-weight metal aluminum in turn grew from obscurity by supplying the tubes, wire, and engineered parts needed by the new war industries.[18]

If Bethlehem was not part of the emerging technologies of aircraft, electronics, and missiles, it did achieve something that confounded the skeptics: it did not stumble. The breakdown in steel production feared by Gano Dunn and other experts in 1941 did not materialize. Picking up their figurative sledgehammers, Bethlehem and other producers did the job, meeting the unprecedented challenges placed upon them. Between January 1, 1941, and the end of 1945 the U.S. industry pounded out a remarkable 427,050,950 ingot tons of steel, besting the production of 1914–18 by 2.5 times.[19]

By manufacturing a commodity so essential in making military supremacy possible, the steelmakers provided the margin of superiority that permitted America and her Allies to repulse the Axis powers and undertake offensives halfway around the world. U.S. output in the war years was 3.6 times that of Germany's Saar region, the world's second-biggest steel production center; it soared ninefold above Japan and her colonies. America supplied not only its own needs, but had enough steel left over to export over 12 million tons a year to hard-pressed Russia and Britain, whose outputs were, respectively, one-quarter and one-sixth that of this nation.[20]

*

Of the vast outpouring of steel, Sparrows Point contributed some 17 million tons between January 1941 and August 1945, cresting with 3.8 million ingot tons in 1944. Production at Sparrows Point, largest of the Eastern plants, came out to an average of 302,000 tons a month for 56 months straight, or about 10,000 tons a day.[21] To produce that amount of steel required incredible levels of exertion. The plant's finest hour was a triumph of conventional steelmaking—nothing fancy, nothing newfangled, just the disciplined application of longstanding practices by an experienced workforce. Deferring many of the issues of wages, seniority, and justice that had been salient in the union drive prior to Pearl Harbor, the men dedicated themselves to the awesome and dangerous task of making steel. When every minute lost meant 7 tons less steel, it was obvious how much depended on the men directing the operation.

A good place to observe the production front behind the war front was the juncture where steel was refined from its mineral ingredients. The open-hearth department at the Point sprawled across some 100 acres of ugly flatlands west of town. The layout of the three main shops, No. 1, No. 2, and No. 3, were similar. In each shop there was a row of two-story-high furnaces that were flanked by a wide platform known as the charging floor. Handling equipment ran overhead on bridge cranes, railroad tracks dropped below, and beneath the whole thing were the pipes and brick chambers for the reuse and discharge of hot gases.[22]

Refining a batch, or "heat," of steel required 9 to 11 hours, tap to tap, and while machinery expedited the process, every phase of the cycle rested on the skill and stamina of an elite cadre of workers known as "first helpers," "second helpers," and "third helpers." The first helpers were responsible for furnace operations. They manipulated the fuel throttles and back pressure wheel that kept the liquid boiling through the chemical reactions. The second helpers prepared bags of river coal and metal alloys (called stock) that regulated the final properties of the steel. The third helpers aided the second helpers and teamed up on the task of opening the taphole for "the blow."

"The blow" was steelman's talk for those first few seconds when the steel burst from its furnace confinement, showering molten steel across the pouring platform. The spectacular sight of white-hot steel rampaging down the runner to the pit ladle was the photographic symbol of American steel. Just as the 20-ton Bessemer blow represented the wonders of 19th-century metallurgy, the 265-ton open-hearth blow displayed the might of industry under the strain of all-out war. For the men who made steel, though, tapping meant one final chore before machinery took over again. The crew ran onto the iron-splattered platform to throw in the stock. This set off new geysers of flame. A crane then lowered its hooks. The hooks grabbed the ladle and, within minutes, carried away the pot of metal to the teeming pits where it was poured into ingot molds for transport to the soaking pits and, thence, to primary rolling.

The men who made steel wore the red neckerchiefs of railroad engineers and the tinted goggles of welders. They walked on shoes soled with rubber tires to keep their feet from getting burned on the brick floor. In summer as well as in winter they cloaked themselves in protective long johns, denim coveralls, flame-retardant jackets, and by day's end what they wore was soaked in sweat. They drank steaming black coffee to keep their bodies hot and chewed Brown's Mule Plug tobacco to keep the dust out of their throats. They ate sparingly to avoid heat cramps.

The men who made steel were on intimate terms with the peculiar slang without which the mill could scarcely function. When open-hearth men wanted a "drink" for their furnace, they threw up a hand signal to the crane operator and got 15,000 to 20,000 pounds of iron back. Furnaces were invariably "she"s and, just as invariably, their behavior elicited the same words that were spoken by old-time iron puddlers: "She's got a bad bottom," "She's got dead metal" (molten steel that is quiet and not giving off gas). To steelmen a ladle the size of a car was a "bug" and a furnace kicking up too much slag was "boiling at the breast."

They were unique among the mill workers: Theirs was a job that remained impervious to assembly-line mechanization. Open-hearth men had to know their metal as no Ford workman need know about an automobile. With the closing of the hot mills in 1938–39, they were the last of a breed and were paid accordingly. The top man on the furnace got the plant's top wage, as much as $25 a day based on tonnage.

The men also knew something that connected them to the soldiers in the field and set them apart from people in other circumstances—they knew that at any moment the machinery might reach out and snatch them from the comparative safety of the charging floor. To actuaries they earned a reputation for higher-than-average death rates. Serious injuries came swiftly in accidents and slowly from bad hearts, dust-scarred lungs, and arthritis.[23]

*

Mike Howard and Ben Womer were among the hundreds of helpers who worked in the shops at Sparrows Point as the mill roared out of the Depression and made 1/25th of the steel produced in the United States between 1941 and 1945. "I felt good," says Howard. "When you're in your prime, physical work is a challenge, and if you're working with other people, there is a certain pride in being on top of the job. The demands of the work bred a kind of man who was fiercely committed to his work." Mike makes an important distinction between two kinds of physical labor. There was "brutalizing work," whose drudgery and tedium take away a man's dignity ("The quicker technology eliminates these jobs the better"), and there was "challenging work," which tests those who undertake it and unites men in a common cause. "In the open hearth," he points out, "you had to cope with problems that involved metallurgy, that involved blacksmithing, that involved judgment. These were not guys who were a bunch of punch-drunk wrestlers. They were bright. There were split-second decisions that had to be made. There were a lot of challenges that kept you on your toes."[24]

Mike's good feelings about the job were improved by his union work. He had entered "the hidden abode of production" because of his political beliefs (he was a member of the U.S. Communist Party) and he was eager to exercise his ideology in practical ways. Not long after the NLRB election, Mike was elected in a departmental vote to the post of zone committeeman. He was the chief union representative for the 2,400-man force in the open hearth, the biggest department in the mill. Settling grievances became his stock in trade. "But I stayed honest," he cautions. "The union work was unpaid. I worked the same shifts, did the same work, as the other guys. Sometimes I had to take off a few hours to go to meetings, but for the most part I squeezed in the union work on my own time."

As zone committeeman he began chipping away at years of autocratic company rule. He did it by becoming a master of procedure, by framing grievances in ways that were procedurally impeccable. He started with the basics.

"Nobody knew how they were getting paid because of all the different rates. If your furnace was down for a scheduled repair, you got a 'down rate,' but if the furnace roof caved in, that was 'an act of God' and the company wouldn't pay you. I wanted to know exactly how I was being paid and nobody would tell me. I don't remember if I filed a formal grievance or just kept at it, but finally I received the entire breakdown of what constituted my pay. I posted it on the bulletin board. For the first time the guys could actually calculate whether they were being short-changed or not.

"As it turned out, the company's pay system was accurate and pretty fair. The problem was their attitude. The thinking was, 'the less you knew about it, the better.' It was foolish and it was counter-productive, but that was the mentality. The war opened up communications a bit, but it was still difficult. It really never got through to the top guys that if you leveled with the men, that would carry a lot of weight. Allow labor to participate in decisions and you'd frankly have some power over labor. Scorn them and, hell, they'll scorn you. So if you want to know what did make the men work hard during the war, it was, first off, patriotism, but it was also little reforms like the pay-rate disclosure that made a difference in shop morale."

Another issue that dogged his tenure was the department's crazy hours. The men changed shifts every two days. A man who was scheduled on 7:00 A.M.–3:00 P.M. on Tuesday and Wednesday switched to 3:00 P.M.–11:00 P.M. on Thursday and Friday, then 11:00 P.M.–7:00 A.M. on Saturday and Sunday, before taking two days off. Mike determined that the 2-2-2-2 schedule was a remnant of Depression cost cutting, a way developed by the superintendent to get around the 40-hour week and related overtime provisions of the code period and Walsh Healey.

"The way they had it figured out, you worked 48 hours over a six day period, but, due to the way the shift hours changed, at least one of those days was on another calendar week. The exception was when you started work on Monday and the sixth day would be within the same calendar week. So in that case the company usually knocked off the sixth day and you had a three-day weekend." Mike arranged to have a departmental referendum. The majority voted to retain the schedule and democracy prevailed. "We kept the schedule. The men were used to it and it did have one good aspect to recommend it—you never had to work for the same foreman for more than two days in a row."

During and immediately after the war the union appointed Mike to various posts. He chaired the "steel side" grievance committee, which met with the personnel department on unresolved Step Three grievances. He sat on the union-management production committee established by the War Production Board. He sat on the USWA* committee responsible for equalizing wage rates. But he always considered himself foremost a rank-and-file man and rejected salaried union jobs. By staying active on the shop-floor level, he gained acceptance in the special fraternity that made steel. From men like Willie, a first helper in No. 2 shop, "I learned that there were all kinds of things which had nothing to do with my arguments about the union that carried a lot with these guys. They were separating men on the basis of considerations, well, to put it in larger terms: integrity, doing your fair share of the work, standing up against the pressure of the bosses. Personal bigness, I guess you would say."

In the summer of 1943 Mike's feelings about the mill and men spilled out in a series of watercolors. "Mill scenes" he called them—

*In 1942 SWOC voted to rename itself the United Steelworkers of America at its constitutional convention in Cleveland.

private visions of a special world. "They kept me out of trouble," he laughs, saying he had been painting since childhood. One of them depicted a second helper who, exhausted from a night of wheeling stock, was slumped over asleep after the furnace tap. Another was a study of the face of a first helper, gaunt and self-composed against the bright lights of the open hearth. His favorite, though, showed a crew of men ramming a long steel rod against a clogged taphole.

"Now that rod weighed something like 600 pounds and you had so many minutes to get it through the hole. It was an emergency situation because that steel is gonna skull [harden] if you don't get it out, and damage the furnace. Everybody's cursing and sweating hard. And the melter foremen—well, some of them—get excited and start yelling, 'Get it up! Get it up higher!'

"So I painted this. I had the first helper at the end holding the rod and determining its slope through the wickey hole. He's got the main weight of it so his torso is falling in a sort of S. Then I showed the group of men, each like a batter at the plate. The whole thing is, well, it's like a Greek athletic event except that everyone is dirty and in heavy underwear. And the furnaces—I had a lot of highlights coming from the furnaces. Reds from the wickey hole. Reds and sharp shadows across the men."

*

In the war years Ben Womer's body took on the contours that would last through his life: not an ounce of excess flesh could be spotted anywhere. His face, sculpted to a Modigliani oval, was burnished red—"sunburns," the men called the small hemorrhages that developed under the skin from exposure to the heat. Ben was 31 when the Japanese bombed Pearl Harbor.

"The bigger you was, the harder it was on you. You take a man that weighed 200 pounds and take a man that weighed 150 pounds. The first guy had 50 pounds more flesh he was carrying around that the heat could get to.

"To this day I don't know how we stood it. Because there was no ventilation. You just worked it. When you'd work in back of them furnaces in the summertime and especially if you had a bad taphole, the sweat would be squirting out between your shoe laces! And don't you think swinging that sledgehammer your legs get like rubber in just a very short while. We had troughs where the furnace-cooling water ran and it was nothing for you to come back and sit in that trough 'cause you couldn't get any wetter than you was.

"Now to speed up the times on the heats in the war, they'd blow the metal in the Bessemer and then bring it up and pour it in the open hearth. And you did, you made much better time, but you had a wild time. The metal be so hot that the runner you set into the furnace would burn through, and then you had an awful time getting the metal in.

"It was hard work, dangerous, but it was surprising the low accident rate we had. When you work around molten steel, that's something you got to have a lot of respect for. You were never careless. But things happened. This one guy went up, this was his first day mind you, and he went to shovel in his spar [fluorspar] when she was reversing and he got it. A flame hit him. And that damn fool ran out on the platform where the wind was and it ignited his clothes. But somebody grabbed a woolen blanket right away and we caught it. I don't think he got burnt bad, but that was the last we saw of him.

"I worked as second helper and in the wintertime, man it was bad. Because where you brought in the stock there was big openings and that wind come whistlin' in there off the harbor. Bring in the snow or rain. Many a time you'd be in back of the furnace and get all sweated up, and you'd come around front, that cold blast take your breath away. How we stood it I don't know. But we did. Ha. Because we was a fightin' a war!"[25]

Combat, that was it. America had been menaced and had to be defended. In Ben's mind he was fighting, lobbing bombs of 60- or 70-carbon steel at the Japs. He was out to redeem something, the disgrace he believed fell on the plant for having supplied the enemy with scrap and pig iron back in the '30s. "Yes, we did ship a lot of steel to Japan, Mitsubishi Company, and it was a foregone conclusion that someday we was gonna get it back. And, lo and behold, we did." He spits out his words in a gravelly, agitated voice. He was always suspicious of the Japanese. "We had a ballfield on B Street and them Jap sailors used to come over and play a game as their ship was getting loaded. They put on a wire mask and they had these bamboo sticks in clusters, and they'd holler and beat the hell out of each other. We used to watch them. They'd never bother us and we'd never bother them."

Steel was part of Ben's house. His dad and all four of his brothers worked in the mill, and the guys he knew best lived out their lives in the act of vulcanism. "My first helper was about as colorful a guy as you would ever want to meet. He was a very small guy, tiny, in fact." Ben first laid his eyes on tiny Taylor Springer at DeLucas Bar in Dundalk, where he had come in to throw back a couple of bolstering drinks after his shift. He called over to the bartender, "John, give me a fifth of that whiskey."

"Well," Ben says, a grin creeping up on his cheeks, "the bartender reached over—'fifth of whiskey'—and put it on the counter and this guy picks it up and starts to walk out. John says, 'Hey, you haven't paid me,' and this guy says, 'John, what did I tell ya? I said give me a fifth of whiskey. So long.' That," Ben chortles, "was the type of comical bird he was. But he was a hell of a steelman. He could tell if a furnace heat was right 50 feet away by lookin' at the flames puffing out of the wickey hole."

On his furnace, a 250-ton monster called No. 84, Taylor Springer taught Ben how to make the fast heats, and one day saved Ben from suspension for "burning the roof." Ben was practicing on the furnace (there was no formal training program) when he lost control of the heat. Misjudging a boil, Ben added too much ore to the brew, causing the hot metal to rise up to the furnace breast menacingly. Just then Springer, decked out in his trademark straw hat, strolled up to the control box. "You sure did do her today, Ben." He then signaled the crane man for some additional fluorspar and manipulated the fuel lines. That cooled down the boils. "He done straightened her out," Ben says, "and the melter foreman never said nothing."

*

The day crew handed it to the night crew and the night crew handed it back. At any moment about 4,000 tons of iron and scrap bubbled in the refractory caves of the 28 open hearths of Sparrows Point. At the end of every day, an average of 10,000 tons of liquid steel had been produced by the crews. From the furnaces the steel was handed to other mills and to more thousands of men. The pipe mill rolled the steel into gun barrels for the army; the rod mill made barbed wire and wire for field telephones. About 20 percent of Sparrows Point steel was routed directly to the loading docks, where cranes swung the ingots into the hatches of merchant ships. This was the most expedient way to ship the metal to Soviet Russia and Britain, where it was urgently needed for armaments. Soon the corporation's own ore-carrying boats were requisitioned by the War Shipping Administration and the orefields of Cuba and Chile shut in favor of domestic supplies from Mesabi.[26]

In meeting the single largest demand for steel—that for ship plate—the continuous hot-strip mill at the Point proved of critical importance. John Butler Tytus's invention was the secret weapon of U.S. steelmakers. Neither Germany nor Japan had access to these $20-million monsters that rolled out miles of highly uniform steel. Engineers quickly found a way to convert the mills from making light steels for tin cans and automobile bodies to making plate several times thicker for ships, airplane parts, and munitions. According to one estimate, the old hand mills could have produced only about half as much plate as the continuous mills turned out. Altogether the U.S. industry made nearly 50 million tons of plate in the five years of war; in the five years before, it had made 13 million tons.[27]

Among the many industrial users of the strong, comparatively light sheets turned out by Sparrows Point was the Glenn L. Martin aircraft plant at Middle River. The Martin plant was the second-largest defense contractor in Maryland and boasted the largest aircraft assembly line on earth—1,100,000 square feet of floor space. Among its wares were the PBM Flying Boat and B-26 bomber. Sheets cold-rolled for ductility were shipped to the plants of American and Continental Can, whose can machinery had been pressed into service for the production of shrapnel, aerial bombs, ammunition boxes, and 4½-inch rockets. Fabricated steel was consigned to the Seabees for temporary bridges and bomber hangars. Most directly related to Sparrows Point was the emergency shipbuilding program across the harbor at Fairfield. The 100th Liberty carrier was launched from the yard in March 1943. The S.S. *John Gallup* was named after an early American mariner. Remarkably the ship had been completed in just 35 days after the keel had been laid.[28]

A Liberty cargo ship consumed 4,500 tons of steel, or a third of its deadweight tonnage. Over 7,000 tons of steel went into each of the aircraft carriers built at the Quincy yards. Forty tankers under construction at the Sparrows Point yard were slated to consume about 200,000 tons of plate. Steel requirements for other items were as follows:

Bomb forgings	1.2 tons
Fighter plane	3.5 tons
Anti-aircraft gun	14 tons

4-engine bomber	15 tons
Medium tank	38 tons
16-inch battleship gun	576 tons
Portable bomber hangar	1,500 tons[29]

"This is a war of steel," said James V. Forrestal, speaking at Sparrows Point on November 6, 1942. The undersecretary of the navy had traveled from Washington to preside over the award of the coveted Army-Navy "E" for production efficiency. At 4th and C Streets a crowd estimated at 5,000 gathered to hear his words. Also present at the ceremony was Herbert O'Conor, governor of Maryland. Forrestal congratulated the workforce for exceeding its production goal for 1942. "Without the work you are doing here, our armed forces would be fighting under handicaps that might prove fatal in our will to win this war. No words and no speeches will win this war. Speeches will die before nightfall. What will not die is our determination to end a war we did not start and to end it for the good of all here and for our posterity." General manager Cort accepted the pennant from the undersecretary. It was raised on a flagpole in front of the general office.[30]

Ben Womer was part of the cheering throng. He flashed a V sign before padding his way to the clockhouse. In the locker room he saw other evidence of the government's push for tonnage. There was a poster of a U.S. soldier bent forward as his machine gun zeroed in on an enemy position. HE'S A FIGHTING FOOL, GIVE HIM THE BEST YOU'VE GOT. MORE PRODUCTION. On another wall was a workman holding his lunchpail and walking into the rays of the dawning sun. STEEL FOR VICTORY was the kicker.[31]

Ben didn't mind the hoopla in the least. It appealed to him to be a member of FDR's "arsenal of democracy." But it didn't distract him from the matter at hand. However well-intentioned, an undersecretary of the navy didn't have to confront what Ben did on a daily basis: that life was tenuous in the open hearth, especially when everything was pushed to its limits. In front of the furnace he shared with first helper Taylor Springer there was a slag-flushing hole. It drained off slag that spilled out in the boiling cycle. The crane operator was supposed to cover the hole with a plate after the operation, but sometimes he didn't. In any case the hole was hard to see and one day Taylor Springer stepped right into it.

The scrap chaser saw him disappear. He raced to the hole and looked down. "Nothing," says Ben. Horrified, he looked around for help. "I was comin' out from the back of the furnace and here the scrap chaser is fainting. He's going down on his hands and knees on the dolomite pile."

"What in the name of God's the matter with you?" Ben asked.

"Springer's fallen into the slag pot."

Ben had a shovel and chucked it against the furnace. There wasn't anything to do but get down to the basement, though at the temperature at which the slag was drained out of the furnace there wasn't much hope.

"So I starts runnin'. The steps are down at No. 85 furnace and I hits the head of the steps flying. Then I damn near fainted 'cause there's Springer coming up the steps."

"Hey, what's the rush, Ben?" he asked.

"Well, nobody never heard of that before. Falling into the pot and coming out alive! Springer never did say what happened, but me and the guys figured it out. He fell down that hole sure enough, but the slag had hardened just enough to make a crust. When he hit the slag, he was able to jump up and grab a ladder on the side. He crawled right out. That slag was hot. But it was solid enough for such a small guy."

Ben was convinced he had witnessed a miracle. And no one disagreed with him. In an occupation where tradition dictated that a man who fell into liquid steel be buried with the full pot to get rid of the curse, Taylor Springer became a legend. He had escaped incineration in a volcanic bath. It was a sure sign of grace to Ben, and 40 years later the rise of Taylor Springer from the slag pot was told not only by Ben, but by other open-hearth veterans, men with broad callused hands and pacemakers in their chests, who would gather at the union hall to share old times.

They also recounted the flip side of Springer's good luck—the incident that seemed to express their worst fears. Pop Weston was a melter foreman who was scheduled to retire during the war. Before his retirement, though, he was asked if he couldn't work a little longer because of the manpower shortage. Pop agreed, remembers Joe Hlatki.

"I was working on No. 72 furnace and Joe Straka was at No. 73. It was quittin' time. Pop Weston, he's out in the middle of the floor, fixing his pipe. He's stuffin' in the tobacco and cussing like always. He's standing there and a crane had a magnet on and they say that he two-blocked it—pulled it up too high—and broke the cable. The magnet, the block, everything fell and it happened that the melter foreman was right there."

In the confusion of sirens that followed, the stink of flesh got stronger and stronger. "The only thing we found was part of his arm. It flew over by the scales," says Hlatki. Splattered like a stepped-on bug, a man supposed to be enjoying his retirement lay crushed under the magnet, half-baked on the furnace floor.

Five blocks away from the mill screams were heard. It was Pop's wife. She lived in a company house with their son. "I've never ever heard such horrible sounds in my whole life," recalls Mrs. Elizabeth McShane, who was walking near the Weston home when the news was broken to Mrs. Weston. Elizabeth was also shaken up, since the accident had occurred in her husband's mill. Colegate McShane had seen a lot in his two decades in the open hearth, but this bothered him terribly. "He was a furnace man, he lived for his work, but when he got home that night he was deeply upset," Mrs. McShane remembers with a shudder.[32]

Reporting to work on the next shift was Mike Howard. He helped Hlatki and Straka clean up the mess. Later another employee found bits of Weston's teeth in the dolomite pile. What could be said? A settlement would be worked out for the family. There would be an investigation and almost certainly a suspension for the crane operator.

Mike had taken up painting as a diversion from his job; now his job intruded upon his hobby. After carrying the scene around in his head for a couple days Mike took a 14-by-22-inch sheet of paper and charcoal sketched his feelings about Pop Weston. "I tried to give the picture the feel of the Guernica painting. Here was a disembodied crushed figure and in the background I had a woman and child grieving. I remember I drew the magnet that crushed Pop Weston like it was really an evil force."

Did the magnet represent the company?

"No," he answered, "I didn't think of it in terms of the company. The whole thing to me was the factory. I was trying to express the feeling that it takes human effort, it takes blood, to make steel."

*

On December 8, 1944, three years and one day after Pearl Harbor, Boeing B-24 Liberators strafed Iwo Jima with 500-pound bombs. The shelling continued for 10 weeks. Between February 16 and 19 the U.S. Navy brought in the battleships *Washington* and *North Carolina.* Their 16-inch guns pounded the island stronghold with 8,000 tons of steel and explosives.

At 0902 hours February 19 the first wave of amphibious LVT's set down their tracks on the beaches. Sixty thousand men from the 4th and 5th Marine Divisions clambered ashore. They were met by strong resistance under General Kuribayashi, the island commander. After violent combat and heavy casualties, 546-foot Mount Surabachi was taken on February 23. AP photographer Joe Rosenthal's shot of four marines raising the American flag on the mountaintop became one of the best-known photographs of the war.[33]

Iwo Jima proved that with sufficient steel even the strongest defensive fortress could be overrun. With a supply line reaching back to the States, the island was secured in the next month, and a landing site for bombers installed only 660 miles from Tokyo. The United States was closing in. On May 7, 1945, Germany surrendered unconditionally. On May 25, 1945, Sparrows Point received for the fourth time the Army-Navy "E" Award for outstanding war production. Three stars now were affixed to the pennant flown over the plant.[34]

The nation's top military contractor was beginning to wind down. Conquest of Japan was only a matter of time. Bethlehem had produced 73,421,622 net tons of steel between 1940 and 1945, approximately one-third of the armor plate and gun forgings required by the U.S. Navy. In the same six-year period the ship division delivered 1,085 vessels, the largest and most diverse shipbuilding program ever attempted by a private corporation. In addition it converted, repaired, or serviced 37,778 vessels ranging from the *Queen Mary* to navy destroyers damaged by kamikaze suicide hits. In 1935, the company's gross sales had been $195 million; in 1945 they stood at $1.33 billion. The company's success mirrored that of the industry as a whole. Between 1940 and 1945 steel revenues amounted to over $32 billion, more than the business had earned during the preceding 15 years. In the final year of the war, as German and Japanese mills lay destroyed, the United States produced nearly two out of every three tons of global steel.[35]

CHAPTER 17

BRAVE NEW WORLD

Demobilization. As Washington steeply curtailed military orders in 1945, Bethlehem management was faced with the classic problem of how to respond to reduced demand. There was, as always, a number of options available under the broad framework of the "three Ms" of manufacturing—machines, men, and markets.

One possibility would have been to reduce the capacity or number of existing facilities. Among the advantages of planned plant shrinkage would be reduction of the industry's notoriously high overhead costs as inefficient practices or obsolete equipment were identified. At Sparrows Point, an area of improvement would have been a reduction in the characteristic lag time between mill processes—for example, the week-or-more delay that normally occurred between the rolling of strip in the continuous mill and the beginning of tin plating.

Another course lay in using the company's immense resources to introduce new or improved steel products onto the market. There was even a niche available for such commercial development—the National Emergency (NE) steels. These had been developed to save on the scarce supplies of imported alloys like chromium, tungsten, and vanadium. Small amounts of two or three alloys were used to replace larger quantities of one, and boron was sometimes added as an intensifier to make the alloy more effective. The new products met the needs for stronger and tougher steels, with improved corrosion resistance. The next step in metallurgical development was super-alloy and stainless steels that could meet the requirements of the Atomic Age which had suddenly exploded at Hiroshima. Wartime research also promised commercial applications for by-product coke chemicals in plastics, for waterproofing compounds, aviation gasoline, and drugs.[1]

But except in a limited way, Eugene Grace chose to ignore all high-tech alternatives, sticking to standard carbon-steel products, and approached the task of demobilizing his corporation by simply downsizing his labor costs. Shortly after Hiroshima he announced that, owing to the reduction of orders, all overtime and premium pay was to be canceled. The men at Sparrows Point, who had been averaging $56.32 a week in April 1945, based on a 47½-hour week, found their paychecks shriveling to $43.38 with the cutback to 40 hours.

This was scarcely the bright future that Uncle Sam's posters had promised. The announcement rubbed a raw nerve. "Hey, are we supposed to go back to the Depression now that we beat Hitler?" asked Mike Howard caustically.[2]

On September 11, 1945, Phil Murray, speaking through the USWA Wage Policy Committee, demanded substantial restoration of war wages. He said a $2-a-day wage hike was required because of the upward revision of the cost of living during the war. In 1945 the Consumer Price Index was 22 percent above 1941 levels, and the food-price component of the index had gone up 32 percent. (Because of low base wages, steelworkers had labored for less during the war than aircraft workers, shipbuilders, and men in the converted automobile plants.) According to Murray, steelworkers were caught in the worst price squeeze since the Depression.[3]

The dollars and cents of the situation were spelled out in a report prepared by the union with the help of the University of California's Heller Committee for Research in Social Economics. The average family in steel country lived under a regimen of enforced austerity. According to the Heller Committee, earnings from the mill alone did not provide a family of four with enough to maintain a minimally "decent" standard of living. Average weekly wages of $56 were $3 short of the committee's recommended weekly minimum of $59.15. However, with an average of $10 a week coming from other sources, such as wages from a second worker in the family or from a boarder, the family was able to exceed the minimal standard by about $7 a week and even to save a little. The typical family had stashed away about $600 during the war—$300 in cash and $300 in war bonds. With the resumption of the 40-hour week, though, it was only a matter of time before these savings would be exhausted and a family would be forced to lower its standard of living.[4]

Impossible, Eugene Grace retorted to the proposed wage scale. A different set of figures sat before him, not "social economics," but rows of cost-profit figures. His notations were emphatic: In 1916 his company had made a quarter more money after taxes than in 1944 when it made six times more steel. Profits had not kept pace with recent billings, and he blamed the government's price freeze and overtime for the shortfall. Accordingly, when his vice president of industrial relations, J. M. Larkin, met with the Steelworkers' negotiating delegation in October, he read a carefully crafted statement from headquarters which concluded that he did not have authorization to proceed with talks based on the union's proposed wage scale. Negotiations were severed.[5] The statement signaled a return to the days of confrontation when the company had refused to bargain in good faith. Little Steel was flexing its muscle. But Phil Murray was a step ahead of the graying fox of Bethlehem.

*

When Grace pleaded poverty as his reason for bypassing union calls for bargaining, Murray's office released a report of the company's finances. It called attention to the fact that in the third quarter of 1945, Bethlehem had written off $53 million in amortization of war-related facilities. This would entitle the company to a $24-million tax windfall from the federal government. In addition, in a period when workers had their hourly wages frozen, Grace and his boys had paid themselves large bonuses as management incentives. Between 1939 and 1942 bonus pay for Grace and the top 14 Bethlehem officers shot up from $319,301 to $1,192,407 a year.

While not so spectacular as in 1929, when he got $12,000 in salary and $1.6 million in bonuses, Grace's pay in 1941–42 had gone up by a factor of three from his average Depression pay. In both years he pocketed $537,724 in annual compensation. This made him the second-best-paid executive in the United States, according to a wartime survey by the Securities and Exchange Commission. (He was outranked only by Hollywood's Louis B. Mayer, who made $697,048 in 1941.) Grace had received compensation 21 times above the $25,000 that President Roosevelt told Congress in April 1942 ought to represent the upper limit of individual private income during the military crisis. "Nothing has ever been invented, in wartime or peacetime, that would make men work as hard as money," Grace solemnly told *Fortune* magazine.[6]

Although Bethlehem's president took a voluntary pay cut in April 1943, reducing his intake to $237,629 a year, the wealth going to management was a source of irritation to the average mill worker. To Joe Hlatki, who had risked his life in the open hearth for a grand sum of $3,701.91 in 1943, the Grace bonuses were galling. "Why do the big men always need so much," he wondered.[7]

Bethlehem was not the only steelmaker whose headstrong leadership seemed compelled to act out the clichés often ascribed to the U.S. profit system. A managerial shift at U.S. Steel had brought to the fore a brass-tacks operating man. He was Benjamin Fairless. He had assumed the boardroom reins of U.S. Steel following the departure of Edward Stettinius to Washington to head Lend-Lease. Born at Pigeon Run, Ohio, in 1890, Fairless had been a protégé of Boot Straps Girdler's, serving as executive vice president of Republic Steel under Girdler between 1930 and 1935. He was later tapped to run the Carnegie-Illinois subsidiary of U.S. Steel. Technically, he shared power at U.S. Steel with two other members of the executive committee, bankers Enders M. Voorhees and Irving S. Olds, but owing to their lack of forceful personalities, the bankers were shunted aside in matters of operations and labor policies.[8]

Fundamental to Fairless's get-tough labor policy was to bring about a rapprochement with the Little Steel faction. In November 1945 Fairless and Grace reached an agreement to present a united front in the pending labor negotiations. U.S. Steel would do the negotiating with Murray, setting the wage pattern, and Little Steel would follow its lead with duplicate pacts. In this way none of the principals would gain an advantage over the other in wage costs.[9]

The sequence of events moved rapidly as management and labor began jockeying for advantage. On November 13, Fairless announced that he would not discuss wage increases with Phil Murray unless the Truman administration approved the industry's application for a $7-a-ton increase in steel prices. In May 1945 the Office of Price Administration had granted increases of $2 to $7 a ton for 14 basic steel products, the first price revision since 1941. But the industry had been sorely irked that sheet and strip products were excluded. Since then it had been battling for a general upward revision of the price schedules.[10]

On November 28 steelworkers at Sparrows Point and other unionized plants voted 5 to 1 in favor of a strike to reinforce their wage demands. The union's strategy committee called a strike for January 14, 1946. Acknowledging the industry's united front, the union called for a nationwide shutdown rather than a selective strike at U.S. Steel. "If there's a strike, it'll be their doing," Grace responded in a radio interview on December 4. When Christmas passed without negotiations scheduled, President Truman named a fact-finding board to sift through the evidence and determine what the industry could afford to pay on the basis of current prices. Meanwhile, a strike by the United Auto Workers over pay restoration had closed General Motors.[11]

On January 8, Truman blinked—or partially blinked. He announced the government would permit a $4-a-ton price increase if the industry would settle the wage issue with its employees. The first negotiations between Fairless and Murray since November 13 were resumed on January 10, but broke down a day later. Answering a presidential summons on the 12th, Fairless and Murray conferred at the White House. The president won a week's postponement of a strike from Murray. Truman then asked the disputants to work out a compromise, and set Thursday, January 16, as a reasonable deadline. On that day Fairless and Murray conferred in the White House again, but reported to the president that they had made no progress. Truman warned them that if they could not adjust their differences he would propose his terms for a settlement.[12]

Friday, January 17. Major shifts were undertaken by the parties. The union scaled down its demand from 25 cents to 19½ cents a hour, the amount suggested by a government fact-finding board to settle the GM strike. Fairless raised his company's offer of 12 cents an hour to 15. At 4½ cents apart, the sides deadlocked again. Truman made his proposal, later revealed to be 18½ cents. This would raise average pay for a 40-hour week back to $50 and would cost the industry $185 million a year, according to the president's figures. He said the government's offer of price relief amounted to $224 million.[13]

Saturday, January 18. Murray accepted the president's proposal. Fairless rejected it. Murray declared to the press that Fairless was willing to accept the president's proposal, but backed away under pressure from Grace and other steel heads. A statement by Fairless said Murray's assertion was untrue. Truman released a last-minute plea to the industry asking for reconsideration of his proposal.[14]

Sunday, January 19. Henry J. Kaiser, maverick owner of the small Fontana steel plant outside Los Angeles, accepted the president's proposal and signed a contract with Murray at the White House. He unleashed a stinging attack on steel management (i.e., on Fairless and Grace), saying they weren't "smart enough" to negotiate a labor contract. According to Kaiser, the productivity gains that would accrue from settling on the president's terms would more than offset the losses from higher wages. "I cannot conceive that a sum of 3½ cents should be permitted to retard or destroy the possibility of real peace and prosperity for the nation," he said. Murray characterized the industry's rejection of Truman's proposal as "rebellion if not actual revolution" against the government. Steelmakers "have deliberately set out to destroy labor unions, to provoke strikes and economic chaos and hijack the American people through uncontrolled profits and inflation," the unionist proclaimed.[15]

*

From its onset at 12:01 A.M., Monday, January 20, the walkout was a dramatic show of unity. All told, 550,000 workmen stayed away from the mills, bottling up 95 percent of the nation's total production. Many metal-fabricating shops were shut as well, as the strike spilled over to USWA workers organized at Continental and National Can and at major railroad locomotive shops.[16]

At Sparrows Point the strike was an unqualified success. Picketers appeared at the plant gates at the stroke of midnight, carrying placards and croaking out-of-tune union anthems. They built bonfires the first night as temperatures dropped into the teens and snow fell on the peninsula. Under a prearranged plan between the USWA and general manager Cort, the mill had undergone an orderly shutdown. On Sunday union members had banked the furnace fires and removed molten metal to prevent damage to the equipment. One battery of coke ovens was kept on fire to produce gas needed for general maintenance, and a corps of repair workers, plus supervisors, was allowed to pass through the picket lines unchallenged. In exchange for the union's cooperation, Cort agreed not to attempt to make steel or ship steel products for the duration of the national strike.[17]

Elizabeth McShane worried from Bosses' Row at the same time Mike Howard crossed his fingers at No. 1 clockhouse. On the first day of the walkout the tension was palpable, but there were no untoward incidents. Howard could happily report back to the union that an orderly picket formation was maintained at No. 1 clockhouse, one of six picketing sites on the peninsula, and Mrs. McShane could relax in the knowledge that her husband was safe from harm. Like most of the superintendents, Colegate McShane was inside his office at No. 2 open hearth, passing the days in idleness. Though less than luxurious, it was better than the picket line outside when the thermometer dropped down to 10° F.[18]

In the meantime public opinion was running against the producers. Ben Fairless's rebuff of the president's compromise raised the cry of "business lockout" in Congress. Even the staunchly anti-Democrat

Daily News faulted the industry for its arrogance. Even U.S. Steel, it said, "can't buck the President of the United States." *Life* magazine declared in an essay headlined **MR. FAIRLESS IS WRONG** that steel should pay the wage increase and get on with the business of supplying steel to the Allies. With civilian populations in Europe and Japan starving "for want of farm implements and steel railway equipment," the magazine said that the strike was causing havoc and hurting America's vital foreign interests. The only public comment offered by Eugene Grace did not seem inspired to win friends. Laughing aside reports that he had been "pulling the strings" in the negotiations, he said, "It's a pity that our workers, who have been making good money for five years, should have to use it up this way."[19]

Whether it was a strike or lockout, the industry was taking a beating. With no sign of weakening resolve on the picket line (in Pittsburgh strikers hoisted signs vowing to stay out "till Fairless freezes over"), the hapless Enders Voorhees was given the task of presenting steel's amended position in Washington. In early February Voorhees put forward a proposal that the 18½-cent settlement might not be so unacceptable if steel prices were allowed to increase "in excess of" $6.25 a ton. At the White House and more emphatically at Chester Bowles's Office of Price Administration, there was little inclination to give the steel masters such a lucrative way out. But as the strike dragged on in February, political resolve lessened and a solution was worked out among government officials, Fairless, and Murray. On February 15, Federal Reconversion Administrator John W. Snyder and Lewis Schwellenbach, Frances Perkins's replacement in the Labor Department (she had tendered her resignation after Roosevelt's death in 1945), announced the settlement of the strike. The terms were that steelworker wages would increase by the full 18½ cents recommended, and steel prices would be boosted by $5 a ton, or roughly 8 percent, effective February 18.[20]

Truman had ended the deadlock, but only at the cost of a general price boost that undercut his own price stabilization program. The direct matchup of wages and prices was worrisome, in that it might diminish the incentive of management as well as labor to improve productivity. But such misgivings were cast aside in the euphoria that followed the settlement of the crippling walkout. Per their agreement with Fairless, Bethlehem and the other operators of Little Steel signed identical pacts with the USWA in the next two days.[21]

The steel contract set the pattern for other settlements. The bitter

General Motors strike ended on March 13, 1946, on the basis of the Steelworkers' 18½-cents-an-hour wage boost. Unionized employees in aluminum, canning, and locomotive shops won the same hourly increase. In the wage-earning precincts of Sparrows Point, the settlement was cause for jubilation. By withdrawing their labor the workforce had won the single largest upward revision of wage scales in plant history.

*

During the winter of labor discontent, a second dynamic was at work. Years of war rationing and forced savings had created an enormous reservoir of consumer demand at the same time that victory on the battlefields had made America the undisputed leader of the Free World. The higher wages of unionized workers only increased pent-up purchasing power. In June 1946, *Fortune* made it official—America was experiencing a boom. "There is no measuring it; the old yardsticks won't do," the magazine said in a special report. "There is a powerful, a consuming demand for everything that one can eat, wear, enjoy, burn, read, patch, dye, repair, paint, drink, see, ride, taste, smell, and rest in."[22]

The steel trade found itself in the middle of the boom. At no other time—except wartime—had steel been in such heavy demand. From consumer goods to capital construction, steel was the building block of the land. Eighty-five percent of all manufactured goods contained steel in one form or another (compared to under 20 percent for aluminum and plastics); furthermore, the amount of steel used per unit of output in many industries had grown. Increased car weight and the universal acceptance of Budd's "all-steel" body had advanced the amount of ferrous metal in automobiles to an all-time high of 85 percent of total weight. This compared to under 60 percent in 1925 when automobiles were lighter and many car makers used wood for body frames.[23]

Detroit stepped up its orders for flat-rolled steel by 55 percent in 1946 as it projected a record 7-million-unit year. The main impediment to car manufacture was getting enough steel. Shipments of structural shapes, piling, and reinforcing rods for construction also raced forward, increasing to 5.7 million tons in 1946 from 4.4 million in 1943. Bridges got longer and buildings got higher, heightening demand. Bethlehem soon won contracts to roll 22,000 tons of building beams for an industrial plant in Iowa as well as to export 11,350 tons of plate girders for 154 railroad bridges to be constructed in China.[24]

Tin-plate sales zoomed as everything from soup to nuts to beer was packaged in cans. During the war the humble tin can had been drafted into service as a container for a thousand and one uses. From 1942 to 1946, for example, nearly 2 billion beer cans were shipped overseas to American soldiers. The introduction of electrolytic tinning (another war priority financed by the federal government) had reduced the amount of scarce tin needed for can coating while improving food preservation. This permitted canners to offer the public such delicacies as chicken stew with dumplings and beef with gravy in canned form. New "dry" foods were developed; one of the most successful was canned dog and cat food. By the end of 1945 the civilian population was opening, discarding, and replacing 45 million tin cans a day—a total of about 17 billion cans a year. "Verily," wrote Charles H. Hession, "the can-opener has become a *sine qua non* of modern urban living and the tin can, as an advertising writer might put it, as much an emblem of our country as the American eagle."[25]

The only major steel-consuming sector that lagged behind was ship construction. A victim of wartime overbuilding, it was becalmed in surplus ships. There was talk that the shipyard at Sparrows Point might temporarily close or convert to navy repair contracts. Then the tide turned. Aristotle Socrates Onassis wanted six ocean tankers from Sparrows Point for his international shipping empire. Orders for tankers of record size followed from Standard Oil and Gulf—28,000 tons and 615 feet from stem to stern. United Fruit called on Sparrows Point for nine boats for its Central American trade, beginning with the refrigerated cargo-passenger ship S.S. *Yaque*. Once again the shipways were alive with the sounds of riveting and incandescent with welders' flashes.[26]

The "economic miracle" of 1946 transformed the political debate of the preceding years. Under congressional pressure the Truman administration was forced to abandon government price controls, closing out the Office of Price Administration and formally lifting the control of steel prices in November 1946. Fear of a '30s-style bust was buried beneath the gushing materialism, the parade of Fords and Chevys and the pursuit of nylon stockings. So were the prescriptions for a planned economy that only a year before had been popular in Washington.

Phil Murray, as head of the CIO executive council as well as Steelworkers president, had advanced the notion of a planned full-employment economy. It was known as the "Murray Plan" or "ever-normal granary." Just as the federal government purchased surplus crops to stabilize agricultural prices, the union proposed an ever-normal supply of steel. Output would be calculated in advance and if the tonnage failed to sell on the market according to projections, the government would underwrite the losses as part of coordinating steel operations with overall economic policy. The aim was to avoid the destructive peaks and valleys of demand.[27]

Even in the best of times, consuming industries bunched up their orders on a seasonal basis, railroads in the fall and construction and cars in the spring. Murray and his allies argued that if such demand were flattened out over the year, steel management would be the beneficiaries because they would not have to maintain expensive capacity that was used only in peak periods. By the same token, with leveled-out production, labor would cease to be the safety valve of convenience, overused in periods of high demand and furloughed when orders dropped off. Murray penned his ideas in pamphlets and magazine articles. He talked about them on Capitol Hill. He put together a book mostly written by Morris L. Cooke, an efficiency engineer, called *Organized Labor and Production: Next Steps in Industrial Democracy.* It called on industry to enlist labor's support to root out wastefulness and achieve higher productivity. While Murray disagreed with Cooke on some matters (they concluded their book with a fascinating give-and-take called "Dialogue between Authors"), he was passionately interested in the business of steelmaking and how it could be related to a progressive labor movement.[28]

Other officers of the USWA were thinking hard about the future. The popular notion that unions did not produce intellectuals—that they thought only in terms of their narrow constituency and of dollars-and-cents wages—was refuted by the ferment of the 1940s. There was a wide range of economic and social ideas expressed on the pages of *Steel Labor.* Clint Golden, formerly of the Brookwood Labor College, and Harold J. Ruttenberg were respected intellectuals as well as seasoned union activists. With a literary flair, they had laid out in *The Dynamics of Industrial Democracy* a blueprint for a more humane and efficient workplace. Their calls for worker participation and partnership have since returned to high vogue as advocated by management gurus such as Thomas Peters *(In Search of Excellence)* and polished up with the "Japanese-did-it" argument. The USWA's research department also ran a consulting program for steel companies under the idea that there were ways to save on postwar manufacturing without cutting into wages or employment. The union had on its payroll time-study men to analyze factory production problems. The program was used by many small companies and was credited with success by Henry Kaiser, who enlisted the union's help to improve productivity at his Fontana works.[29]

As a whole, though, the industry was not impressed with Murray's call for new approaches to labor and production practices. It was far more willing to bear the cross of excess plant capacity than to submit to a state-regulated granary or the affliction of an interfering union. In testimony before Congress, U.S. Steel's Enders Voorhees voiced objections that ranged from the impracticability of regulating customer demands to worry that stockholders might be asked to guarantee labor wages during a slump. In a telling aside, he said, "At the best, these demands, if granted, would result in a kind of steelworkers' WPA." The fact that some manufacturing companies, such as Tom Watson's International Business Machines, had adopted a no-layoff policy to secure employee loyalty made no impression on mainstream steel managers. From their point of view, it was labor's fate to rise and fall on the business cycle.[30]

Voorhees's remarks were a not-so-subtle repudiation of the conciliatory path taken by Myron Taylor and Edward Stettinius before the war. Now with private capitalism returning to claim its place in American society and with the Cold War making a state-steered economy politically unpopular, steel officers were more dogmatic in asserting their feelings about government, unions, and the linkages between economic and political life.

Rising above his peers to expound his agenda for the future was Eugene Grace. "A one-man Praetorian Guard of the U.S. profit system" was how one writer described the razor-thin, hawk-eyed mogul, now 71 years old. Appropriately, Grace chose to present his case as speaker at the first annual Charles M. Schwab Memorial Lecture of the Steel Institute. The date was May 21, 1947. In the audience at the Hotel Pierre was a consortium of familiar faces: Tom Girdler, Ernest T. Weir, L. E. Block, and Frank Purnell of Little Steel, and Ben Fairless and Enders Voorhees of toughened-up Big Steel.[31]

"As we gather here today to do him honor," Grace told them, "we ask ourselves what would be Mr. Schwab's appraisal of the current economic outlook for our country. While he was a natural optimist, nevertheless I believe he would agree with us that his optimism could only be justified by the correction of the restrictions that have been leveled at management and business over the last decade." Grace then proceeded to enunciate to his colleagues the standards for which his company stood. Chief among them were a "revision of the country's labor laws," a reasonable tax program, and "elimination of other uncertainties." But above all Bethlehem stood for the principle of big rewards for officers. Bonuses were essential for encouraging the ultimate in management achievement. In reference to the controversy swirling over his own compensation, Grace noted evenly, "You have cycles of restraints and criticisms; and every age faces a change, for that is the law of life; but the fundamental principles continue to apply."[32]

Grace had made certain bureaucratic adjustments in accordance with his strategy. He had taken over the board chairmanship and installed Arthur B. Homer, vice president of shipbuilding, as president. Homer had spent his whole career at Bethlehem. Joining the ship division in 1919 at Quincy, he had become manager of diesel engineering and sales in 1921. He was named manager of sales in 1931 and assistant vice president of shipbuilding in 1934. His big break came in 1939. Informed of navy plans for a two-ocean fleet by Secretary Charles Edison, he got Bethlehem in shape for the influx of orders even before President Roosevelt signed the enabling legislation in 1940. He oversaw the Liberty shipbuilding program, where he won a reputation as a numbers cruncher. Except for an interest in sailing ships, he was short on color; he had little sense of humor, and was circumspect beyond his age. "He sticks to the business at hand with a singlemindedness that overlooks no pertinent detail," was one appraisal. "People who do not like him call him ruthless. Those who like him call him determined."[33]

In the other important executive slot, that of vice president of steel operations, Grace promoted a nuts-and-bolts technician. He was Stewart Cort, Sparrows Point's general manager, replacing Quincy Bent, who retired in 1947. Cort, 66, had worked 30 years for Bethlehem; he was typical of the unimaginative production types bred under the company's internal management program. He kept a low public profile, reviewed the tonnage figures reverently, and kept costs down. The *Sun* described him as robust and well-respected in engineering circles. His former position at the mill was filled by his deputy C. E. Clarke, while L. F. Coffin, plant engineer, was moved up a notch to assistant general manager.[34]

With his number-2 and number-3 men selected, Grace confided to the audience at the Pierre that he planned to take some time off to finish a project that was of longtime interest. He wanted to establish at corporate headquarters a memorial library to house the keepsakes and speeches of the late chairman. "If Art Homer will look after his new job of being president, I intend to work at this memorial personally and see that it comes to pass." In the past 43 years there had been only two persons in the chairman's seat at Bethlehem, and Grace conscientiously sought to keep the cult of the founding patriarch alive. Pledging his fealty to the old ways, Grace said soberly, "Mr. Schwab passed away on September 18, 1939, but his influence remains with us as a guide and inspiration; an inspiration for the tasks of the future, and there will be plenty of them."[35]

*

When Sparrows Point rolled out its first steel rail in August 1891, its rated capacity was 350,000 tons a year. That was more than the output of Sweden and Italy combined. When Schwab spoke of making the mill the world's biggest in 1916, he was thinking in terms of about 4 million tons a year. This was based on his experience at Carnegie Steel and from the Gary works of U.S. Steel, which approached this level of giantism during the First World War. Ironically, Schwab was on the verge of his goal when the Depression clobbered his capacity-raising exertions in 1931. The plant did not recover for a decade, and then it was only a world crisis that put into operation the final 538,000 tons of capacity that Schwab had first envisioned.[36]

In 1946 the plant was the possessor of 3.9 million tons of annual ingot output, a mammoth agglomeration of tonnage by any standard of measurement. As flagship of the Bethlehem chain, it outstripped the capacity of the whole National Steel organization. In other words, this *one* plant of the nation's second-largest producer could make more steel than *all* of the Middle West operations of the nation's fifth-largest producer. Sparrows Point was exceeded only by the two biggest mills of the world's biggest producer, the 6-million-ton Gary works and 5-million-ton South Chicago works of U.S. Steel.[37]

For Bethlehem, as for many extractive and transportation companies, new applications of power and material handling made an expanded scale of production feasible. A case in point was the transformation of railroads to the diesel age. The cost savings of diesel engines, introduced en masse during the war, were considerable. A lash-up of four of them could pull 100 or more coal cars over most

Appalachian grades; it had taken two or three steam locomotives to pull a coal train of half the size. The diesel engine became to railroads what the jet engine was to become to aviation—a new phase of mechanical development. Equally important to management's thinking was that three or four diesels coupled together were a "unit" that required a single crew as opposed to crews for each steam locomotive in use, a factor that led to many railroad junctions turning into virtual ghost towns as employment vanished in the '50s.[38]

In ocean shipping, too, the concept of bulk transport changed. The oil tankers under construction at Sparrows Point bespoke the possibilities of a new generation of vessels. The tankers on order for Gulf, Standard Oil, and Aristotle Onassis involved record deadweight tonnages. The largest of these vessels were 195 feet longer than the Liberty cargo ships of 1943. They nearly tripled the deadweight tonnage—28,000 versus 10,000 tons. This greater size led to lower unit costs, an investment point that Bethlehem convincingly demonstrated in its ore hauls from South America. The eight vessels it completed in 1946 and 1947 for its Ore Steamship Corporation subsidiary carried 24,100 tons, nearly 5,000 more than '20-era ships. What's more, the carriers were faster, cruising at 17 to 18 knots rather than 10 to 11. This clipped off full days in travel time. In its inaugural round trip between Sparrows Point and Cruz Grande, Chile, for example, the S.S. *Venore* was clocked at 28 days compared to the 36 to 42 days it took the older ships.[39]

Increasing size to increase unit earnings—a principle as old as integrated steelmaking—held enormous attraction to a steelmaker that had reclaimed its bulk during the war. In-plant growth seemed the best strategy to management. Construction of a fully integrated plant would cost about $225 per ton of steel produced, or around $400 million for a plant of 1.8-million-ton capacity. On average, in-plant expansion went for only $112 a ton. Post-war expansion at Sparrows Point, though, was contingent on a number of factors, the most important of which was assuring a supply of fresh water. Like all major steel mills, the Point had a raging thirst. An average of 157 tons of water was needed for the production of every ton of finished steel. Water circulated constantly through the plant for use in cooling furnace boshes, washing blast furnace and coke oven gases, and generating steam. Since Frederick Wood's time the plant had gotten its fill from the harbor and freshwater artesian wells—200 million gallons a day of the former. The salt content and mineral impurities of the harbor water, however, had been causing problems for the rolling equipment used in the hot-strip, cold-reducing, and continuous-rod mills. The salt corroded the rolls, forcing frequent replacement, and damaged the finished strip and tin plate with streaking and scar marks.[40]

When the hot-strip was started up in 1938, the company relied on fresh water in the underground streams first tapped by Frederick Wood. Constant "mining" of the underground streams, though, had dropped the water table, making it more difficult to get the water. The draining of the reserves also caused contamination of the remaining water by salt and acids, a problem that spread during the war across 60 square miles below the harbor. The Army Corps of Engineers issued emergency instructions to local industries to limit the intake from artesian wells.[41]

The deficit of water could be made up by tapping into the freshwater supplies in Baltimore County, but the price would be stiff. So the company hired Dr. Abel Wolman as its water consultant in 1940 to explore possible alternatives. A member of a prominent Baltimore family, Wolman was professor and chairman of the department of sanitary engineering at Johns Hopkins. Over the prior 20 years he had held a number of important state jobs, including chief engineer of the Maryland Health Department and chairman of the state planning commission. Widely acknowledged as the "dean" of Maryland engineers due to his involvement in publicly-financed sewage and road projects, Dr. Wolman did not limit himself to technical matters; in an interview, he boasted of his role as the unofficial "rabbi" for Albert Ritchie and other Maryland governors.[42]

In his initial report to the Bethlehem Steel Corporation, labeled "confidential," Wolman confirmed that use of city reservoir water would be an expensive proposition. He estimated it would cost about $5 million in capital outlays for pipeline and filtering facilities and another $500,000 in annual water and operating charges. But there was another possibility: water from the municipal sewer plant located several miles from the mill at a whimsically named offshoot of the bay, Bread and Cheese Creek. Because Wolman served as Baltimore's sewer consultant, he was in an excellent position to evaluate his own proposal. This he proceeded to do with the blessings of George E. Finck, city sewer engineer, and George L. Hall, his former health department deputy and now chief engineer of Maryland.

His thinking was that if the sewage water could be made usable through re-treatment, the savings to the state's largest employer would be considerable and the supply long-lasting (since it seemed highly doubtful that sewage water, unlike fresh water, would ever drop in quantity). He assigned a promising graduate student, John Geyer, to run a battery of tests, then concluded that the sewage, properly flocculated, treated with alum, given a good airing, and doused with more chemicals, would be machine-compatible.[43]

Consequently, Wolman drew up a contract between Bethlehem and the Department of Public Works, assisted by Charles Evans, the city solicitor. The municipality agreed to supply Sparrows Point with a maximum of 50 gallons of effluent a day, guaranteeing certain standards of treatment, and the company agreed to pay a small sum for the liquid. ("I told Mr. Cort, you pay the city a little something and then a city councilman or a newspaper reporter can't come back at you and say, 'Look, the city's giving something away for free,'" Wolman said in a 1980 interview.)[44]

Approved by Mayor Howard Jackson and the Board of Estimates, the agreement worked as follows: If Sparrows Point used an average of 25 million gallons of retreated sewage a day or less, Bethlehem Steel would pay a fee of $1,000 a month. If intake exceeded 25 million gallons a day, the fee would go up to $1,500 a month, and for over 37½ million gallons, $2,000. Counting the cost of building the pipeline and related facilities, Bethlehem got a virtually inexhaustible source of water for 1.73 cents per 1,000 gallons. This was a savings over the former cost of obtaining artesian water, which averaged 3.142 cents per 1,000 gallons. Stewart Cort and his superiors in Bethlehem expressed delight.[45]

For slaking the mill's thirst Wolman didn't do badly either. His contract with the steelmaker specified a consulting fee of $250 a day for the first 25 days of service and $200 a day thereafter. In his first 13 months of part-time work he was paid a whopping $19,450, or more than $130,000 in present coinage. In fact, his first-year fee amounted to more than the city realized by providing the sewage and treatment facilities in the first place. Queried about this arrangement, the professor said the fees were based on a fee schedule for consultants of the American Society for Civil Engineers. A review of ASCE records, however, indicated that no such fee schedule was in existence in 1940, the year when Wolman first entered into his contract with Bethlehem. Instead, the society had published a survey of consulting fees and other "ethical guidelines" which were described as recommended for its members. An expert of Wolman's long experience and top caliber would receive a per-diem rate of $20 to $50 a day, according to the survey. Questioned further about his consultancy, Wolman expressed resentment that his fees might be construed as excessive and abruptly terminated the interviews.[46]

A number of cooperative officials of state and city government also benefited from the engineering-cum-political decisionmaking of the project. Among these were Ezra B. Whitman, chairman of the Maryland State Roads Commission, whose engineering firm was appointed general engineer of the project by Bethlehem, and Baltimore Solicitor Charles Evans. As the city's chief legal counsel, Evans had negotiated the original sewage contract with Wolman. Several years later, entering private practice at a prominent Baltimore law firm, Evans was named Bethlehem's counsel. In this capacity he represented the company in its renegotiation of the sewer contract in 1947, which retained the same extraordinarily low fee structure. Evans stayed on as company counsel in governmental matters through the '50s.[47]

The only objection to the sewage project came from the local union. Union officers questioned whether the effluent might not cause health-related problems. Employees found that the liquid was emitting an obnoxious odor because of its high urine content. "The stench was something awful," recalls Bill Forbes. "Some days, like in summer, the whole damn sheet and tin-plate mill smelled like an open sewer." The liquid ran down beside the rolling equipment separated from the working area by iron floor grates. Sometimes the hot, agitated water foamed up on the floor. Sometimes the spray guides clogged up with condoms or fecal matter, requiring a repair gang to root around in the mess.[48]

The men denounced Dr. Wolman's brainstorm as "shitwater," but their hands were tied. Union officers were informed none too delicately that their objections were irrelevant. So long as approval was granted by the Maryland Health Department, the effluent system was there to stay. And Dr. Wolman had made sure that state approval was granted. As early at 1941 he talked privately with Chief Engineer Hall and other health department officers to gain their support. He argued that the effluent was safe because it was segregated from drinking water. Delay "threatened seriously to hinder the work of the Company which holds large defense contracts." Chief Engineer Hall approved the effluent system without any serious review of it. Practically everything was okayed on the basis of the expertise and good judgment of his former boss at the health department. Similarly, when the system underwent a 50-percent expansion in 1947, which raised the daily intake of re-treated sewage to 75 million gallons, approval was granted—almost automatically—by Robert H. Riley, director of the department.[49] The system was certified as "safe and sanitary" under permit No. 3755.*

Air was another commodity needed in massive quantities by the expanding mill. In the blast-furnace department alone, 4.5 tons of air were used to refine 1 ton of pig iron. Mixed with coal and natural gas, air produced the combustion that refined steel in the chambers of the open hearth. The problem was not obtaining air but in discharging it at the lowest possible price. Luckily for the company, the plant was outside the boundaries of Baltimore's smoke ordinances. Still, assurances were needed that state officials would not take untoward action against the steelmaker as it increased its postwar capacity and smoke emissions.

Again the company's relationship with Abel Wolman proved invaluable. Wolman was a specialist in particulate matter and other technical aspects of air pollution. Before joining the Hopkins faculty he had had responsibility over statewide pollution issues as chief engineer. If Bethlehem was to be approached about air pollution in the metropolitan area and over the Chesapeake Bay, the person who would have to take the lead was Dr. Wolman.

As he recalled in an interview, he was privileged to be in the position to probe unofficially about company plans to deal with smoke emissions and perhaps to make some recommendations. ("I want to avoid conflict," he said in an interview. "What I want is a solution to problems.") Regarding the steelmaker's discharges, which were beginning to cast a semipermanent cloud of smoke over the peninsula, and which would get worse under the expansion plans, Wolman said he broached the subject one day to plant engineer Coffin. Walking near the blast furnaces he pointed to the reddish smoke belching from the stacks. "I said, 'You know, you could water-wash those iron particu-

*The same health department permitted Bethlehem to discharge 450 million gallons of water used by the plant daily in outfall sewers that lined Baltimore Harbor and Bear Creek. The water that flowed into the harbor contained various mill discharges and chemical residues, including sulfuric acid (pickling liquor), suspended iron particles, lead, cyanide, benzene by-products, phenol, chrome, and oil. Toxic wastes from these and other discharges from local industry were beginning to impact severely on local waterways, forcing the closure of a number of public beaches and causing deformed crabs and tumored fish.

lates before they got out of the stacks, put in two simple door tanks, settle it out, and pump the [iron] sludge back into the furnace for reuse.'"

"Your idea is a good one," Wolman recalls Coffin replying, "but I'll tell you why it wouldn't be done. When I come to the board of directors with your recommendation for recapturing the iron ore particulates, the first question they are going to ask is, what will it cost to do that? I would tell them what two door tanks would cost. And then they would say, 'Well, if we invest in that procedure, the return would be about 6 percent, but if we use the same money just to make steel and not to worry about Dr. Wolman's stacks our return will be 20 percent.' Our board of directors, I can tell you, will make the decision that way."

Wolman accepted Coffin's explanation because he said he agreed with the underlying assumption. It was not fair to ask Bethlehem to finance outlays in pollution control devices if the investment return did not compare favorably with the return on other capital projects. Net return was a fact of business life. "I have to remind people that industry is not a philanthropic institution. Some people mix them up."

His rendering of a corporation as a simple engineering tool by which managers maximize stockholder profit, devoid of public or environmental obligations, was all the more startling given his government positions as scientific expert and his self-description as an environmental "advocate." Apart from his role as advisor to Maryland's governors and Baltimore's DPW and Board of Estimates, Wolman served on the National Research Council's committee on chemical agents, was consulting engineer for the TVA and U.S. Atomic Energy Commission, member of the Advisory Committee on Nuclear Reactor Safeguards, consultant to the U.S. Surgeon General, and chairman of the World Health Organization's Expert Committee on Environmental Sanitation.

When Dr. Wolman talked about a clean environment, he meant using more chemicals in public reservoirs. And while he defined the behavior of business in the self-interested terms of cost-effectiveness, he sat on government committees that, directly or indirectly, in the postwar era or in the Depression, had their hands full of environmental problems caused by the promiscuous use or foolish neglect of the environment.

"An engineer is a person who can do for $1 what any damn fool can do for $2," was the motto he recited to the author. And at $200 a day Bethlehem had gotten itself the best damn sanitary engineer; one who would not be so indiscreet as to raise the matter of air pollution again with Mr. Coffin, even after Sparrows Point's blast furnace H went into production in 1948 and began spewing out a record 12,000 pounds of iron particulates and sulfur dioxide emissions per day.[50]

*

What went up, at least partially came down. When the stacks "blew," everything would turn red, remembers Bob Eney. "You'd be walking along C Street by the executives' houses and here would be these big Victorian houses and shade trees, but the sky overhead would be pink from the smoke and the houses would glow a rusty red. You'd pass these nice houses, round a corner and see a big cloud of dust coming right at you."[51]

The red dust powdered the streets, the roofs, shrubs, grass, everything. Cartwheeling down the sidewalks and eddying in front yards, it blew into porches and penetrated windows and doors. Cars were coated with it and street signs obscured by it. And it did not respect social boundaries, raining down on the executive side of town as heavily as on the workingman side.

"If you looked out in the morning there would be soot all over your windows," Elizabeth McShane recalls. "Sometimes it was kind of blackish-green, but mostly it was red, and soon it got so that houses looked like they were painted red."

"Oh my," adds her sister Katherine, "our little town was becoming so dirty. We couldn't sit outside any more. On some days it looked like it was raining red peppers outside."[52]

Bulging with more steel than ever before, the peninsula was becoming a grimy, foul-smelling place, "a factory town" rather than a model town. While the natives rued the passing of a community they had regarded as clean and green a decade before, it was something they insisted couldn't be helped. It was the price of progress. "You didn't think about it," Elizabeth says, expressing a sentiment heard repeatedly around town. "You accepted it because you knew what it meant. It meant that the mills were running full. That's what people thought about. If someone said something about the dirt, you just said, 'Yup, it's raining dollars.'"

CHAPTER 18

UNION BUSINESS

The speed with which Sparrows Point was rushing forward was astonishing. In a matter of four years (1946–50) the peninsula had been stuffed full of additional plant and machinery. More than 500,000 tons of new steelmaking capacity had been anchored into place along the south shore at the megacost of $40 million—six times the plant's original 1890 price tag. The second continuous hot-strip mill, opened in 1947, doubled the output of strip, and tin-plate manufacture was enhanced by the start-up of a new cold-reducing plant which darkened Humphrey's Creek with an unbroken ridge of brick, steel, and ventilating stacks. Including intermediary equipment, the tin-plate division alone comprised a roofed area with a circumference of three miles. Enough canning plate was made there each 24 hours to span the distance between Baltimore and Trenton, 125 miles, with a ribbon of steel two feet wide.[1]

Fat intestinal pipes, the outgrowth of Dr. Wolman's industrial water system, wound through the body of the plant, discharging a torrent of treated sewage into the bay. Once upon a time Rufus Wood had lit the town's Fourth of July fireworks in Pennwood Park. Now, with the erection of a coal-fed power plant on the last piece of parkland, enough kilowatt-hours were generated to supply the needs of half the residential customers of Baltimore. According to the company, Sparrows Point consumed 1/500 of the nation's total electricity and burned up 11,200 tons of coal daily.[2]

Keeping Sparrows Point sated with iron ore had exhausted the Cuban orefields that Frederick Wood had discovered in 1882. Less than 9 million tons of ore lay at the old Juragua mines near Santiago, and they were of poor quality. The Mayari ore on Cuba's north coast, although abundant, had a high content of moisture which made it too costly to ship at a favorable rate. The Cuban mines that had played such an important role in Sparrows Point's early history were closed down, and Bethlehem looked at Venezuela to fill the gap. A fully owned subsidiary, the Iron Mines Company of Venezuela, had been organized for the purpose of mining the recently discovered El Pao lode in the Orinoco basin.

The property was centered around a mountain whose top was found to be almost pure iron. In all, 60 million tons were on or near the surface of a 20,000-acre plot, a jungle Mesabi located 195 miles down the Orinoco River from its confluence with the Atlantic Ocean opposite Trinidad. Construction was underway for a rail line, mining equipment, and port facilities to tap the new deposit for shipment to Baltimore 2,115 miles away. The army takeover of Venezuela's civilian government in 1948 was a stroke of good luck for Bethlehem. One of the first acts of junta strongman General Marcos Pérez Jiménez was to declare a policy favoring U.S. investment in oil drilling and mining.[3]

For the interim the mines at El Tofo, Chile, kept the plant running at full tilt. Rich hematite and magnetite ores were being delivered to Sparrows Point for under $6 a ton, an excellent price, and nearly 20 million tons of ore lay untapped on land owned by Bethlehem north at Romeral. In 1949 Chile agreed to put up $2.5 million of the $9 million needed to exploit Romeral and contracted with Bethlehem to sell a portion of the ore to its new state-owned steel mill, Acero del Pacifico.[4]

Bethlehem also looked overseas for another basic material for Sparrows Point. This was scrap. Over the years, scrap had become increasingly valuable as an ingredient for open-hearth steelmaking. By mixing cold scrap in a 1:1 ratio with hot pig, steel could be made with about 400 pounds of coal added per ton for fuel.

Abundant supplies of scrap lay rusting in the battlefields of Europe and the South Pacific. After purchasing the rights to the salvageable wreckage, Bethlehem opened the largest metal reclamation facility in the world in Baltimore. Between 1948 and 1950, 270 ocean freighters tied up at the wartime shipyards in Fairfield to unload 1.1 million tons of scrap. At one point 1,800 army tanks, allied and enemy, were piled up in the yard awaiting their turn to be dismantled and ultimately coverted to new steel. Another 800,000 tons were reclaimed by cutting up 384 surplus war vessels, including battleships, aircraft carriers, troop transports, hospital ships, and submarines. Scrap from the yard was barged across Baltimore Harbor, where it was melted down to feed the open hearths of Sparrows Point.[5]

Employment on the peninsula had jumped to 23,000 steelworkers

—2,000 above the wartime average—and there were 5,000 employees in the shipyard. Through 1947 and 1948 orders averaged 98.2 percent of capacity, the best record ever in peacetime. Three times a day a mass of humanity tooled up North Point Highway. Used-car lots, taverns, trailer parks, and restaurants had sprouted everywhere. One could dine on hard crabs and T-bone steak, play shuffleboard, listen to Joe Luhrman & His Trio, or sit back and contemplate the foam uncoiling from a cold tin can of National Boh. *Ships, Steel, Tin is all you hear / Above the din of the glass / At the bar of Hillegas,* shouted the advertisement of one popular watering hole.[6]

*

While generous pay kept management's eyes peeled on the company and production, rank-and-file workers were glimpsing hope of a better material and social existence—the longstanding promise of the CIO union struggle. In the late '40s the United Steelworkers of America not only represented over 90 percent of the 500,000 production workers in basic steel, but were spread across other building-block industries. Steelworker country encompassed the iron-ore mines of northern Minnesota and the aluminum mill of Alcoa, Tennessee. Members were found in the railroad shops of Schenectady and Eddystone as well as the canning factories of East Baltimore, Chicago, and South San Francisco. Fully a national union, with more than 2,000 locals, the USWA was integrated at the top with a unified leadership. National policy was propounded from Pittsburgh's Commonwealth Building where the international officers resided on five floors with their staffs. The key officers, forming a de facto board of directors, were President Murray; two vice presidents, Van A. Bittner and newcomer James G. Thimmes; secretary-treasurer David J. McDonald; and six department heads with a total of 80 underlings.[7]

"Nothing remotely like the confusions and rivalries of the United Automobile Workers exists in the Steelworkers," wrote John Gunther in his 1947 best-seller, *Inside U.S.A.* "Their union is run, and run well, from the top; it believes in what I heard called 'centralized decentralization.' " The international was foremost a banking authority; it controlled the locals by controlling their purse strings. Income from monthly dues of $2 per member was checked off at the worksite and sent directly to Pittsburgh for deposit. The international kept half and returned the other half to the originating local. This gave the international unusual clout in the labor movement. It had the potential to cut off funds to locals that incurred the leadership's wrath. On the other hand, it gave the Steelworkers an enviable reputation for financial probity. An independent audit of union funds was published every six months and the books of the locals were examined by a traveling band of accountants assigned by secretary-treasurer McDonald. Try as they might, the many congressional critics of unionism in the '40s could not tar the USWA with charges of racketeering and manipulation by the criminal underworld.[8]

Between the international and locals, serving at the pleasure of the former and supervising the affairs of the latter, was the district director. There were 36 district directors; District 8, Baltimore, was in the hands of Albert Atallah. He was a snub-nosed organizer from Aliquippa, Pennsylvania, brought in by the international to replace Nick Fontecchio during the war. Atallah was the man in the middle. The international paid his $6,000-a-year salary and approved the costs of his office and small staff of organizers. Below him were the two locals formed to cover Sparrows Point employees: Local 2610 which represented "steel side" workers, and Local 2609 which represented the finishing forces. The president and other local officers were elected under surprisingly vigorous competition given that the work they did was notably unremunerative. The best pay offered was a $20-a-month salary for the local's president.[9]

The rank and file elected the stewards in their immediate departments as well as zone committeemen who represented major divisions in the mill. While not paid for their work, they were recompensed by the union when their duties resulted in time lost on the job. The zone committeemen were responsible for coordinating the activities of the shop stewards and handling the "meat and potatoes" of daily unionism—not strikes and picketing, but grievances.

Under the contract, employees had the right to file complaints over seniority, scheduling, pay, and working conditions which were "grieved" through a five-step procedure. First Step involved taking the complaint to the immediate supervisor to seek a resolution of the difficulty. If this failed, the complaint was appealed to the superintendent on Second Step. The majority of cases were resolved at this level, but if there was still a disagreement, the union was entitled to move the complaint up a notch to the union-management grievance committee. From there, it could go to Fourth Step, a formal review by the plant personnel director in consultation with the union. If still unresolved at this juncture, the grievance was removed from the jurisdiction of local officers and placed before the district office, where the international staff then presented the case in a hearing before an impartial arbitrator. This was the Fifth (and final) Step.[10]

For the average worker unionism was a source of considerable pride and loyalty. The struggle was still much too close to be taken for granted. "What the men themselves wanted the union for," recalls Joe Hlatki with strong conviction, "was to have everyone respect your seniority and your ability, and not have these here bosses' pets and brown-nosers get the best jobs. We got more money, that was important, sure, but we got respect more, that was number one." Hlatki's words expressed the consensus of a rank and file that was conservative, sober-minded, and didn't trust management to honor its collective existence.[11]

Prosperity threw a curve into an organization born in an era of scarcity and depression. The "agitators" of a decade before were now the pin-striped administrators of the nation's then-largest industrial union. In the words of one activist, the USWA had gone "from being a cause to being an institution." Was the institution of unionism fulfilling its original promise? The experiences of Charlie Parrish and Mike Howard pointed to how the union handled two of the most volatile issues of the period—racial discrimination and Communism.

*

Charlie Parrish had picked up, and believed, the words of the union's organizing literature: picked it up from Oscar Durandetto in 1940 and forged the message on the anvil of his own experience. For a man beset with the daunting problems of the "color line," the union offered a fighting opportunity for improvement, and that's what Charlie says he was "hunting for." Like any black person living south of the Mason-Dixon line, he was painfully aware of the persistence of racism. When he took the streetcar to work he was required to sit in the back or ride in an attached "colored-only" car. He was forbidden to eat at many lunch counters in the company town. His wife Alice could not try on hats at the local department stores. The list went on and on.

While Charlie figured he didn't have a chance to defy the social customs of his time, he and several other men were determined to chip away at job discrimination at the Point. Their tool was the union contract and its guarantee—cosigned by the company—that seniority and workmanship were to be the determining factors for job advancement. Charlie's first effort was typed into the grievance form on November 9, 1942:

> Employee: Chas. Parrish, BB 315, Repairman B
> Foreman: Geo. Martin
> Steward: Hural Thompson
>
> Statement of Employee:
> The reason I want more money because I am doing the work satisfactory. I lays off material and I fit it up. I am a first class Burner, and a copper Burner and teach men that don't know the job. I am asking for repairman A.[12]

A reply was made the next day by the chief of furnace mechanics:

> Disposition of Foreman:
> Employee Chas. Parrish, Ck. No. BB 315, is now classified and paid as a repairman B. I oppose his being promoted to repairman A as he is not qualified to do the type of work we expect our repairman A to do. This grievance was presented to me on Nov. 9th, 1942.
>
> J. C. Royston 11-10-42[13]

Charlie's grievance was appealed to Second Step by shop steward Hural Thompson. On November 17 Superintendent C. F. Hoffman refused to reconsider the proposal. Charles Parrish, he wrote, while "a very good worker," cannot "take a print or sketch from his foreman and go ahead with the layout work, for he cannot read a drawing." Two months later Thompson (who was black) and Mike Howard (who was white and headed the 2610 grievance committee) disputed Hoffman. They said the blueprint matter was a smokescreen since Parrish was fully capable of doing the layout work based on his 15 years of experience in the crew. The matter was appealed on Fourth Step to plant personnel director Francis Wrightson. Wrightson reconsidered. Conceding that the union had a winnable case, he relented and put through the order on February 5, 1943, for Charlie's promotion. Charlie thus became the first black repairman A at the Point.[14]

Over the next five years he worked in the stack-and-bin gang and kept his sights on his long-range ambition—to advance to millwright. In 1948 a vacancy opened up. A millwright named Patton filled in for a sick foreman and another repairman A named Tricinelli moved into the millwright's slot. Tricinelli had 9 years less seniority than Charlie. When Tricinelli's position was made permanent, Charlie filed a grievance.

This time the main office was adamant: in Fourth Step Wrightson refused the union's demand of promotion. This set in motion the first case where racial discrimination was broached openly at arbitration. While the union did not attack the ingrained pattern of discrimination at the Point, where white and black workers were channeled into separate and unequal job classifications, it did question the disqualification of an unusually qualified black to a higher job. The union and Charlie seized on what amounted to a historical accident—his appointment to an all-white mechanics gang back in 1927—to claim that under seniority rules he was in a position to claim a job that was at the top of the hourly ladder.

Case No. 2099-3860, *In the Matter of the Arbitration between United Steelworkers of America (CIO) and Bethlehem Steel Company,* was held on March 29, 1949, before B. M. Selekman, a Boston attorney and contract arbitrator. In attendance on behalf of Bethlehem were James C. Phelps from the vice president's office, lawyer Gerald J. Reilly, personnel director Wrightson, George C. Little, assistant master mechanic, and four other plant supervisors. Representing the union were Al Atallah and Glee Smith of Baltimore District 8.[15]

Charlie was not in for an easy day. On the witness stand he was questioned searchingly by Reilly as to his exact qualifications, his education, and his ability to read blueprints. On the final count Charlie pleaded no contest: With an education that ended in fifth grade in Goochland County he had not learned to read blueprints; nor was such information provided him by the company, whose one technical school, the so-called "Gas School," never managed to have room available for black employees.

"I didn't know blueprints, but I know how to do the job," Charlie says. "If you know something, the millwright don't need no blueprints over in the blast furnace."[16]

George Little, assistant master mechanic, begged to differ. As chief witness for the company, he rattled off jobs he said Charlie did not have the qualifications to perform. These included the repair and assembling of lorry cars, fitting of bearings and cables on the skip hoists, and adjustment of the control valves of the pneumatic bell cylinders. However, he conceded that Charlie had never failed to perform a job he was asked to do and that during the war emergency he had worked directly with millwrights on several of the jobs cited above. "During the war we were shorthanded and anybody that could do anything was called on to do it," Little explained blandly.

In response, the union introduced evidence concerning the limited opportunities which the company had permitted Parrish. Atallah asked Little why formal training had been offered to Tricinelli, who had 14 years experience as a repairman, but denied to Charlie, who had over 20 years of repairman's experience. Little answered that from what he knew Parrish had not applied for training. In another exchange Glee Smith, an assistant for Atallah, pressed Little on the opportunity issue. From the transcript:

> MR. SMITH: What I am trying to get at is that this man never had the opportunity to advance that Tricinelli had. How many colored employees do you have in the pipefitting gang?
>
> MR. LITTLE: We don't have any.
>
> MR. SMITH: How many do you have as Repairman A besides Mr. Parrish at this time?
>
> MR. LITTLE: We don't have any as Repairman A.
>
> MR. SMITH: You have none in the millwright classification?
>
> MR. LITTLE: That's right.
>
> MR. SMITH: How many have you got in Stacks and Bins?
>
> MR. LITTLE: We have one, Mr. Parrish.
>
> MR. SMITH: Mr. Parrish had 22 years of service, and you have isolated Mr. Parrish and the rest of this [black] group by itself and have not given them an opportunity to advance...
>
> MR. LITTLE: We have assigned him to jobs within his ability.
>
> MR. SMITH: How do you weigh the ability?
>
> MR. LITTLE: By observation of the man's workmanship.
>
> MR. SMITH: Well, each time that a man performs his job correctly, or as nearly correctly as can be done, don't you think that he should be given an opportunity to show something a little better in the way of ability?
>
> MR. LITTLE: One of our principal objectives in the Blast Furnace is to develop good men.

MR. SMITH: You said Parrish is a good man.

MR. LITTLE: Parrish is a good man within his field.[17]

After several more exasperating go-rounds with Little, Al Atallah summed up the union's case. "We are forced, under the circumstances, to make the statement that Parrish was discriminated against because we have found, and the testimony here shows, that people with much less seniority than Parrish came into the plant and to the department in the same way that Parrish came in, as general laborers, and that they have been promoted to different jobs and given a chance to familiarize themselves with all phases of the operations."[18]

Not true, countered Reilly. "Actually, what Mr. Atallah's statement amounts to, if I may characterize it, is that Mr. Parrish should have been moved up into the millwright position regardless of his qualifications for the job. In other words, the Umpire is asked to direct the Company to discriminate in favor of Mr. Parrish without regard to the provisions of the [contract] agreement." Given his lack of "all-around qualifications," said lawyer Reilly, "to have moved him into the higher-rate job would have been unfair to him, it would have been unfair to supervision, and it would have been unfair to the department."[19]

Mr. Wrightson, the personnel director, then asked for permission to speak. "I would like to say a word or two about this matter of racial discrimination. It gets under my skin." The fact that Charlie Parrish was appointed repairman A, a position second only to that of millwright, was clear evidence that he had not been a victim of any type of discrimination.[20]

Arbitrator Selekman agreed with Mr. Wrightson. On May 25, 1949, he dismissed the union's charge of racial discrimination. Management, he said, was within its legal rights under Section 1 of Article X of the USWA contract to exercise discretion in appointments of personnel so long as they conformed to objective criteria. "The Company marshalled an impressive array of objective data to combat the allegation that racial discrimination has operated in any decision involving Mr. Parrish," Selekman wrote. While the charge of violation of Section X requiring the posting of shop vacancies was upheld by the arbitrator, he stated that because the violation "did not substantially alter consideration of the claim," no further action other than affirmation of the violation was required.[21]

What it got down to were those blueprints. According to the findings presented by the arbitrator, the best course for Charlie lay in attempting to enroll in the Gas School—"Management is willing to help Mr. Parrish qualify if he should so desire," Selekman wrote hopefully. But Charlie wasn't listening to the arbitrator; his attention was turned to the union and his own apprehension of what was right and wrong.

"The arbitrator is not supposed to like you or the company neither," he says. "Whoever was right by the contract. But the laws then gave the company so much footsteps to do the things they were doing. That was wrong. That arbitrator had to play ball 'cause we had no equal rights laws then. No other millwright had to go through what George Little fought me on. He says, 'You a good worker for what you can do.' That gets me angry. Oh, the white man down there! He have all the education and no common sense."

He continues, "I took it from that point. I knowed I was being tested and I knowed that Al Atallah was behind me. I had patience. Oh, I had. I said nothing till the next opening come up. It come up for millwright for the cranes [a year later]. I walked in the master mechanic's office and told him I wanted that millwright's job. The master mechanic said, 'Charlie, who sent you?' I said, 'Nobody sent me. I come for the opportunity.' He said, 'Well, come back Wednesday.' Wednesday I'm standing in his office at 8:00 in the morning. He said, 'Why you doing this, Charlie?' and I said, 'I want the money.' Now that wasn't what I was huntin' for, but that's all they understand. So I said I want the money.

"The master mechanic goes and calls the foreman in. And he says to me, 'Now, Charlie, if you tie this machinery up and stop up production, we're gonna take you off.' I said, 'I know that.' Gracious, when I first took hold of that burning rod and made that first burn [in 1928] I knowed if I made a mistake I'd be back in Big Jeff's gang. I knowed that all my life. The master mechanic just looked out the window. Long time he looked out, then he turns to the foreman and says 'Well, Charlie's going down to the ore docks. He's going to be made a millwright.' "

Four years after Charlie made millwright at Sparrows Point the Supreme Court ruled that racial segregation in public schools was illegal in *Brown* v. *Board of Education.* Five years separated his promotion in 1950 and the Montgomery bus boycott. It would be 10 years before the first sit-in demonstration at Greensboro, North Carolina, protesting segregated eating facilities, and 12 years before James Meredith enrolled at the University of Mississippi. On April 14, 1964, at the annual meeting of Bethlehem Steel, a stockholder asked Edmund F. Martin, chairman of the board of directors, why no blacks could be found in management positions. "We would like to employ Negroes in top jobs, but we can't get them," was the executive's answer.[22]

*

The ideological tempests of postwar America would have a profound impact on Mike Howard, chief grievance officer of Local 2610. In the Depression many union activists had been Communists or socialists depending on how they viewed the problem of replacing capitalism with an improved economic model. Mike did not regret having joined the Young Communist League in 1932.

From his perspective, progressive unionism was passing through a critical test between 1945 and 1948. Uppermost in his mind in these years was Truman's nationalization of the coal mines in 1946. Mike interpreted the move as an attack on labor and John Lewis, its towering leader. "That was the start, Truman calling out the troops on Lewis and the Miners. I was convinced that Truman was out to break organized labor." In 1947 the Taft-Hartley Act was passed by a Republican-dominated Congress. The law signaled the beginning of an alliance between white segregationists and big business to beat back CIO organizing in the South, according to Mike. "Union-busting, white supremacy, and the rise of Cold War—all these things convinced me that the advances made since the '30s were going to be rolled back. It was a real crisis, and I thought the Steelworkers had to take a strong stand and not just back Truman because he was the Democrat available."[23]

Many in the CIO ranks wanted to break with Truman in the 1948 race in favor of Henry Wallace, the former vice president who was running on the Progressive Party ticket. But at the 1948 annual convention of the Steelworkers, held in Boston on May 11–15, Phil Murray put a damper on the effort on behalf of Wallace. While not specifically endorsing Truman, the international sponsored a resolution passed by the delegates which condemned the "third-party" movement, accusing its supporters of "providing aid and comfort to reactionary candidates for public office and to the enemies of organized labor." Furthermore, by voice vote the USWA constitution was changed to deprive members found to be "consistent supporters" of the Communist Party from holding a position in the union.[24]

About the same time Mike found out that the leadership of the American Communist Party was distancing itself from the Progressive ticket. He was astonished. "Either a very devious or a very stupid game was being played by the CP. People like myself were going all out for Wallace, while the leadership was making back-door deals in support of Truman." Something important snapped in Mike. "I had been involved with the Party up to the point where I suddenly realized they were a bunch of bush hitters. The people I was looking up to, well, I lost confidence in them. It had nothing to do with my Marxism. I'm still a Marxist. It was that I sized up the CP leaders as people not worth following, so I didn't."

Withdrawing from CP affairs ("I stopped going to the meetings because I was so out of sympathy with what was going on"), he redoubled his efforts to persuade his local to support the Wallace candidacy. On "Candidates' Night, 1948," held at the union hall in East Baltimore, he spearheaded the pro-Wallace faction.

Al Atallah was there to speak on behalf of Truman. "The first time I met Al Atallah he was very Left," Mike recalls. "That was in 1936. Now all Leftists were fellow travelers and dupes of the Kremlin. He was without doubt a solid union guy who always kept his eye on one thing, namely his job. So when the leadership jumped from Henry Wallace to Truman, he was right there, a political chameleon. But the thing was, the guys didn't think much of Atallah. We had a local organization in place before Pittsburgh put Atallah in as director of District 8. He hadn't done as much as a lot of local guys for the union."

Following Atallah's speech Mike gave his pitch. "I remember I had a long list of arguments why Truman didn't deserve the support of the labor movement. I guess," he laughs, "it must have been a pretty good speech." Local 2610 voted to endorse Wallace. "Well, that was it," he continues. First he says he noticed some "sniping" behind his back. Then a formal complaint was filed against him. "As chief of the zone committeemen I was responsible for grievances in the 12 zones at the Point. Well, some little flunky of Al's put in a complaint against me for having failed to appeal his grievance. The complaint was nothing, but it served their purpose. Since they couldn't vote me out of office, they had to get rid of me another way." It was the new regional director in Philadelphia, Hugh Carcella, who obligingly removed Mike from the zone committee post and chairmanship of the plant grievance committee for malfeasance in office.

Mike found it painful to observe how closely the union's tactics matched the devious ways of the Party. He had been expelled by the CP after he was tried in absentia. Now he found himself ousted from an elected union post without any opportunity to defend himself against his accusers. He decided to take the matter to Phil Murray. "I had known him since the Canonsburg convention in 1936. I went to see him and he was nice to me. 'Mike,' he said, 'I'll straighten it out.' But he never did. It was disillusioning. Goodness knows, I didn't want to be a union bureaucrat. I wasn't bucking for Atallah's job. But I did want to work in the local. I liked it. I had figured I was going to spend the rest of my life at Sparrows Point, doing rank-and-file union work and getting the experience to become a first helper."

*

Since the breakup of the U.S.–Soviet wartime alliance, Communists had been hunted for and discovered at the State Department, in Hollywood, in the army, in literary circles. On June 11, 1951, they were hunted for and discovered at Sparrows Point. The instrument of exposure was a Virginia housewife and ex-beautician, age 29, named Mrs. Mary Stalcup Markward. She had a bizarre story to tell. Speaking as a government witness before the House Un-American Activities Committee, she said that she had been recruited as a "confidential agent" of the FBI while working at Mrs. Ethel Meyer's beauty salon on Massachusetts Avenue in Washington.

In 1943, she said, an FBI agent approached her and asked her to join the Washington-Maryland Communist Party organization for the purpose of "furnishing information of what I found out to the Federal Bureau of Investigation." Agreeing, she took part in Party activities until 1949, paid by the FBI while acting as the treasurer of the D.C. Party office and signing the Party checks.[25]

In her testimony concerning "Communism in the District of Columbia–Maryland Area" she read the names of 297 persons whom she said were members or sympathizers of the Party. Five were from Sparrows Point, and one of those was Mike Howard. "Mike Howard—he was particularly active at the 1945 convention when the Communist Party was reformed. He was a member of the Leadership Nominating Commission, of which I was also a member. His subsequent activity has been less," said the Virginia housewife.[26]

A week later Howard was hauled before HUAC, where a host of stock questions awaited him:

REPRESENTATIVE JOHN S. WOOD: Do you know the name of the president of the Bethlehem Steel Corporation?

MR. HOWARD: You got me. It used to be Mr. Grace and I don't recall the man who replaced him.

REP. WOOD: Do you know Mr. Grace?

MR. HOWARD: I never had the pleasure of meeting Mr. Grace.[27]

When he refused to name his immediate supervisor, Chairman Wood quipped, "This might be doing him [the supervisor] an injustice." When Mike gave the name of the plant general manager as "Mr. Clarke," Representative Clyde Doyle demanded to know why he hadn't refused to give his name. "I must insist that my reasons for refusing to answer on grounds of self-incrimination are sufficient to me; I beg your indulgence," Mike replied.[28]

Then came the ritual confrontation based on the Markward dossier. Asked to identify a long list of individuals as members of the Maryland or Baltimore Communist committees, Mike took the Fifth Amendment. He protested to Chairman Wood the way in which the hearings were conducted, saying the release of his and other names to the press two days prior to the hearing was a "clear invitation for employers to fire employees."

"I hope you do this committee the fairness to refrain from the inference you have been badgered into making a statement you prefer to make or not to make," replied Chairman Wood.[29]

Thus, a man who had been expelled from the CP in 1948 for his free thinking stood revealed in 1951 as a Communist subversive. "The odd thing was," Mike recounts, "they were throwing names at me I really didn't know. I said to my lawyer, 'They're throwing fictitious names at me.' And he said, 'Oh shut up and take the Fifth.' The lawyer, you see, was very unhappy to be up there with me."

GROUP GIVEN NO CLUES ON SUBVERSIVES was the page-one headline in the *Sun*. It was coupled to an even bigger headline—PROBERS TELL OF COMMUNIST CELLS IN PLANTS HERE. Over and over the article noted grimly, "Howard would not answer when asked if..."[30]

However, it was not such dishonest insinuations which got to him ("it doesn't take much to add your little tin can to the tail of someone being whipped"); it was the dagger thrust he got when he reported to work the next day. "I go to the locker room in the open hearth and there's about 800 or 1,000 men in it. It's noisy and usually it's 'Hi,

Mike,' 'Hi, Mike,' 'Hi, Mike.' Well, I come in that day and you could hear a pin drop. Nobody's sayin' a damn word. I walk to my locker and I felt—I don't want to get too romantic about it, but here are guys I worked my ass off for. They know it. I have settled grievance after grievance after grievance. For whites and for blacks. You'd think they'd rise up in a mass and say, 'We don't care what those idiots in Washington are doing!' But I think that they felt, well, there's something going on they don't understand. They're taking stock. They didn't rush to my defense and in a way I didn't expect them to. But I didn't expect quite the silence. You know, it's hard when you've been well liked to suddenly find yourself an anathema."

Mike, pausing, turns analytical. "Really, I thought I was fighting on a different level. I was fighting on a level which went to my Marxist beliefs, and I was working for them on a level which represented only their particular interactions with the company. And perhaps it shows what a poor job I did in radicalizing and politicizing the people in my department. Perhaps I should have done a better job of bringing the two together. I'm sure I could have done a better job than I did. Well, I sort of washed out all around. Everything—the Party, the union—was flying into orbit."

The aftermath of HUAC was a whirl of disconnected events. Mike remembers forcing himself to attend the local union meetings, hating every minute of it, and talking with Francis Wrightson, the personnel director, and feeling the better for it. The company man expressed sympathy about Mike's plight and said that as far as he was concerned his job at the mill was secure. "Mr. Wrightson was a gentleman to me. I remember saying to him, 'If you think I'm capable of sabotage I'll quit right now. I'm having a rough time as it is.' And he said, 'Don't leave.' From all the talking we had done across the table over grievances, we had built up a rapport and mutual respect. It struck me that in a way I had more in common with him than with the men in my own department. For a fleeting moment, we had human contact without the posturing and union-management stuff coming in."

The silence in the mill seemed to last an eternity. In real time it may have been about six weeks before there was a crack in the wall. He recounts the circumstances. "I was working in the manganese bin. I'm working next to a guy who is in the National Guard Air Reserve. Very patriotic. He and I are breaking the chunks of ferromanganese with our 16-pound mauls. We're both wearing our goggles and we're sweating like crazy. And he's not saying a word. Finally he turned to me. 'You know, those bastards are crazy.' I said, 'What are you talking about?' 'Those guys in Washington. No spy would be taking a job like this! There's a hundred jobs around this goddamn mill where you'd never break a sweat. They must be crazy!' And that broke the ice with him."

Around the No. 2 shop some of the men started to open up to him. Mike heard things that taciturn men tend to keep to themselves. An elderly worker confided that he was a fugitive—he had abandoned his family and town after he was blacklisted in the Gastonia textile strike in 1932. In all these years he had lived in fear of being found out, or coming face to face with his family. With men like the National Guardsman Mike became something of a folk hero. "I'd find myself being introduced to the new hirees on the floor with, 'Hey, this is the guy those crazy bastards in Washington were after!' "

While he regained the respect of at least some of his coworkers, it was not enough. "To me my work in the mill and my work as a union person were part of a single package. To work without having any purpose to it—I didn't see myself pushing a wheelbarrow, to spend my life working, penned in, that way." Mike was doing time in the open hearth. He did it out of principle—"to have quit would have been an admission of guilt of some sort"—but also because he couldn't think of what else to do. "I was in a very strange frame of mind, very muddled." All he knew as he continued to tend his furnace in 1951 and 1952 was that McCarthyism was at its height and that for some unfathomable reason the company hadn't fired him outright.

In 1953 he quit. "I just wanted to get away from the local scene. So I went to upstate New York. My sister was up there and she invited me to look things over. Her husband's family was involved in a big dairy. So I worked for them for a while. Then I got interested in a chicken farm. When I was a boy Shangri-La was having your own little chicken farm. Everybody dreamed of retiring to a chicken farm. Anyway, I finally pulled myself together and came back to my family. Talked to my wife about my plans to go back to school, to get a certificate in electrical engineering. I was 38 years old. I thought, well, I have a fighting chance of rehabilitating myself, of getting into a new line of work. And she was delighted. She was afraid I was going to stay in upstate New York."

Over the next two years he took engineering courses at Johns Hopkins University and worked in an electronics repair shop. In 1955 he joined a small instrument-making firm outside of Washington, D.C.

where he would spend the rest of his working life in research and development, experimenting with metering instruments for space research and materials analysis.

"I had first seen metering instruments in the open hearth and I was always a little shy of understanding how they worked. So I did this. I made the decision I never was going to be dependent on a show of hands. I wanted to study things that could be physically measured."

When he looks up his eyes are like polished silver. They are glazed with tears. "Damn it. I wanted to quit those mills when Phil Murray, my union president, backed Truman. In '48. After Truman called out the troops against the miners."

*

The late '40s and early '50s did mark at least a psychological turning point in the history of the Steelworkers. A new agenda had emerged and to those who were not its beneficiaries it seemed hostile. President Murray could talk all he wanted about industrial democracy, but to members like Mike Howard it was obvious that the union leadership had become entrenched and self-protective. Al Atallah was the local monarch for life, his power absolute so long as he toed the line from Pittsburgh. Machine-style politics had emerged as the dominant institutional framework of the union. Reformists wanted the union to address a progressive, if not socialist, agenda. But many workers consented to, if not actively favored, a political framework that retreated from the controversial items of the day. Like their counterparts in big-city machines, union officials demanded a don't-rock-the-boat conformity and believed, with some justification, that internal dissent weakened labor's hand against a tough-guy management. Compared to the organizing days of John L. Lewis, the union had backtracked from its position as a leading force in the battle over ideas, especially if those ideas strayed from the Democratic Party platform.

As it lost its edge as an organizing force (a post-war attempt to organize Southern factory workers was a dismal failure), the USWA took great pains to service its own constituency. It was still massive —even cab drivers belonged to a Steelworkers' local in Pittsburgh— well trained, and disciplined; in buoyant economic times, the delivery of annual hourly wage increases wasn't all that hard to come by. The 18½-cents-an-hour victory of 1946 was followed by wage increases of 12½ cents in 1947 and 13 cents in 1948. A major priority of the union was the establishment of uniform wage rates through the trade. This was done over several years through joint union-management committees that established 30 general job categories based on the skill required of an employee, physical exertion, experience, and level of possible danger. The management-union committees then agreed to a pay schedule for these jobs. As part of the agreement, the "Southern differential" that had kept wages in Birmingham low was phased out.

Bonus premiums also were simplified and made more uniform. The old Schwab incentive plan had become an administrative nightmare for Bethlehem; with little fuss, the company agreed to scrap the program in favor of a more limited bonus scheme which was agreed to by other large producers. In a final yet important matter, the USWA secured an agreement after a four-week strike in 1949 that broke new ground on employee pensions. By one estimate, more than half of the steel men retiring from the mills were disqualified from company-administered pension plans. Keyed to Social Security, the USWA-negotiated pension was the first in private industry and made a big difference in retired life in steel towns.[31]

"If that is bread and butter unionism, I am for it," David J. McDonald said. McDonald epitomized the "business union" wing of the USWA, which gained strength as the organization expelled members who were advocating principles that were in vogue a decade earlier. As secretary-treasurer, McDonald had consolidated his power through control of the organization's dues and finances. A handsome, well-groomed unionist with a taste for the high life, he gladly served as point man for the Red-baiting which became a hallmark of steelworker country.[32]

Phil Murray died of a heart attack on November 9, 1952. He was 66. The executive board of international officers and district directors named McDonald acting president. Without significant opposition McDonald was elected president and I. W. Abel, director of District 27 in Canton, Ohio, became secretary-treasurer in a membership vote. Business unionism ("pie-cardism," in Howard's terms) was in full sway, as indicated by the tone and substance of McDonald's inaugural address given on March 11, 1953, in Pittsburgh. "From the very start of the Steel Workers Organizing Committee," McDonald asserted, "we rejected Marxism in all its forms whether it was called Socialism, Fascism, Nazism, or Communism. Down through the years we have brought constant economic improvements to our members.... If this is bread and butter unionism, we can all be proud of it, because it means helping the people whom we are privileged to represent, helping them in a practical, substantial, day-to-day fashion, and not offering them pie in the sky through some crazy idealism on a basis that is unsound."[33]

CHAPTER 19

WOMEN'S WORK

At the tin mill tens of millions of dollars depended on the touch of a woman's hand. After the laborious process of rolling slabs of steel into paper-thin tin plate was completed, the resulting sheets had to be checked for defects prior to shipment. This was the task of women who worked at tables stacked high with freshly tinned steel and earned the nickname of "tin floppers" from the sound of their work.

Flopping tin called for coordination of hand, eye, and wrist. The hand ran along the rim of a sheet, thumb and forefinger gauging whether the edge was straight, the surface unwavy. The eye patrolled for pinprick holes, black spots, scale pits, and other signs of machine defects. The wrist arched the sheet from front to back and into one of several classificatory bins, the trademark *flop* coming from sheet hitting sheet.[1]

A dozen tasks were compressed into this circular sweep of touch, look, and throw, carried on at a top rate of 30 sheets a minute. Such speed, it was believed by management, could only be achieved by women; so women were welcomed to the sorting room.* Their workplace was an ugly and functional building deep inside the three-mile circumference of the tin-plate division. Skylights let in smudgy sunlight during the day and giant lamps illuminated the tables at night.

About 200 women a shift worked among four rows of tables, peering at the tin plate and deferring to the iron will of their boss, Eliza-

*During the war, about 700 women had been hired as crane operators and machine tenders at the steel mill, part of the corps of "Rosie the Riveters" who also worked in the emergency Fairfield shipyard and the airplane plants. Following V-J Day in 1945, all of these war workers were laid off, leaving tin inspection the only production job open to females at Sparrows Point.

beth Mak Alexander. Under the policies of Grace and Schwab, line foremen had a great deal of discretionary authority over their departments, and Mrs. Alexander wielded her power freely. Strict rules were in effect at all times. Each girl was required to wear a blue uniform of prescribed length—never more than 12 inches above the floor. When the whistle blew at 9:00 A.M., the daylight shift marched two abreast down the center aisle and into a rear dressing room, commencing a 15-minute break. At noon this procedure was repeated for lunch. Talking among women was frowned upon; it was cause for dismissal if one of the parties happened to be male.

Men, considered hopelessly ham-handed when it came to flopping, did the heavy labor. Reckoners counted and weighed the piles of inspected tin. Wirers secured the bundles onto skids. Pile-pushers and pile-pullers pushed and pulled the tin plate around the room and over to the shipping department. The sorters were separated from the men by different job classifications, lower wages, and by 36 assistant fore-ladies who patrolled the floor in bright pink uniforms.[2]

At the turn of the century women had played an important role in Rufus Wood's scheme for a model community; now five decades later a lady with a thick Hungarian accent played a curiously complementary role on the floor of the tin department, in equal measure "civilizing" and bullying her charges. The fierce maternalism of Elizabeth Alexander in the paternalistic setting of the mill made an indelible impression on those who worked for her. To many she *was* Bethlehem Steel.

"She walked up and down that damn floor like a general, and when you saw her coming you'd straighten out, boy."

"She'd protect the girls from them mill men that try to make you."

"Her husband was a reckoner in the same room and they never said a word. Never know they were married."

"Some days her glory was to tear you up."

"Everyone said she had a bad past—ran a 'house' in Ohio."

"She brainwashed those girls. The job meant a lot to them and they buckled down. She was gettin' production out of them, all right."[3]

Mrs. Alexander has her own story to tell, one that dates back to the 19th century and Imperial Hungary. For hours at a sitting she talks as if mesmerized, framed in her living room by a statue of an African prince and photographs of herself, reliving a passionate past. Her handsome good looks—blond hair, unblemished white skin, clear blue eyes—belie her age (in the late 80s). She recalls her victories by nodding her head knowingly, and when a memory displeases, her hand curls up into a fist.[4]

*

Elizabeth Mak Alexander was born in 1895 in a farming village in southeast Hungary. Not long after her birth her father moved the family to Budapest, where he found a job at the royal dairy. The splendors of imperial life sparkle in her mind. She says she attended the very best convent school in Budapest, the one favored by the King Franz Josef's granddaughter, and that by age 12 she had embarked on a career in the theater, debuting as a *huszar* or boy soldier in a Budapest cabaret. "I could sing, I could dance, I could learn anything, one, two, three." In 1908, a pistol shot shattered her childhood. Her father had accused a manager at the dairy of embezzlement, and in the ensuing duel he sustained a bullet wound to the face. The incident caused panic in the family—"dueling was against the law and my poor Daddy could be jailed"—and he decided to flee to America. He made his way to a mining town south of Pittsburgh where he had the names of some relatives. After he took a job in the mines, he notified his family to come to him.

When Elizabeth and her mother stepped off the train near Pittsburgh they were greeted by several miles of dirt road that led to Pricedale, a settlement owned by the coal company. The company houses all looked alike and were filled with Hungarians. "That's where Daddy was, where so many of them got hurt in the mines. I tried to be brave but I hated this country. Because you had outside toilet, you have to carry the water, you have to wash the clothes, and I wasn't used to it." Nor did she like Americans "when I saw them making fun of the other nationalities. They felt above them." The Welsh kids beat on the Hungarian kids in Pricedale, and Elizabeth defended her turf by acting tough and dressing well, "better even than the Welsh." She wouldn't associate with the neighborhood kids. "Some of them don't even have no boots. I was really embarrassed by it." She says a black girl became her best friend and taught her English.

After a year of living miserably, Elizabeth says she took things into her own hands. Dressing especially carefully one day, she walked in to a local factory, stretched her age from 14 to 18, and demanded a job. The factory turned out to be a division of the McKeesport Tin Plate

Company. As a tin-sorting learner in the sheet mill, she received 9½ cents an hour. Although the work was hard and the bosses unpleasant ("they was so strict because they knew the foreigners, we had to take it"), she remembers being very happy because she hoped to earn enough money to return to Budapest.

But things didn't turn out that way. World War I came along, blocking an easy return to Budapest. When her father became sick, she found herself the breadwinner of the family. Life became a constant shift of addresses as the family moved to the ebb and flow of job openings in the flopping departments of the Ohio Valley—from Pricedale to McKeesport, then to Wheeling and from there to Martins Ferry, Ohio. In the last town she met and married a tin reckoner in the mill: Mozes Alexander, an immigrant from Transylvania. They moved to Yorkville, Ohio, where he worked in the mills and she raised their three children and ran a boardinghouse for steelworkers.

She hands over a photograph from this period. It shows her beside a road, a slender, pretty woman dressed in a long dark dress. With jaunty abandon she has raised her left leg on the running board of a car, and her pleated hat crowns a very fashionable face: thin, high-cheekboned, with hair bobbed and tucked into a hat. Over fifteen years had passed since she and her mother had stepped off the train, bound for Pricedale, and what did she have to show for herself? Although she had inspected hundreds of thousands of sheets of tin plate, she still couldn't afford a decent house and didn't have the money to buy pretty furniture. Her husband's earnings were low and the family's savings were depleted during the seasonal shutdown of the factories. "The mills worked rush, rush, then shut down. There was nothing enough to make the kind of living I wanted." Life in America had put a dent in her living standards, but it certainly had not destroyed her reserves of will—she is smiling crisply at the camera and seems ready to go places.

*

If the road to Sparrows Point was long, Mrs. Alexander's climb up the factory ladder was swift. In June 1925, she came to the plant seeking to manage a newly opened restaurant. When a man got the job instead, she began to work in the sorting department. When somebody died in July, she was made a forelady. When somebody else left in December, she became the head forelady. From bottom to top of the department in a matter of months—the ascent, she says, was simply a matter of the company needing her. "I knew more than everyone about tin. I could tell sheet weights just by touch."

There was no doubt that she came to the plant at the right moment, a bit of luck in which her ambition and the aims of Charles Schwab coincided: In 1926 Bethlehem was involved in an expansion program that would increase tin-plate output by 33 percent. While the overall program would prove a costly misuse of capital funds as Bethlehem chose to ignore the Armco hot-strip mill for traditional technology, it was a boon to Mrs. Alexander. In the sorting department a reorganization was necessary to handle the increased output. She convinced her superiors to allow her to change inspection practices. In short order she ended the system in which an inspector handled as much tin plate as the hours permitted and then left the remainder to the next day.

Henceforth, when the girls arrived in the morning they were confronted with premeasured piles of plate that had to be sorted by day's end. These piles varied according to the level of inspection required. For "special" tin plate (top prime quality plate going to canners) the sorter faced a pile about 45-inches high, or roughly 5,000 sheets, to inspect, while for standard sorting where absolute perfection was not required, the piles totaled about 90-inches high, or roughly 10,000 sheets. Those who failed to meet the quotas, or let their eyes stray from the flailing tin, were discarded, and in their place Mrs. Alexander picked the best out of the pile of applicants.

She wanted girls who were young, agile, and not afraid of hard work. "I was a tough boss," she says with open glee. Not only did productivity increase, but wages were stabilized at "a decent level for women." Wildly fluctuating incentive wages which had been the norm in the department were abandoned in favor of the straight hourly rate of 36 cents. While this was the rock-bottom wage at Sparrows Point (black laborers received 1 cent more), the money was high in comparison to wages for women in the oyster-canning or clothing factories of Baltimore, where hourly wages of 15 cents were not unheard-of.[5]

"I won the girls over," she says. With sorting showing constant improvement under her touch, she set herself the task of improving the quality of her employees. For the most part the floppers were first-generation Americans who lived with their parents and answered to surnames like Stachowska, Flippo, Flower, Rohrbugh, and Caralle. Their ignorance of the social graces was appalling. With a shudder

Mrs. Alexander recalls finding a young Greek girl crouched in a toilet stall one day after work.

"Here she is putting her comb in the water. Then she combs her hair. I says, 'My dear, goodness, what are you doing?' In broken English she says, 'Everytime I see in the newspaper, toilet water good.' Now you could imagine some of these poor, ignorant people, how they were."

She was ready to educate them, but first they had to shed their old ways. Odd or objectionable clothing was the garb of some of the floppers. One girl came to work in a ragged, full-length evening gown she had found. Several others wore no underwear. They may have been poor daughters of immigrants, but Mrs. Alexander was determined that they wouldn't *look* that way. So in spite of the Depression around her, she arranged with a department store in Dundalk to stock navy-blue uniforms. The girls were instructed to buy uniforms at the discount price of $3.60 each.

Mrs. Alexander decided to attire herself in tailored dresses of green gabardine, and she chose pink for her foreladies so they could be spotted on the floor without difficulty. As part of her upgrading campaign she offered important tips to the girls on hairstyling and personal hygiene. "I would tell them, 'You'd look so good if you got a little skin cream. This druggist will fix you up. Don't waste money on the doctors.' And they listened."

Now the department presented a pretty picture. Behind rows of tables were rows of attractive young women, their blue uniforms mixing nicely with the shimmering silver of the piled plate, and their busy activity brought to order by the rhythmic flop of the tin itself, tin bound for Campbell's Soup in South Jersey, for meat and vegetable packers in East Baltimore, for lithographers in New York, for Dole Pineapples in Hawaii, for hundreds of overseas factories in Algiers, São Paulo, Bombay, Melbourne, and Yokohama.

*

The scene Mrs. Alexander created was perhaps too pretty, for among the girls were men—wirers, pile-pushers, shearsmen, and reckoners like Mozes—whose eyes also alighted on the blue uniforms and nice skin. She did not want the men to intrude on what she defined as women's space. They distracted the sorters. They'd flirt at the drop of a hat. They also had bad habits, smoking and cursing and sometimes fighting among themselves. But the sorting room needed them to move the many tons of plate to the shipping department. They were needed, but not welcomed. Much of Mrs. Alexander's time was devoted to separating men and women stationed a dozen feet apart, separating them by a ban on nonbusiness conversation, by frequent patrols by herself and her assistant foreladies, by different break times and lunch periods, by female-only lunchrooms and water fountains. "I had to be mean," she recounts. "The fellas come around and then talk to the girls. Sometimes I had to grab them and say, 'Look young fella, you don't get paid for this. Now get away. This is important job these girls are doing.' I had to yell at the men, but the girls, no, never. And another thing, I always teach my foreladies never even put your hand on a girl. Never do that! I always told them not to consider yourself any better than the other person."

The male bosses presented a tougher problem. Distinguishing between legitimate supervision from improper socializing was often difficult. "I had to be sharp," she says. Early on came the discovery that a highly positioned boss was dallying about the floor—"While he's going over the plate, he's making a date!" She confronted the boss and made him stop; the trouble was that the problem never did. It seems that all the men had the same intention. "They want to pick their wife out in the mill." Not only did this run counter to her own plans (which were "to help them get something better"), but she was convinced that the men were motivated by personal hostility. They wanted to get back at her.

The situation came to a head with Mr. Wise, the new superintendent of the tinning and sorting departments. A big man with plenty of muscle, he had a temper so terrible that another boss, no pushover himself, "trembled like a leaf" when Mr. Wise made his way through the mill. From the start Mr. Wise made it clear to Mrs. Alexander where she stood. To everyone in the mill she was "Miss Alexander," but to him she was Elizabeth. "He was the only one that ever call me that," she says. And soon he was barking orders at her.

Mr. Wise had committed an unpardonable sin in her book—"he did work against me"—and "people like that I try to give back what they give to others." One day she came into his office to show him some samples of tin. "I don't even have the chance to put the plate down, he start to hollering about something. I don't know exactly what he holler, but he holler. In a real high voice. I took the whole damn plate and I dropped it right in his office, in the center of the office, and I walk out. And then he sent for me. And I said, 'He's gonna talk right or he's not gonna talk to me and he's gonna let me have my pay.' I said that to the messenger. He got scared and after that he talked to me the right way."

The dropped plate did not end their feud, it just submerged below the surface. Miss Alexander would agree to one of his orders only to disobey it in ways he could not detect, and he would relent on one issue only to explode on another. It did not take long for the battle of wills to focus on wages. He would forget to request a raise for his subordinate and she would resent being paid "women's wages"—about $250 a month in the mid-40's. "If I was a man, I would have got three times more. See, those foremens don't know anything. They bring them in from other places and I have to do their work and they start at $600, $700 because 'they have to support a family!' "

The new boss, though, did serve one useful purpose—he confirmed to her the rightness of her ways. In almost everything she saw him as her opposite. If he couldn't bear to delegate authority to anyone and turned up at the most unexpected moments, causing consternation among his staff, she went to great lengths to include all the girls in the work before them. If he yelled and carried on and his foremen lived in perpetual terror, she used low-keyed persuasion to solve a problem, "always counting to myself, one to ten, to calm myself down." Boss Wise, for his part, did nothing to make himself popular. On the contrary, he seemed to find a perverse pleasure, a secret thrill, in calling forth hatred. If nobody liked Mr. Wise—a point he confessed when he was in one of his despairing moods—Mrs. Alexander felt absolutely secure of her place in her girls' affections. They realized how she was helping them, how she was more than a boss. She had given them pretty uniforms and tips on hairstyling, she wiped the leers off the strutting mill men, and she periodically gathered the girls together on the floor and said, "We must all do good, save a penny or a half penny where we can, because this is where your daily bread come from, from the Bethlehem Steel." The immigrant girl who hated this country had become a booster of everything American because, she says, "this was the country that give you a chance."

*

Mary Gorman's job was to maintain order in the dressing room. When the floppers filed into the room after their march from the sorting floor, bedlam could break loose without a matron like Mary Gorman around. The rest period was short, and 200 young females could cram in a lot of activity: sipping coffee, for instance, or examining themselves in the bank of bathroom mirrors. Mostly, though, the girls would sit on the benches and snack on gossip: They had been standing for a couple of hours and, in a few minutes, they'd begin another round of enforced silence.

Mrs. Gorman walked beside the lockers and inspected the girls. She'd see to it that the floor was not littered with the shells of sunflower seeds that the girls in the "Italian section" were crazy about. In the toilet area she'd listen to the exchange of confidences, soothed by the fact that the whispered words between the stalls sounded a lot like the previous day's latest. And off in a corner she'd frequently spot a girl, usually a new recruit, with the timid expression of a child that had been beaten. The girl, she knew, was recovering from a rebuke from a pink-suited forelady or, worse, a verbal lashing by the Old Lady herself. It was in such cases that Mary quit her role as observer and would dispense some timely advice.

"Never cry. Don't give her that satisfaction. Don't cry in front of her, 'cause she'll know she's got you."[6]

Such advice was not only appreciated by the girls—it made Mary's life easier, for in addition to handling the girls on break she served as Mrs. Alexander's personal attendant. She ordered the boss's meals, ran errands for her, and collected money for social activities. Theirs was a relationship symbolized by Mrs. Alexander addressing Mrs. Gorman as "my little Mary" (although little Mary was over 40 years old), and Mrs. Gorman calling her boss "the Green Hornet" behind her back. Without obeying her own advice it would have been difficult to take the times she was stung by the Hornet. "She always started with 'Ya-ya,'" Mary remembers, and some days were filled with ya-yas—"her givin' me hell for this or that, letting the girls gossip in the dressing room or something. She was theatrical, let me tell you." The matron would meekly bow to her boss's wishes only to mock her with impersonations of her accent. With a giggle she recounts the day she sent down to the kitchen a breakfast order that read "2 sof bol eeg wid doast."

"She and I got along good," Mary concludes after a few more chuckles. Despite the surface tension, the two women shared a lot—a passion for order, for one thing; also an appetite for work that came from "eating a lot of fillin'," as Mary puts it. If Mrs. Alexander was bossy to the girls she gave it to the men too, and Mary had enough bad experiences behind her to value the Hornet's ability to "keep them in their place." She also couldn't help but admire the woman's audacity. "She fought her way through this place," Mary says with admiration, and once she had it made she purchased a beautiful house in the superintendent's section of Dundalk, a sacred ground heretofore not besmirched by a foreigner and certainly not one who started off "low."

Mrs. Gorman could appreciate what this meant because her life had been enclosed by the horizons of the plant. Born in 1908 in a company house, she had been raised near the machine shop, the only girl of the four children of Patrick and Mary Durkin, who came to Sparrows Point in 1895 from Ireland. Living the whole of her life in Sparrows Point gave Mary a cozy view of the town. The furnaces always sighted by outsiders do not loom large in her recollections. Asked to describe what it was like growing up, she speaks of little things like the wooden fence that went around her parents' house. There she liked to sit on the top cross-piece and knock off the hats of passing workmen with her brother John. It was great fun, she says, because the targets of their mischief would rarely look up to see them. Instead, the men absentmindedly bent down to fetch their hats and went on their way. She also recalls an earlier memory—of following the men one morning, her mother holding her hand. A binge by her dad had threatened his position in the rail mill, and her mother had come to the mill to plead with the boss. Mary was present to tug on the boss's heartstrings a bit.

"We needed money bad," she says. In pursuit of making ends meet, her mother cooked the meals of a half dozen other workmen. They were called "table boarders." Every morning at 6:00 sharp her mother began serving up a breakfast of eggs, bacon, soda bread and oatmeal "so thick you couldn't stir it," and in the process wrapped two sandwiches, fruit and cake for each of the men's lunchpails. For dinner she cooked many variations of Irish stew or corned beef. Mary helped out by fetching the water from the hydrant down the street and going to the company store on errands. From the store's flour sacks her mother made underwear for the kids. This routine followed the constant running of the mill—the alternating six- and seven-day workweeks and the shift changes that sent different men on different eating schedules. "It was a big job but my mother was a lion, she could do that."

In 1930, at age 22, Mary married a steelworker, Ed Gorman, and they moved into her parents' house. In the Depression she began working as a midwife for the company doctor, Dr. Farber, knowing "the ropes" from having already given birth to the first of her three children. Dr. Farber would go see the colored women, but the town dentist refused; unregistered midwives worked in some of the outlying shantytowns and delivered children that never got birth certificates. Dr. Farber paid her "peanuts" for spending hours with expectant mothers; "he wanted to be called just when the baby was coming, not a minute earlier, and, bang, he'd be gone right after delivery." Whereas he charged $25 to deliver a baby, he told Mary it was up to her to collect her $5 fee directly from the families. This was often impossible for her to do. Once she delivered a baby as the father sat nearby with pins sticking in his fractured leg. "God knows how long it would be before he'd be working again," she says, and she refused to collect her fee.

When the wartime labor shortage opened up the job of matron in 1942, she grabbed it. "Dr. Farber had fifty fits, but I told him my family needed the money. It was paying $40 a week." The job proved to be great for her personal life and especially for her relations with her husband. "I think Ed felt a lot of me for taking it. He was sick at the time and the money helped a lot. And I thought I was gettin' somewhere. I could see what was being accomplished at home. The kids were beginning to grow up. They were getting their homework done, they had no trouble finding where they were at."

Juggling duties at work and at home was definitely complicated by the weekly shift changes, but "if you have a system it will work out." She arranged her life meticulously: when she would wash and hang the clothes; when she'd cook the meals; when she'd phone in instructions to her husband on how to heat up supper. Ed would agree to help her around the house so long as nobody knew he was doing "women's work." "If he dried the dishes the first thing I'd know he'd be shuttin' the window shades and saying, 'I'm not going to have my neighbors lookin' in at me doin' women's work!'"

A new shift began every Monday. If she liked the 3:00–11:00 turn because of all the housework she could get done, "the killin' part" was the graveyard shift. "I'd be all right til about 3:00 A.M. That's when the girls come in on break (there wasn't the pressure on them like there was on day turn). We had beds there and a couple of 'em would lay across the beds and say, 'Wake me up in ten minutes.' Well, I'd rather lay in bed ten minutes myself, but I kept on the ball and made sure they didn't overstay their time. I dragged through it. At 5:00 in the morning I had a terrible time keeping my eyes open. And that's when I gotta be awake 'cause then they start coming to work for the next turn."

Besides the revolving shifts, the worst part of the job had to be the men. She lingers on the indignities handed out by some of them. "They treated you like a stick of wood. Like we had a haughty man in the dispensary. I knew by his attitude he despised women. Didn't have a kind word to say—all business. I took some new girls from Virginia to the dispensary and he said, 'No vaccination? Did you realize that you were coming here to live among civilized people?' And I didn't like that one bit. You know what I thought: A kind word would have been much nicer than to talk that way. I said to him one time, 'Did you ever have a mother?' "

As part of her matron's job Mrs. Gorman practiced first aid on the floppers who lost a round to the sharp-edged tin. "We had all kinds of cuts and little nicks." The administration of antiseptic cream and Band-Aids broke up her days nicely. In the summers the heat inside the mill sometimes caused spells of fainting—"a girl would heave and fall on the plate." According to the boss lady, women going through menopause were more susceptible to such episodes; hence, she tried to restrict hirings to women younger than 35. According to the whispered consensus, the time of the month also had something to do with these bad spells; hence, the Kotex.

Kotex by the case was stocked free of charge by Bethlehem Steel, but Mrs. Gorman collected 5 cents for each package from the girls. Mary then deposited this money in the bank. These nickels added up and under the close supervision of Mrs. Alexander they became the financing arm of the big social events which the boss lady sponsored every two or three years for her girls. Soberly entitled, "Banquet and Dance by the Ladies of the B.H. Department," they were simply the "Kotex Balls" to Mary Gorman and the other girls: a night of catered food, fancy mixed drinks and wild skits conceived by Mrs. Alexander. "Oh, gosh, it was crazy," she says, shaking her head.

*

At the Kotex Ball no holds were barred—not on the costumes (supplied by a specialty shop) nor the locale (a ballroom in an important Baltimore hotel) nor the clientele (families of the floppers and a who's-who of mill bosses), but especially not on the skits performed. Under the gaze of chiefly male eyes, tin sorters pranced about in diaphanous veils and hula skirts. So popular became several of the skits that they were repeated at almost every banquet. At the 1954 Kotex Ball, for example, there were nine acts, among them:

> ***Sultan Chooses His Favorite.*** Sultry sorters lounge in Sultan Abdul's harem, chained to a life of leisure, until summoned to please the Master by performing the exotic Snake Dance.
> ***Sailor Blues.*** Girls from foreign countries encircle the ballroom stage to the beat of a slow love song for the enjoyment of floppers dressed up as sailors.
> ***Gay 90's.*** A barbershop quartet of mustachioed male reckoners harmonize for the enjoyment of Diamond Jim Brady and Sophie Tucker as blackfaced "Harlem Dancers" and "Can Can Girls" cavort in the background.
> ***Hawaiian War Chant.*** The theme of carefree living returns as the girls engage in some spirited beachside hula-hula dancing.
> ***Grand Finale.*** Sorters, now solemn and wearing the native clothes of Italy, France, Hungary, Poland and other nations, march beside a float carrying the Statue of Liberty and Uncle Sam. Two soldiers unfurl a sign, FOR THE PEOPLE, BY THE PEOPLE. All stand and sing, "God Bless America."[7]

Mary Gorman was in several of the skits. To please the Sultan, for example, she dressed up to look like Mae West ("you know, all bosomy") and belted out "Pistol-Packin' Mama" and "Oh, Johnny, Oh" to uproarious laughter.

But such skits paled in comparison to the perennial favorite of the girls—"Eight Ballerinas." Eight of the heaviest male workers were put onstage, adorned in mascara, blond wigs, and frilly tutus. "When the men in the mill did the can-can and all, it was a scream." What an absurdity to see the men trying to negotiate a high kick or stumble through a flatfooted pirouette. All that brawn could be laughed at now.

One, in fact, cannot talk to Mrs. Gorman or page through the old photo books of the ball without being struck by the relationship between this occasional pageant and routine life in the mill. While the Kotex Ball may have been held in a downtown Baltimore hotel, it seemed to spotlight tensions of the sorting mill 12 miles distant, and reflect back images nobody could handle at work. At the Kotex Ball the company rules were flouted by normally obedient workers, flouted to the applause of normally tyrannical bosses. Here was where young women could shed their blue uniforms for skirts far above the

12-inch rule. Where men and women could freely talk to one other. Where girls who sorted in silence could sing before the glare of flood-lights. Where children of immigrants could feel a part of America. Where on one night every two or three years those who would never cry in front of the boss lady could sniffle to the strains of a love song.

Presiding over the affair was the boss lady herself. It was her show: She (the child *huszar* of Budapest) wrote all the skits; she (a teenage beauty contestant in Pittsburgh) picked the costumes; she (a frustrated actress) rehearsed Mary Gorman and the other performers in the basement of her house. In a sense, Mrs. Alexander accomplished something quite remarkable as well as highly allegorical at the balls; she transformed the "woman's curse" into an affirmation of her own and her sorters' femininity. And she did this despite the hostile male environment of the mill—did it by defying her boss, Mr. Wise, who objected to the banquets as an unholy "mixing" of workers and supervision. Perhaps this explains why during the final event of the night Mrs. Alexander would be summoned to the announcer's microphone. Feigning surprise, she would ask the announcer if it could possibly be she whom he wished to see. "*Yes!*" roared the audience as a bouquet of red roses was carried to the stage.

A photo from the September 25, 1954, ball shows Mrs. Alexander accepting the bouquet. She was wearing an elegant black dress ruffled with lace. Her eyes were arched upward and on top of her blond hair there glittered a five-point tiara that she had rented for the occasion. At 59 but looking much younger, she surveyed her kingdom with a triumphant smile.

*

Not long after the 1954 banquet Marian Wilson staged a coup d'état—she came to work without her uniform on. Accompanied by four co-conspirators, she marched into the sorting room wearing a blouse that was equipped with enormous bat-wing sleeves and a crinoline skirt that was narrow at the waist and balloon-like at the hem. As a final touch of parody, she wore "little ballerina shoes" in place of the regulation rubber-soled Oxfords.

"So here five of us girls go bouncin' down the aisle with our bouncy skirts and the Old Lady, the Green Hornet, she comes runnin' out on the aisle and screams, 'Do you know this uniform has been in this plant for 20-some years? Who do you think you are coming in here looking like this?'

"And I said, 'Well, I guess if we've had something that long it's time to change it.'

"The veins in her neck stuck out, and she said, 'Don't you dare come in here lookin' like that again. You'll be sent home.'"[8]

Marian knew what the contract said—it said the company could not dictate what an employee wore unless it paid for the clothing—but she also knew that a contract was only a piece of paper. To make it stick, the other floppers had to join in the protest, even the old-timers. "The senior girls whispered we were troublemakers," Marian says, "but when they found out she couldn't get on your back any more, you'd see one girl come in with a skirt on, then a couple more and then many started wearing slacks. After a while, even the foreladies started wearing street clothes."

Marian Wilson was a member of the "Generation of '51," a group of hirees who came to the sorting room during the Korean War boom. They were different from the other floppers. Before they would submit to the company's tight leash, they would get the local steelworkers' union to represent them. Learning from the union men, they developed enough internal cohesion to confront management on working conditions in the mill. According to Marian, conditions were "the pits." The water fountains didn't work. It was freezing cold in the winter and roasting hot in the summer. Yet none of these battles with management had really encroached on the Hornet's personal authority, and it was after much talk and trepidation that Marian had decided that she had to be confronted on the uniform matter.

"You *were* scared to death of her," she says of Mrs. Alexander. "People being as they are don't like to be hollered at in front of others and she'd do that. It made you feel small."

The confrontation, though, had another crucial element. Marian and her allies rejected the boss's smothering maternalism. They disdained her banquets ("She wanted to show the girls off to the big bosses and get all the glory," says Marian), and they rebelled against the source of her greatest pride, the blue uniforms. Those uniforms tagged the sorters as "just tin floppers" when Marian and her friends went out to a bar after work. They wanted to look good and wear the latest fashions. "The young girls liked to be glamorous; they were men-crazy," she says. Still in her early 20s, Marian was no different.

*

Tin-flopping changed Marian, first physically and then mentally. She says she was overweight when she started work at the mill. Within a couple months she had lost 20 pounds. "It was all that heat and physically it was strenuous. Before I got used to the job I would come home so fatigued I took a nap and then didn't want to get up and eat. My mother and father got kinda concerned." Her father, in fact, was upset when she took the job. An employee of the steam department, he was proud of Bethlehem Steel, he defended the company and was grateful because they gave him a job when he came up from Virginia in 1936. "He always said that Bethlehem paid our bills." But he didn't want his oldest daughter to go to work in the mill.

One thing that upset her father was the effect the company paycheck had on Marian. She was spending Saturday afternoons shopping, and other reflections of worldliness were creeping into her appearance: lipstick and eye shadow, for instance. Where she worked the men noticed quickly: Although talking between the sexes was officially outlawed, the men never seemed to miss an opportunity to look up from the tin plate and silently appraise the girls, especially the new girls. "When you walked down the aisle—we had to go down two by two—the men stood there and looked you up and down and I guess picked out the one they figured they'd like to have. Really, they'd look at you like you was on an exhibition show."

Marian, though, found she had to watch out for more than roaming male eyes. In her first weeks she learned how the foreladies liked to sneak up on a beginning sorter and pounce. "The forelady come up on me and really yelled at me about my work. She then told me I was stupid. She got me so I went down the aisle cryin' my eyes out." With humiliation ringing in her ears Marian said she had one thought in her mind—to get out. The advice of another flopper stopped her. "She said to me, 'They get a big kick in making people quit. Don't let that forelady think she's made a fool of you before all the girls.'" So Marian stuck it out that day.

For a while she simply fought the exhaustion and tears, but gradually she says she realized that the greatest thing she had to combat was the fear. Fear filled the room. Girls were afraid of their own footsteps, dreading a walk to the water fountain since it might provoke a response from a marauding forelady. "In order to get a break from the tin plate, we'd go over to the fountain to get a drink. But every time the forelady saw you over there, she'd make this hissing, *psst*ing sound to get you back to your table. It was a way to let you know that if you don't shape up, you are gonna be reported. And they threatened you with all kinds of things if you didn't work overtime."

It was not over these issues that Marian first confronted the company—it was over a cigarette. "The women couldn't do a lot of the things the men already done. There was a sign, it's against the fire regulations to smoke, but the men are walking around with them in their mouths." What happened when she lit up a cigarette is recalled by this terse exchange with a forelady:

"You can't smoke in here," the forelady said.

"Why not?" Marian asked.

"'Cause it's a company rule."

"But I see the men smokin' all over. What's the difference in me smoking and them?"

"I don't know, but you can't smoke here."

"Well, if you're big enough to make me put it out I'll put it out, but if not I'll just keep on smokin'."

If you're big enough to make me became the rallying call of the young rebels in the department. Marian kept on smoking and she and her cohorts agreed they would no longer put up with some of the conditions in the mill. "Things started to break," she says. "Some people were fighting the system and the union backed us." First, impromptu meetings were held in the dressing room; later several sitdowns were staged in the lunchroom under the glare of grim foreladies. "If we didn't like what was going on, we just set down and wouldn't go back to work. We never really complained about the wages (a qualified sorter got $1.56 to $1.58 an hour then, and to us that was good pay for a woman to find), it was the working conditions. I felt we were really done wrong. We were not treated like human beings."

Not long after Marian burlesqued Mrs. Alexander's blue uniform on the sorting room floor, the Hornet offered her a job as a forelady. Marian took the offer as an insult, a last-ditch attempt to buy her off and save the Old Lady's crumbling dictatorship. "She was really gruff when I turned it down," Marian laughs. "She's the type, if she asked you to do something, she expected you to do it. But I wasn't about to go management. And I told her we'd still give her a hard time if she kept trying to put the fear in people."

Shortly afterward Mrs. Alexander entered the hospital. She had recovered quickly from a previous breakdown during the war, but this time she did not. She suffered from painful heart-related ailments. After going in and out of the hospital for two years, she reluctantly took her retirement.

Upon her departure many of the rules that specially governed the tin sorters were lifted. Some of the male supervisors appeared positively delighted to banish her memory from the floor. The job of head forelady was eliminated from the mill hierarchy. While there remained several assistant foreladies, they now reported to the superintendent of the tinning operation. The power of a female to run an entire department had been discarded.

The same fate befell the gala evening under the floodlights. A move by Mary Gorman and several old-timers to hold another Kotex Ball was rebuffed by supervision and, under pressure from Marian and her allies, the money accumulated from the Kotex was distributed by the company back to the female employees. From then on, Kotex was free.

About a decade later, Marian ran into Mrs. Alexander at a local shopping center. The chance encounter yielded up a big hug "like I was some long-lost friend," Marian says. Talking in one of the steel-ribbed corridors of the mall, the women presented a study in contrasts—Mrs. Alexander blue-eyed, blond and rather frail; Marian big and tall and dark all over; Mrs. Alexander the former boss lady who still believed ardently in Bethlehem Steel; Marian the shop steward who had married an officer of the steelworkers' union.

A decade's time had drastically changed the tin-sorting operation at Sparrows Point. The sorting process had been overtaken by automation and reduced demand. Now Marian pushed buttons in a machine station as a coil inspector. The rhythmic flop of piled plate no longer resounded in the cavernous room (stripped of its rows of tables, it had been converted to a warehouse), and millions of dollars of orders bound for Campbell's Soup in New Jersey, packers in East Baltimore, and Dole Pineapples in Hawaii depended no more on the touch of a woman's hand.

"Things are different down the Point today," Marian began.

"I hear things has changed very much," Mrs. Alexander replied. "For the better I guess."

Her remark made an awkward meeting even more uncomfortable. "I really didn't miss her," says Marian, "but I did miss what she represented. Because I loved sorting tin plate because you could touch the tin and it felt good. Because that's what I went there to do and the changes made everybody unhappy. Nobody likes to change."

So Marian shook her head and reassured the Old Lady. "I said to her, 'Oh, no, Miss Alexander, don't say that.' And she looked at me and smiled."

CHAPTER 20
COMPLACENT GIANT

In the high noon of '50s prosperity kids followed their dads into the mills. They did so mechanically, because that's what they were trained to do. The strength of steel in steel country, be it in Sparrows Point or Pittsburgh, could not be divorced from the insular climate it created. "The sun used to come up over the blast furnaces and sink over the 40-inch plate mill," remembers Ed Gorman, the oldest son of Mary Gorman. Going to work for Bethlehem, he says, was "going into the middle class."

"That was it. That was the lifestyle. All the people wanted to work at the steel mill. It was the best damn paying job around here."[1]

Bob Eney felt the same gravitational pull. He was part of the postwar generation of raw material for the works. He was the last of four sons of Milton and Viola Eney, country people from rural Maryland. His father punched his first timecard as a delivery boy for the company store. Later he worked at the dairy owned by the store, milking cows. When Bob was born his dad had been promoted to a general clerk by Linwood Ensor, superintendent of the company store. So delighted were his dad and mom with the promotion that they gave Bob the middle name of Linwood.

"When a superintendent family wanted new linoleum in the kitchen and windowshades for the pantry, they told my father," says Bob. "He then took the stuff from the store and piled it on the company delivery truck. And when he was done he brought home a cigarbox full of drapery hardware. It would all be put on the superintendent's store account and forgotten about. I know the executives got all kinds of stuff from the company store because what was left over went in our garage. You see, my dad played along too."[2]

When Bob turned 16 he decided he wasn't going to follow in his father's footsteps. Wasn't going to take a job with the company, or take satisfaction at rising to some predestined level in the hierarchy. And most definitely he wasn't getting near those mills, whose noise and dust filled him with horror. "I was going to be an architect," he resolved. In his senior year he revealed his secret to those gathered in Electa Bimford's front parlor.

Electa Bimford lived alone in one of the mighty houses on B Street. "She was the widow of some bigwig, but she was good for some of us because she took an interest in the art-room crowd at school. She would invite people like Caroline [a friend] and me to dinner. She was very concerned about education. She would ask us, 'What are you going to do after school?'" Now Bob had an answer to Mrs. Bimford's question. But how to go from a steel town to a fashionable architect's studio?

"I knew that you had to go to college and get special training. Well, I always did well in art class and I was interested in history. As a kid I drew pictures of the furniture sold in the company store. But I had some lacks. I was a poor mathematics student. I figured this was something I had to overcome."

Overcome, but how? This was a question he finally decided to bring up to his parents. "My parents encouraged me to think about architecture, but neither my mother or father were well enough educated to offer me anything but hope. They were from the country, you know, where they didn't educate you and the idea of a college was totally alien to them." They ended up arguing with each other. Because his dad had read somewhere that industrial drawing was sought after by prospective employers, he pushed for a trade school. His mother, meanwhile, thought he should go to an arts college. As a compromise it was agreed that he should talk to his homeroom teacher. "I asked her, 'I need some idea of what it would take to become an architect,' and she told me to talk to the shop teacher. And the shop teacher said, 'Forget it.' He said my aptitude wasn't sufficient in that direction."

Bob disregarded what the shop teacher said. Everyone knew "he was paid to train kids for Aunt Beth"—so while the interview was unpleasant, Bob figured he was on the right track. What he needed was the advice of someone not beholden to the company. "Do you know there wasn't a single college counselor at the school? And we had more than 500 kids in our graduating class. I found out from my cousin that the Baltimore County School Board had college counselors that went around to the different schools. At my cousin's school they had gone over all his questions and told the students how to fill out the application forms. So I waited for them."

Late in the spring of his senior year (too late to apply to many colleges) the counselors arrived. "There were about 40 kids who were in the academic track. They were the superintendents' kids and they had hard-and-fast appointments. For the rest of us [Bob was a general student], we were lined up in the hallway outside the auditorium. There must have been 150 of us."

The mood was festive, Bob recalls, like a class outing. Even several kids from the "greaser car crowd" were there and seemed keyed up. Two hours later the hallway scene was grim. Less than half of the academic kids were through yet, and the only sense of movement came from the occasional swish of the door as the secretary walked over to the principal's office for another set of manila folders.

Another hour passed. The underclassmen flooded the hallways and stared at the seniors. It was past 1:00 P.M. when the doors were flung open. A counselor stepped out. He began calling off names, including Robert Linwood Eney.

"They took us in groups back into the auditorium where these people had tables set up and we sat down, two to a table, with the counselors. They sort of read off our grades and names. The counselor I had was looking at my report card when he asked if I had a particular vocation in mind. I said yes, I wanted to be an architect. And he said, 'Well, your grades don't substantiate that.' "

The counselor's words came in a firm, businesslike tone and Bob took the advice at face value. "I figured that I had really messed up. I just didn't have what it takes." But then the counselor said something that shocked him out of his complacency. "He said, 'Well, all is not lost.' Then he indicated very clearly that what I should do is file for employment at Bethlehem Steel. 'They need employees,' he said, and even told me that I showed an aptitude for a good job! Later I found out that they said the same thing to every one of us."

On that spring day Bob learned precisely and in the most personal of terms the tribute the company demanded of the families it nurtured in the bosom of its company town—and that the company's demands were not subject to appeal. "As far as that counselor was concerned, I was nothing but a mill rat."

Seething with anger and frustration, Bob tried to get around the company's summons. For the next three years he tried to educate himself. He enrolled in a special drawing course at the Maryland Art

Institute. But he didn't "make it," failing in the drafting course. For three years he tried to find a paying job in art. The best he could find was as window decorator for a downtown department store. The pay was a miserable $18.50 a week. "I was terribly discouraged because it stayed with me that I would never be an architect, which is what I desperately wanted to do all my life."

Finally, broke and beaten, he succumbed to his father's pleas and filed for employment at Sparrows Point. The company gave him a clerical job in the continuous hot-strip mill. "I was a tracer in the mill and I was making big money—like $100 a week. And how I hated it! How I hated Bethlehem Steel! It was so dirty. Everything was covered a filthy red—even the rats. They'd come into the office when I was working night shift. And because I had a nonunion job, the other employees made you aware you were different, especially during the overnight shift. The work was nothing and the smell—God, it was terrible. So I enlisted in the army."

A hitch in the army proved a temporary expedient; after it was over Bob found himself back in Baltimore, once again fighting the gravitational pull of Sparrows Point. He had a brother in the mill. His father was there after the company store was closed down. His other two brothers had worked for Bethlehem Steel. Nothing was available to free him from the dismal monotony of a pinched factory existence. Going to his high school class reunion only rubbed it in.

"There was such a sameness. All of those people had gone to work for the company. I felt I had nothing in common with them. Except for maybe four or five people, their interests were so limited. All they talked about was the mill—it was either somebody they didn't like or about the money they were making. Like my friend Caroline—she later married a company guard and everything she did she wanted to relate to Sparrows Point. There were all the old school cliques—the car crowd, the sports crowd, the elite crowd—but nothing beyond that. I saw my whole life rolling out in front of me. Nothing except a big car and a little suburban house."

*

The "macro" view of the period did not take into account such matters as youthful expectations dashed. What was considered important was that factories like Sparrows Point got their fill of manpower. The overall sense on Wall Street and Main Street was that it was the best of times. The economy was suffused with growth and steel was the symbol of American brawn. What made this country strong was simply that it could roll 110 million tons of steel in 1953, overwhelming Communist Russia by a 2½-to-1 margin and exceeding the production of the rest of the Free World combined.[3]

Dwight D. Eisenhower stated the political significance this way. "Steel," he said, "is an important factor in the Communist struggle to dominate the world and the steelmaking capacity of Western Europe must be defended" to reduce the chances of Russia winning a long war. Others praised the ability of U.S. producers to squeeze out more of everything during the Korean conflict when production reached 102 and 103 percent of capacity.[4]

America's domination of world markets was supreme. U.S. exports of steel products outstripped imports by 5 to 1. Bethlehem individually had the capacity to make as much steel as the whole of Germany's rebuilt industry. Its production of 17.7 million tons in 1953 compared to 16.9 million in Germany, and was double the size of Japan's industry. Corporate sales of $2.1 billion were five times more than in 1939. Working capital had more than tripled in the same period. "Obviously considerable postwar meat has been added to Bethlehem's bones, and the bones themselves have grown," remarked *Fortune.* "Bethlehem has maintained its capacity position of 15 percent of the steel industry, half of U.S. Steel, during the industry's greatest expansion period."[5]

If the majority of government leaders and opinionmakers were content with the torrent of hot metal pouring from the open hearths of Pittsburgh, Youngstown, Sparrows Point, and Gary, there was a minority that wondered whether the industry wasn't enormously flawed. From the perspective of an odd lot of free-market advocates, liberal Democrats, and small businessmen, the industry was misusing the current strong market for steel to sit back and extract the highest possible profits from the body economic. To them the trade was the spoiled child of the industrial fates—prodigally wasteful of natural resources and hidebound technologically. They questioned the farsightedness of management and faulted the industry's take-it-or-leave-it stance toward customers.

In dealing with the production of iron, why hadn't Bethlehem or U.S. Steel developed research labs remotely on par with those of AT&T, Du Pont or General Electric? Why, in fact, were highly profitable steel companies at the bottom rung of R-and-D expenditures? Did this explain why the last important production breakthrough, now nearly 30 years old, came from little Armco Steel? Physicists reported that the theoretical strength of iron crystals was 1,000 to 10,000 times greater than their actual developed strength. The problem was to get rid of the chemical impurities that ordinarily existed in commercial grades of steel. The demands of jet engines, nuclear reactors, and gas turbines were taxing the limits of standard carbon and alloyed products. Moreover, the stronger the steel, the less of it would be needed for a given application, thereby cutting down on weight and cost. Improving strength-weight ratio was a persistent quest of materials engineers and product designers. If standard steel could be made twice as strong as at present, the weight of steel in car bodies, skyscrapers, and tin cans could be cut in half, bringing great economies in tow.

Traditional steel production was inefficient in many respects. The energy required for heating, cooling, and reheating metal as it went through the steps of processing was expensive. Variations in the chemical composition of steel in the ladle caused high wastage ratios when the steel had to meet precise mechanical and dimensional specifications. For years steelmen had talked about subjecting their production practices to a thoroughgoing scientific analysis, but little had come from this. The industry was dominated by a mindset eager to defend itself from the inconvenience of change. As economist George J. Stigler noted pointedly, "One can be opposed to economic bigness and in favor of technological bigness."[6]

What ignited the strongest criticism, however, was the constant spiral of prices. Since 1945 steel prices had advanced about 80 percent faster than the average of all industrial prices. Why did prices go up so much? One reason undoubtedly was the pressure for higher wages by the USWA. Unit labor costs advanced on average 5 percent a year—overall about 15 percent faster than labor costs in the economy as a whole. But the industry contributed to the wage spiral by its own strategy. This strategy was to increase prices each time wages were increased, but to raise them *more* than the increase in unit labor costs.[7] As a result, prices were not only chasing wages, they were exceeding them.*

*As an example, steel prices rose 66 percent between 1945 and 1950, according to the Bureau of Labor Statistics' Index of Steel Mill Products. In the same period, compensation for all steel employees, union and non-union, worker and management, had gone up by 42 percent. Between 1950 and 1954, steel mill prices continued to soar compared to unit labor costs, rising 31 percent, or double the pace of labor. (From price and wage indexes of the United States Department of Labor)

The pattern of these increases struck a raw nerve among trust busters. Every summer like clockwork U.S. Steel would announce new "published prices" for steel products. Then another major producer, often Bethlehem, would raise its rates by the same amount. This in turn would be matched by the other integrated operators, usually within days. The behavior smacked of collusion, a follow-the-leader syndrome that perverted the principles of free enterprise and free competition. "Steel," wrote economist Walter Adams in 1954, "must go out and play the capitalistic game according to its naked, shameless, yet vitalizing and invigorating rules."[8]

In testimony before Emanuel Celler's House Subcommittee on Monopoly Power, Dr. Stigler said that price rigidity was one of the baneful consequences of the dominance of too-large firms that had gained their market power through monopolistic mergers. "Not one steel company has been able to add to its relative size as much as 4 percent of the ingot capacity of the industry in 50 years by attracting customers. Every firm that has gained 4 or more percent of the industry's capacity in this half century has done so by merger." He advocated the breakup of the so-called "tropoly" of U.S. Steel, Bethlehem, and Republic into a considerable number of independent companies. "Such competitive organization of the industry seems vastly superior to both the existing organization and to detailed regulation."[9]

The Celler subcommittee heard testimony that concentration of raw materials in a few private hands inhibited technological change and reinforced the pricing policies of the dominant firms. For the most part Midwest steelmakers were dependent on U.S. Steel for traffic leases, docking privileges, and ore-mining arrangements, a legacy of the Carnegie-Rockefeller alliance of the previous century. Only U.S. Steel controlled sufficient domestic sources of ore to sustain its blast furnaces indefinitely.

The raw-material sinews of Bethlehem, of course, lay in a different direction. Owning active mines in Chile and Venezuela, and holding undeveloped ore properties in Oriente and Camaguey, Cuba, Bethlehem was nearly alone in its command of foreign ores coming to the U.S. "I do not think it would make any difference whether you have five companies in the steel business or 500 companies in the steel business, if one or two companies controlled the source of raw material. The monopolistic features would be there in the controlling point," tesified Oscar Chapman, Truman's secretary of the interior.[10]

Aside from the potential for monopoly, there were pages of testimony about actual monopolistic covenants foisted on the independents by the large producers. Frank Smith, an ore producer in Georgia, showed the Celler panel the contract he said he was forced to sign with Republic Steel. It committed him to sell all his ore to Republic, but permitted the steelmaker to suspend its purchases on 30-day notice. According to Smith, Republic had an understanding with U.S. Steel whereby it had a clear field to obtain Georgia ores, which were then shipped to its Midwest mills. Because of railroad freight differentials in the East, Smith said it was impractical for Southern ore miners to seek contracts from Bethlehem or other operators.[11]

On the other end of the spectrum—that of finished products—the panel heard allegations that Bethlehem and U.S. Steel were putting the squeeze on independent firms competing with their product lines. Gustave Koven, president of L. O. Koven & Brothers of Jersey City, said his company could not get sufficient boiler-plate metal from the integrated steelmakers, whose subsidiaries competed with his business. As a result Koven said he had to forgo some contracts and meet others by paying premium prices on the steel "gray market." The large companies, he charged, were acquiring facilities in the fabricated-products business in order to build a safe market for their products and to avoid aggressive competition that existed in the sale of finished steel-mill metals. "The problem as I see it is this: How can we keep the small and medium-size manufacturer from extinction either through merger or sale of assets to major steel companies or by becoming discouraged and closing up?"[12]

Samuel Wasserman, president of American Pipe & Equipment in Baltimore, a small wholesaler of pipe, said he was being put out of business because he had not toed the line on prices orchestrated by Bethlehem Steel's pipe salesman. He said his troubles began when he undersold local jobbers and escalated when he lost a valuable contract with Bethlehem Shipbuilding. Although he was willing to buy his steel from Sparrows Point, he found he couldn't place orders with the steelmaker. Another businessman told him he was being squeezed out because he wasn't "a member of the fraternity" and "I had to be on my own." Testified Wasserman:

> Baltimore used to have a pipe club, where the jobbers, as you gentlemen are sitting here, would meet once a week in the hotel at lunch and discuss prices of pipes and what was being done with pipe in the city. We weren't a member of that pipe

> club, but we would hear of the luncheon after it was held. Mr. McGuire, who was then salesman for Bethlehem Steel in Baltimore, was the unofficial spokesman for the pipe club. Any jobbers that stepped out of line, Mr. McGuire would come around, he would know them personally, he would talk in a friendly manner. A number of times he stopped in and saw me. He would say, "How are things going?" and I would say, "All right, Mac." "What are you doing with prices?" "I am meeting the competition." Or, "What are you doing in the shipyard, are you behaving yourself?" All in a friendly way he would be making these inquiries.[13]

In world trade James S. Martin, a former officer of the American military government in Germany, gave evidence of collaboration between U.S. and European makers to limit shipments to American markets from abroad and to fix quotas in other markets. Documents which he said were removed in 1947 by a U.S. military team from the confidential files of the European Steel Cartel in Luxembourg indicated that representatives of U.S. Steel, Bethlehem, and Republic discussed world steel prices with the cartel and agreed that development of steel manufacturing in Third World countries should be opposed. Martin read a letter from a British steel representative which praised Eugene Grace and Benjamin Fairless by name for their efforts to revive the steel cartel, which had operated before the war as the American Steel Export Association (and prior to 1913 as the International Rail Pool).[14]

Advancing his own view of the situation, laying stress on the domestic market, Chairman Celler voiced the worry that U.S. producers were neither big because they were efficient nor efficient because they were big. Rather they were adept at using their economic clout to intimidate their smaller competitors. Celler expressed lively doubt at the big firms' commitment to technological advances, and echoed the opinion that the firms spurned advances in steelmaking that threatened their hegemony.

As an alternative Sidney Williams, a Warren, Ohio, consulting engineer, testified before the committee. He unfolded a scenario of a "new frontier" of a rationalized, low-cost industry. He cited a number of new developments in Europe, among them the Krupp-Renn smelter of Germany, the Wiberg-Soderfors sponge-iron process developed in Sweden, the modern top-charged electric steelmaking furnace, and the

Tysland-Hole electric smelter developed by Norway Steel in conjunction with the Norwegian government.[15]

Such processes could liberate the industry from the high costs associated with building conventional plants and coke ovens, he said. Top-charged electric furnaces could circumvent the tremendous capital expenses associated with open-hearth shops, while the Krupp-Renn unit, a rotating kiln, by direct-reducing ore to pig, eliminated the intermediary step of coke-making. Such new methods, Williams conceded, were commercially untested and would require cheap electrical energy as well as a plentiful supply of scrap. But because they depended less on primary process and could use a poorer grade of ore, they augured a day when it would be economically feasible to locate steel facilities outside of the Pennsylvania-Ohio-Chicago axis, with small facilities spotted in rapidly growing areas of the West and South.[16]

*

Reaction to the testimony was quick in coming. "All of it," Eugene Grace informed newspapermen testily, "is aimed at forwarding the socialistic state and nationalization and bureaucratic control of business." By his reckoning it was part of the old conspiracy that lurked in Washington—"the theories of the New Deal."[17]

Relations were never easy with the "Star of Bethlehem." As the commander of a company committed to stamping out anything that did not conform to its quarterly bottom line, Grace was increasingly unwilling to countenance criticism—or even to hear it. His isolation had increased since the war. Having deeded to Ben Fairless the position of industry spokesman that Schwab once held, Grace stayed behind in his Pennsylvania mountain retreat, venturing out for trips to a favorite golf course in South Carolina or business meetings in New York. An officially sanctioned portrait in *The New York Times*—arranged to coincide with his 80th birthday—gave a glimpse of Grace's life as a sheltered deity:

> Mr. Grace spends at least three hours a day reading newspapers from Bethlehem, New York, Washington and elsewhere. He's particularly interested in the financial and sports pages, and can rattle off major league batting averages as readily and accurately as the vital statistics of steel plants. He attends at least one game of every World Series.

> But golf is his favorite hobby after steel. He used to say, "I have shot my age beginning at 67, and that's becoming easier all the time."
>
> And asked recently why the company did not split its stock, he replied, "It took us some time to get into the blue chip class. We're enjoying it."[18]

His underlings mirrored these parochial sensibilities. The 15-member board was still wholly "inside"—nobody but officers sat on the board and none of the officers was permitted on anybody else's board. With the return of stockholder passivity in the '50s, the board could do pretty much what it wanted. The lineup now included, besides Chairman Grace and President Homer, the ten departmental vice presidents, comptroller Frank R. Brugler, and two officers from the treasurer's office. Except for the vice president of shipbuilding, all of the executives lived in the executive compounds around Bonus Hill and Weyhill. It was a cushy life. On Wall Street Grace's company was called "the biggest little company in the world."[19]

As in the old days, the officers lunched daily in the corporate dining room, occupying leather chairs with their names affixed to gold-plated plaques, and on Saturdays they socialized at the Saucon Valley Country Club. While not seeing fit to pay for a decent R-and-D laboratory, they didn't stint when it came to their own luxuries. The executive country club had undergone a multimillion-dollar refurbishing courtesy of corporate funds. The main clubhouse had been redecorated with chandeliers and other fixtures. A three-pool swimming complex had been built, and indoor squash courts added for the executive heavy hitters. "Even the bridge crossing little Saucon Creek was a showpiece, built for an estimated $100,000 in 1950 dollars," wrote John Strohmeyer, former editor of the *Bethlehem Globe-Times*. "Bethlehem Steel paid all the big bills, sparing members the capital assessments that normal country clubs would impose."[20]

The privileged existence was exemplified not only by the continuance of Schwab's famous bonus system, but by the "ingots and bourbon" parties renowned on Bonus Hill. Key dates on management's calendar revolved around these affairs, where a discerning Bethlehemologist could pinpoint subtle shifts in the pecking order of staff. Other highlights of the calendar were Alumni Day, when Grace's surviving Lehigh classmates made speeches about what they had done in the past year, and Christmas Night, when he and his wife Marion hosted a buffet supper for 100 steel executives and their wives. Yet another ritual was to take the incoming class of Loopers to the Schwab Library for a peek at the keepsakes of the founding father.

Shorn of its capitalist frills, the organization had much in common with an Eastern-bloc Communist country. A single dominant party fused ideology and company politics in a way that made conformity paramount and smothered originality. Officers were groomed in an elaborate internal management program that began immediately after college and continued through years of slow but steady promotions within the bureaucracy. It was a closed system, fundamentally antidemocratic, that attracted and rewarded good soldiers. Getting a salaried job at Bethlehem meant security for life. White-collar employees were protected from cyclical downturns, unlike their blue-collar brethren, and were paid during union strikes.

Even Bethlehem wives had a role to play, according to insiders. "Her wardrobe, topics of conversation, general appearance, drinking habits, skills as a hostess, and certainly her devotion to her husband and his career [were] carefully scrutinized," wrote Strohmeyer. "Certain taboos—such as a wife with a personal career—were rarely tolerated. An inclination to the Democratic Party was considered a sign of disloyalty; and if politics was discussed at a dinner party, it was a man's topic and women were expected only to listen."[21]

The same inbred spirit permeated the seat of production at Sparrows Point. With local management engrossed in honing its skills in daily production, one had to look long and far to notice anyone questioning the basic premises of integrated steelmaking. They were too busy speeding up tap times and labor productivity.

No. 1 open-hearth shop boasted the best average heat times, tap to tap, in the country; No. 2 shop was second. It was such evidence of managerial prowess that made the pulse quicken. In the blast-furnace department, furnace H was racking up record production totals. In 1951 it twice broke the world record for monthly pig output, producing 55,835, and 56,010 tons. Overall Bethlehem's labor costs were almost 3 percent less than U.S. Steel's, while the company produced 2 percent more metal per worker.[22]

According to Joe Hlatki, the mental pressure was going up in direct proportion to management's efforts to speed up production. As first helper, he was responsible for timing his furnace heats down to the minute, while other crew members logged in the charging times, pouring times, and teeming times. These steelmaking facts were handed up the chain of command for scrutiny. Once when his crew broke an in-house record on a heat, he got a bottle of liquor. For the most part, though, his reward was in his paycheck, which had gone up to about $500 a month. "They got the work out of you all right. You'd get a melter foreman down there that when you try to say somethin', he'd go, 'No, no. no! We ain't got the time. We'll lose too much time!'"

With the supreme test of steel manufacturing defined as squeezing the most out of men and machinery, Sparrows Point reigned supreme in the business. Says Hlatki's shopmate, Ben Womer, "We had the steel mill that the other people wanted. Hell, none of this business about Japanese steel or German steel. People came to visit us. We were making steel and making it faster than anybody else."[23]

The same good opinion was offered by C. E. Clarke, general manager of the Point. "I'll never see the day when Bethlehem stops growing," he boasted in an interview with *Fortune.*[24] To underscore his point he invited the writer to take a good look from the window at the general office. Here through dusty panes was the company's view of itself: a ponderous collection of pipe-wreathed furnaces and conveyances bristling with enough liquid steel to frame out a 30-story skyscraper every day.

*

The mood among the 1,200 executives who assembled at the Waldorf-Astoria in May 1955 for the annual meeting of the Iron and Steel Institute was similar to the instinct of 1929. A giddy confidence ruled. Steel was in the midst of its best year ever: It was projected that the industry would put out a record 115 million tons of metal in 1955. (The actual amount was more—117 ingot tons.) That was fully 31 percent above the wartime high of 89 million tons in 1944. Many mills were booked solid into the next year. For the immediate future Eugene Grace told reporters that he'd be disappointed if operations sagged below the 100-percent mark. "It would be hard to imagine a high-placed steelmaker saying anything more startling than that except during an all-out war," *Business Week* reported. "The steel business [has] burst through its own built-in sonic barrier and assured itself that the future is probably as good as it looks."[25]

But in two days of champagne and self-congratulations, it was Arthur Homer, heir designate of the Bethlehem chain, who rallied the steel fraternity with the bubbliest talk of all. In a keynote speech, Homer predicted that steel was on the verge of its greatest growth period ever. Overall he predicted that the trade would have to increase capacity by over 50 percent, or 60 million tons, by 1970 to keep pace with demand. "As someone has said, the American people are 'wanters.' Their wants are going to require a great deal of steel," Homer said, noting:

> It is estimated that the total national production of goods and services may be expected to grow from the current level of some $370 billion to an annual average Gross National Product of perhaps $570 billion in 15 years. We know that the growth in the demand for steel in the past has closely paralleled the growth of the Gross National Product. Therefore, if the projections for the economy as a whole are realized, the steel industry will need an ingot capacity of about 180 million tons in 15 years to handle peak requirements.[26]

Such expansion would add hugely to the nation's capital spending stream. Steel had the largest stock and equipment of any industry—$12.3 billion in 1955 dollars—and shared first place with automakers as a capital-goods purchaser. At an average price of $150 per ingot ton, the 4 million new tons per year cited by Homer would translate into the expenditure of $600 million a year, without counting the replacement of obsolete equipment. If replacement was done over a 30-year cycle, that would require another 4 million tons of steelmaking equipment a year. Using the same $150 tag, that would add another $600 million to steel's capital spending, or $1.2 billion per year for new and old plant over the next 15 years.[27]

Homer's assurances that steel was poised for its greatest spurt in history was based on his corporation's reading of the past. It was significant, though scarcely surprising, that Homer assumed that steel's growth would be impaired only if the national economy stumbled. While the steelmen knew that some steel lines would experience declines in sales (rails, for example), overall demand would continue to equal, if not exceed, the rate of the GNP, they reasoned. Happily Homer recited statistics to demonstrate that per capita consumption of steel had risen every decade with the exception of the '30s. Use of steel in the United States was at its historical peak—1,500 pounds per man, woman, and child. There was little worry expressed about imports for they amounted to under 1 percent of the national market. Japan, for instance, shipped only 96,371 tons to the United States in 1955, less than 1/1000 of the output of the domestic industry.[28]

Having replaced wood, glass bottles, and cast-iron products over the years, steel was one product, management assured itself, that could not be replaced by substitutes. Consider the "war baby," aluminum: It was still a drop compared to the titanic bucket of steel. The biggest single use of the light metal was for residential sidings and window-frames. Colored aluminum exterior panels made their debut that year on an office building in Cincinnati. Otherwise the metal was primarily an engineered material used in engines and frames of airplanes, in automobile components, and for special extruded parts. Over a few years it had also been tried in experimental passenger trains. The latter had proved unsuccessful because the trains bounced too much and didn't trip safety signals properly. Except for pots and pans, aluminum was not a big item in the U.S. kitchen. Stainless- and enameled-steel products were less expensive and more popular. The Reynolds Company was experimenting with ways to die-roll aluminum for the can trade, but the process was not in commercial use. As steelman figured it, the aluminum can was likely to go the way of lightweight trains.[29]

The plastics industry was similarly in a period of laboratory gestation. Output of plastics and resin products stood at under 2 million tons in 1955. Significant interest was centered on various heat-resistant polyethylenes, which were under experimental testing. The year did prove important for the introduction of thermoplastics by American Cyanamid and Hercules. Products molded from thermoplastics were able to withstand the heat of boiling water without distortion. This discovery would portend a breakthrough in kitchenware. For the most part, though, plastics were concentrated in industrial uses, chiefly in the building trades, in such forms as vinyl resins for flooring, adhesives, and paper-treating. Phenolics and other tar acids were used for lamination, adhesives for brake lining, molding, and coatings. These tar-acid plastics were derived from the by-product coke chemicals made by steel operators. For Sparrows Point, the plastics industry promised to be a steady customer of coke chemicals.[30]

Even if plastics and aluminum managed to make inroads in building products and household goods, how could they possibly replace steel in the giant capital-construction projects of the time? Bethlehem's fabricated-steel division, the former McClintic-Marshall organization, was at work on many bridge and building contracts. The division had handled all steelwork on the 4.35-mile Chesapeake Bay Bridge, which opened in 1952. Steel consumed: 60,000 tons. The division currently had contracts to build the 666 Fifth Avenue building in New York, the 60-story Chase Bank headquarters, a new Senate Office building in Washington, and the Mississippi River Bridge at New Orleans.[31]

In his talk Homer could take further pleasure in how steel had grown as an integral part of the auto business. On V-J Day there had been 25 million cars registered in the U.S., down from 27.5 million in 1940. By 1955, registration exceeded 50 million. Not only had motor-vehicle ownership doubled in ten years' time, but the amount of steel in cars had gone up to a record 3,000 pounds per unit. "Unisteel bodies" were the rage of Motor City. They consisted of heavy welded steel bolted to the frame, with other sheet-metal components like hood and fenders attached. The popular chrome grill was actually steel coated with a thin layer of chrome. General Motors touted the extra rear-end weight of its Chevy Bel Air as the "style which moderns definitely prefer." Heavyweight cars, in combination with V-8 engines and 4-barrel carburetor power packs, made for "sure, constant control under any conditions," a GM ad informed the buying public. "In it you laugh at curves and hills that make others strain and sway!" The ad showed a Bel Air streaking up Pikes Peak "in record-breaking time."[32]

By the end of 1955 Detroit's consumption of steel topped 18.7 million tons—almost three times the 1946 total of 6.6 million. One out of every 4 tons of steel output went into the manufacture of the internal-combustion machine, a rate not unlike that for railroads before the First World War.[33]

*

"With respect to labor matters, we can also point to progress," Homer went on at the Waldorf gathering. "There has been a heartening improvement in the relationships between steel companies and the labor officials representing so many of their employees. It has been gratifying that union officials have shown an awareness of the importance of complying with agreements—not only with regard to slowdowns and wildcat strikes, but also in other aspects of the contractual relationship."

Labor did seem becalmed under the regime of David J. McDonald. Since his election as president after the bitter 1952 strike and Truman's nationalization effort, McDonald had started a publicity campaign to mend the breaches between management and labor. To this end he had convinced U.S. Steel's Ben Fairless of the wisdom of touring the corporation's properties together. The tours eventually covered all of the firm's plants, with McDonald going through the mills shaking the hands of the rank and file and taking local officers aside for pep talks. In the same cheerleader spirit, Fairless announced that labor and management were "inseparably bound together in a state of economic matrimony." According to *Iron Age*, "His relations with David J. McDonald, president of the United Steel Workers, has [sic] blossomed into a 'Ben' and 'Dave' friendship."[34]

The peace pipe was smoked during the 1955 round of bargaining: Under the wage pact signed by U.S. Steel, Bethlehem, and other operators on July 1, wages were raised by an average of 15 cents an hour; they went from $1.57 to $1.685 for the bottom mill jobs, and graduated up to job class 30 where pay was boosted from $3.275 to $3.545 an hour. The industry then passed on the higher labor costs to the steel customer, as it had done every year since 1950.

U.S. Steel led the announcements of general price hikes, effective in July 1955. They averaged $7.35 a ton for more than 50 heavily used carbon-steel and forged products. A sampler from U.S. Steel's price list: standard structural shapes up $7 a ton from $128 to $135; hot-rolled strip, up from $81 to $86.50; galvanized sheets, from $109 to $117; concrete reinforcing bars, from $86 to $93; and standard T rails, from $89 to $94.50.[35] The size of the boost was a surprise to many analysts, who had forecast a rise of no more than $6 a ton. But the prices stuck and the ramifications of the increase were felt across the country in the four out of five products that used steel.

CHAPTER 21

THE EDIFICE COMPLEX

The steel companies glided contentedly along the highway of established practices, the heavy-metal, high-octane Chevy Bel Airs of U.S. industry. As the producers rolled up profits in a tight market, they figured that everybody would simply get out of their way as they divided up orders among customers. "Our salesmen don't sell steel, they allocate it," gloated a Bethlehem official.

The trade had even invented a vocabulary to justify its trust-based mode of organization. "Meeting the competition" was management talk for the matched prices that the ten important steel operators instituted nationwide. "Unfair competition" was anything that might undercut these uniform prices. "Prudent investment of capital" was investment in open hearths and other conventional equipment. "Inelastic demand" was the ostensible reason why steel was outside the laws of supply and demand and why the trade could advance prices with impunity.[1]

Arthur Homer added to the store of circumlocution when he was called before a congressional panel to explain why his company never varied its prices more than a nominal dollar or two from the published prices of U.S. Steel. "If we should lower our prices," he asserted, "then it would be met by our competitors, and that would drop their profit so that we would still be right back to the same price relatively." His dumbfounded interlocutor, Democratic Senator Estes Kefauver, wondered whether Homer wasn't proposing a new theory of price competition—that of safeguarding the profits of the least efficient manufacturer:

> SENATOR KEFAUVER: But the trouble is, Mr. Homer, you are operating under the umbrella of those who are not as efficient as you, as indicated by the profit figures. If you would lower your prices, which might bring your profits down a little bit, not much, then the inefficient would have to get more efficient or they would fall by the wayside, and that would be true competition.
>
> MR. HOMER: I cannot see that you would get anywhere at all by a process such as you are suggesting, because one company is more efficient than the other. You are, in effect, saying that to be efficient is a bad thing, because you want to turn around and give a lower price, and then your competitors meet that, and then the next time you give them another one, and pretty soon you end with with a zero proposition.[2]

Eisenhower did not have the stomach to deal with steel and prices. Lulled by prosperity at home, the president found ways to ignore the upward march of steel prices during his two terms. He was at his Gettysburg farm, celebrating his 39th wedding anniversary with Mrs. Eisenhower, when U.S. Steel announced the $7-a-ton increase in 1955; the president could not be disturbed, press secretary James C. Hagerty said. In another episode the president said confusedly to reporters that, while "I do stand firmly upon the idea I advanced, which is that government alone cannot preserve a sound dollar," he would not ask the Justice Department to look at the merits of steel price increases. "Very naturally, I don't have the exact knowledge that would allow me to make [such a request]," he explained.[3]

The policy of "sales reciprocity," pioneered by Bethlehem, strengthened the hand of the big producers in obtaining a share of orders regardless of the cost or merit of their product. Schwab had summed up reciprocity succinctly in a 1926 speech: "Be friendly with us, and we will be friendly with you. Help us and we will help you." Bethlehem's Reciprocity Department carefully reviewed the "friendliness" of suppliers. An example was provided by Samuel Wasserman in his testimony before the Celler committee. He said that after he had won a contract to supply valves to Bethlehem Ship, he got a call from a Bethlehem representative who asked him for the name of the manufacturer of the valves. It was Mercer Tube. Checking the ledger books, the representative found that Mercer had not made equal or equivalent purchases from Bethlehem. Because of this infraction, Wasserman said he was stricken from the company's list of eligible contractors; this punishment, he said, had forced him to go out of business.[4]

If Bethlehem and other steelmakers cooperated on pricing and market allocation, there still was one arena where members of the fraternity butted horns. Scale, size, tonnage—notions that had mesmerized the industry since its formative years—loomed large as ever in the macho imagination of management. Steel as a tale of progress to greater brawn appealed enormously to steelmen. The whole urge of the trade was to keep on building, to put earnings right back into existing mills. In this way each firm's capital structure was preserved as rigidly as its prices and market share.

On January 26, 1956, Eugene Grace appeared with Homer in the office at 25 Broadway that adjoined the boardroom where Charlie Schwab had maintained his private quarters. He confirmed to awaiting reporters that an enormous building program was slated for Sparrows Point. The board of directors had authorized $300 million for the expansion of steelmaking capacity from 20 million to 23 million ingot tons. Of this expenditure, $200 million was to be spent at Sparrows Point to enlarge its ingot capacity by 2 million tons to over 8 million tons a year.

The numbers were important. Schwab's goal of 40 years' duration was about to become a reality—Sparrows Point was to become the single largest steel plant in the world. The expansion would boost the plant's rate of production from 720 to 930 tons an hour, or greater than 22,000 tons per day. It would add approximately 3,000 workers to the mill, bringing the number of employees to 30,000, and raise the weekly payroll from $2.25 million to $2.8 million. Truly, promised Grace, rattling off these enormous figures, Sparrows Point would be "the greatest show on earth."[5]

At the same press conference Grace astounded financial analysts with the disclosure that Bethlehem would bankroll the entire capital expenditure without venturing outside its Pennsylvania fortress for capital. "The money will come out of our bank accounts," Grace exclaimed. Capital reserves were at a record high of $801 million, including $690 million in cash and marketable securities. With operations averaging 97.9 percent system-wide—and 104.6 percent at the Point—word was out that the new capacity must be put on stream as rapidly as possible.[6]

The scope of site preparation was illuminating as to the size of the project. In Frederick Wood's time the underlying strata of Arundel clay, judiciously mixed with granite blocks and boatloads of oyster shells, had been able to hold up the steel works. By now, however, heavy construction had evolved to a state where the clay had insufficient shear strength to withstand the cumulative weight and thermal shock of the furnaces; steel pilings were necessary to support the structures. At the excavation pits a squadron of steam pile drivers started hammering, banging down steel piles to a depth of 125 feet. At the open-hearth site alone 5,548 piles were sunk.[7]

Above these sturdy footings Bethlehem's structural workers pieced together the largest open-hearth shop in the world. Stretching the length of two football fields, the colossus was named No. 4 shop. The Boswell of No. 4's attributes was George C. Little, assistant superintendent of the mechanical department (and company witness in the Charlie Parrish arbitration). Flowing from his pen onto the pages of *Iron and Steel Engineer* were the furnace "stats" and flow capabilities that excited steelmen. Final specifications called for seven identical furnace chambers 104 feet, 3¾ inches long, 25 feet high, and 28 feet,

10½ inches wide. Combustion air was delivered by a forced-air draft fan from basement heaters at a rate of 50,000 cubic feet per minute. The main burners fired 1,200 gallons of Bunker C oil per hour into the furnace maws.

And there were more superlatives. Each furnace had a horizontal fire tube boiler, with superheater, that could produce 300,000 pounds of waste gas per hour at 560° F. The waste-gas flues were big enough to permit maximum firing around the clock. The smoke and red dust were gathered and discharged through a smokestack that topped all previous stack dimensions—240 feet high and 12 feet in inner diameter.[8]

The output of the new furnaces exceeded the limits of existing handling equipment. The solution: more gargantuan ladles and pit cranes. Little happily described the steel ladles and the pit crane as "the largest ever built." The crane was designed to handle 500 tons on its main hooks. The weight of the crane lifting a loaded ladle across the floor comprised a moving load of 1,400 tons, the same as lifting 900 Chevy Bel Airs. Little noted:

> This unusually heavy moving load has made necessary a new design of end trucks and runway girders. The crane is carried on 24 30-inch diameter wheels on each end. These are mounted on 12 double-wheel axles and operate on two 175-pound rails, which are carried on two runway girders. Each girder is about 12 feet deep and 120 feet long. The physical dimensions and weight of these girders are very near the one-piece shipping size limit of a structural member.[9]

Stunning yet also highly self-referential. In designing the mill, mechanical staffers strayed little from accepted practice. Uppermost in their minds was the avoidance of anything that might upset target output—or elicit disapproval from their superiors. The fact that they were reporting to a vice president of operations (Stewart Cort) who had started his career in the open-hearth department guaranteed that nobody was going to stick his neck out. "The thinking was, if Mr. Cort trained on it, then it was fine, except maybe getting more tonnage out of it," recalls a supervisor assigned to another mill.[10]

In designing the tenth blast furnace for the mill, the engineers again gazed back reverently at earlier equipment. With evident pride Little characterized the new furnace as "a near duplicate" of the H furnace built in 1948. The Russians may have had success with high top-pressure techniques and the Germans may have developed a direct-reduction blast furnace, but such methods were not for Bethlehem. "We don't want to invest in a facility unless it will return, on the average, 20 percent before taxes, operating at 60 to 70 percent capacity," Comptroller Frank Brugler instructed his subordinates.

According to Little, the biggest challenge that faced the engineers in designing the No. 12 coke oven was constructing the battery over open intake drains. Otherwise the battery was "a near duplicate" of the No. 11 battery. Additional installations erected in conjunction with the steelmaking facilities were all of World War II vintage. Prominent among them were an auxiliary ore storage field, a desulfurizer plant, sintering plant, an enlarged pier, seven blocks of soaking pits, a 45- by 90-inch universal-slab mill, a 160-inch sheared-plate mill, a galvanized (zinc-coated) sheet line, two electrolytic tinning lines, and an electric-resistance pipe mill.[11]

Whether these facilities were very efficient or merely very productive was a question that Little did not address in his article. He was too busy totaling up the materials used for the facilities: 79,000 tons of structural steel, 59,000 tons of steel pilings, 122,000 cubic yards of poured concrete, 15,000 tons of reinforcing rods, 70 miles of pipe, and 21 miles of double-track railroad. In all, about 179,000 tons of steel were used. The Golden Gate Bridge had been built with 100,000 tons of steel and the Chesapeake Bay Bridge with 60,000 tons. In effect, the engineering department had succeeded in cramming the two famous bridges into an already crowded peninsula.[12]

*

At this juncture Bethlehem was making crucial decisions not only about its own future growth and technology, but about the environmental conditions around the Point and the Chesapeake Bay in general. As the plant doubled in size between 1945 and 1957, its thirst for water rose from 450 million to 640 million gallons a day. One hundred million gallons of industrial effluent from the city sewage plant were now circulated through the complex and 540 million additional gallons were drawn directly from the harbor.

With the aid of its $200-a-day consultant, Dr. Abel Wolman, the company still used the city's treated sewage at the same minimal rate negotiated during the war. Bethlehem Corporation now had use of 100 million gallons of water a day for the grand yearly fee of $48,000. By contrast, in Amarillo, Texas, local industry paid the city government $100,000 a year to use only 8 million gallons of treated effluent a day.[13]

Every day the same Niagara of 640 million gallons rushed out of the mill from 20 outfall sewers and artificial lagoons. The liquid contained corrosive pickling liquors, steel chips and flue dust, oils, heavy metals including lead, copper, and nickel, coke-tar chemicals, and an array of other hazardous substances. In a report Henry Silbermann, district engineer for the Maryland Water Pollution Control Commission, estimated that about 590 tons of waste pickle liquor were discharged from the plant daily. On average the liquor was composed of 17.6 percent sulfuric iron and 5.6 percent sulfuric acid.

"No treatment is attempted at this time of waste pickle liquor," Silbermann wrote in his 1954–55 report for the pollution commission. "The most extensive pollution results from the discharges of waste pickle liquor which has caused pH of 4 in Bear Creek and floating patches of iron hydroxide. The bottom of Bear Creek is also coated with iron precipitate."

Silbermann cited other pollution hazards in his inspection report. He noted that steel chips and iron particles drained from the rolling mills were placed in settlement basins known as scrap pits. While an overhead clamshell scooped the iron sludge from the basins, the leftover material was washed down the outfalls. Waste palm oil from the tin-plating lines was another problem spot. A recovery plant took in waste water containing 500 to 600 parts per million of the oil. After a separation process and skimming, the water was discharged into Humphrey's Creek, mostly in an emulsified sludge form. It had 70 to 80 parts per million of oil, which formed oil slicks on the surface of the creek. Some of the oil was collected by a skimmer at the mouth of the creek and removed manually. And the rest? So-called "overflow waters" conveniently carried the debris into the harbor.[14]

For a half dozen years Inspector Silbermann had dutifully filed his reports about mill pollutants and for half a dozen years the steelmaker had done as little as possible to abate the problem. As Dr. Wolman advised, industry should involve itself in what was "feasible" and "practical" and ignore indirect or little-understood chemical-biological reactions. But Wolman's strategy of avoidance was increasingly difficult—the chemicals in the discharges were misbehaving, causing an environmental crisis around the Point.[15]

One problem was the outbreaks of algae growth. They formed when the organic effluents in Wolman's industrial water system mixed with mill pollutants like sulfuric acid. The algae formed in the polluted bottom waters came to the surface in ill-smelling, greenish mats that sometimes grouped into floating "islands" as large as 100 feet around. What's more, the sludge on the bottom gave off gas in the summer heat, making the water look like it was boiling. In daylight hours the algae produced oxygen which lured schools of fish into the area. At night the algae gave off carbon dioxide, depleting the oxygen supply and sending the fish gasping to the surface.

Fish kills (called "catastrophes") were reported around Sparrows Point at Humphrey's Creek, which flanked the tin and sheet mills, and at Back River, where some Bethlehem wastes were discharged with the city's normal flow of sewage. In 1952 the Back River Improvement Association accused Bethlehem of responsibility for a fish kill. The company declined to comment. Abel Wolman was not so rash as to call attention to the matter. He advised the company to pour more chemicals into the creek. By utilizing "newer algicides" he was of the opinion that the algae growths could be controlled at Humphrey's Creek. Definite action was delayed by the company and Wolman didn't press matters. In the meantime the seasonal run of herring up the creek ceased. A small but ecologically significant piece of the estuary ("the immense protein factory of Chesapeake Bay" in Mencken's coinage) was slowly dying off in a bath of pollutants and fermenting water.[16]

Such high-volume releases of plant wastes, however, could not be contained indefinitely in the local waterways. In his report to the polution commission, Inspector Silbermann packed a bit of punch. He concluded under "Needed Improvements" the following: "Eliminate the major source of pollution produced by waste pickle liquor. Improve treatment of acid plant effluent. Maintain all treatment devices at top efficiency." It was time for more environmental "handlers" to get to work.

They were John C. Geyer and Charles E. Renn, sanitary engineers at Wolman's Department of Sanitary Engineering at Johns Hopkins. Among their jobs was to examine the coloration problem of the pickling discharges. "The color was striking against the surrounding water," Dr. Renn remembered in an interview. "It was a deep, slaughterhouse red from the suspended iron molecules in the outflow. You could see it from the air. It caused a sharply defined bloom. The locals called it 'blood water.'"[17]

Engineering controls had been developed to reduce such pickle outflows. One technique called for settling tanks to capture the iron-acid wastes before the liquid was discharged. As an alternative, the pickling liquor would be treated and returned to the mill for reuse. The latter course had been followed at the Kaiser works in Fontana, California, where fresh water was at a premium. The Kaiser plant used only 1,400 gallons of water per ton of steel produced, or about one-tenth of that used at Sparrows Point. "Can other steel companies do the same?" asked Edmund B. Besselievre, a sanitary engineer who helped design the Kaiser system. "Of course they can.... It would cost something to rearrange their mills for this purpose, but the result in lessening the demands of the available water supply would be tremendous."[18]

But Renn and Geyer didn't consider conserving water or reducing the overall volume of pollutants viable options for money-conscious Bethlehem. Early on Dr. Geyer had written to C. E. Duffy, fuel engineer, suggesting that the company conduct its own waste and creek surveys. The company could then propose a "reasonable" program to the Water Pollution Control Commission "prior to a hearing" before the public. In this way groups like the Back River Improvement Association would have no say in the final recommendations.

Soon thereafter, a colleague of Geyer's at Hopkins, Donald W. Pritchard, had a better idea. He suggested that Bethlehem finance a study to see whether the acid wastes couldn't be discharged from pipes placed under the harbor, thereby diluting the wastes and getting rid of the telltale red coloration. This was a cosmetic fix and Bethlehem gave the go-ahead for a feasibility study through Pritchard's Chesapeake Bay Institute, a research group cofunded by the state government and Johns Hopkins. The upshot was "Controlled Discharge of Acid Waste Through a Submerged Pipeline," a curious little report that was submitted to state regulators. The report stated:

> An experiment involving the introduction of an acid waste (spent pickle liquor) into the harbor waters from a temporary rubber pipe line, with measurement of the dilution and dispersion of the waste, was conducted by the Chesapeake Bay

Institute of The Johns Hopkins University and The Sparrows Point Plant of the Bethlehem Steel Company.

Figure 1 shows the area in which the experiment was conducted, and gives the location of the pipeline. The release point for the wastes was located some 200 yards off shore on a relatively flat bottom having depths of about 15 feet. This depth was representative of an area extending several hundred yards outward from the release point, in which the slope directed toward the channel was very small.

The bottom in this area was composed of a surface layer of soft mud having a high percentage of powdered coal and probably some unidentified coal products.

Observations of the background pH of the receiving waters in the vicinity of the pipeline were made on 3 October and on the morning of 4 October. Considerable variation in pH was noted between the two days, with the pH on the 4th being considerably higher than on the 3rd. These observations are tabulated in Tables A-1 and A-2 in the Appendix.

Information regarding time of initiation of pumping of acid wastes, rate of discharge, and periods when pumping was interrupted due to some mechanical difficulties, will be supplied in a separate insert when this material is made available by the Research Division, Bethlehem Steel Company.[19]

From the measurements taken, the researchers could not determine with any degree of accuracy the chemical interactions between the bottom waters and the pickling acid. But that did not deter Dr. Pritchard from leaping to the conclusion that harbor waters could withstand much greater loads of pickle liquor, either through underwater pipes or surface outfalls. This was the argument sanctioned by Dr. Wolman's sanitary engineering department—that the "assimilative powers" of the harbor were great and, hence, evidence of localized deterioration had to be balanced with the harbor's overall health. This premise was taken up later by the director of the Water Pollution Control Commission, a bureaucrat named Paul W. McKee, who stated haughtily, "Where I sit I have a responsibility for proper economic development in the state. I can't simply keep everything the way it is, say we have to preserve everything as John Smith saw it when he came up the Bay."[20]

Out of this confluence of vested interests there emerged a pickle-acid "limit" for Sparrows Point that raised permissible discharges by some 65 percent. The pollution commission, "guided by the statements and expert testimony" of the Hopkins staff, permitted an equivalent of 180 tons of pure sulfuric acid pickling liquor to be discharged daily. Academic science had helped the state's biggest employer evade its environmental responsibilities. And as fish catastrophes became more and more commonplace in the 1960s and early '70s, Director McKee joined Bethlehem in decrying the "mysterious malady" which was killing so many harbor fish.[21]

*

On October 2, 1957 the crown of "world's largest" passed to the peninsula from the Gary works when the first heat of steel was tapped from No. 4 shop. Sparrows Point now towered above Homestead and the other famous mills of western Pennsylvania's Mon Valley, above Johnstown, Birmingham, Youngstown, and Lackawanna. Some 28,000 men and 600 women were at work at all hours of the day and night on the peninsula. From their labors 1/15 of the nation's steel could be produced.[22]

There was nothing in the Free World that could match the muscle of the Point. Japan's major plant, the Yawata works, had an operating capacity of under 3 million tons. (The rapid progress that would characterize Japanese steel in the 1960s was not yet in evidence.) The largest mill in Europe was the Dortmund-Horde mill in Germany, with 2.5 million metric tons of capacity. Port Talbott was the largest in Britain, with slightly less capacity. The Russians also had not reached the level of magnitude of Sparrows Point. Their showpiece, the Magnitogorsk works in the central Urals, was believed capable of pouring 5 million tons of steel annually.[23]

Sixty-seven years after it had first lit up the skies of the harbor, Sparrows Point was the hub of a manufacturing wheel that clacked out some $2.8 billion worth of goods in 1957. Blue-collar Baltimore was second only to New Jersey's Chemical Alley in terms of the intensity of industrial development and the dollar volume of business on the East Coast. If Kansas was flat and Chicago was tall, then Baltimore was a dingy industrial city, marked by block after block of look-alike rowhouses, a fading downtown, and traffic jams that tried the patience of citizens motoring through from Washington or New York.[24]

More precisely, Baltimore could be thought of as a tidewater West

Virginia, a port of call for the import and assembly of bulk commodities for large conglomerates headquartered elsewhere. It was the place where superphosphates were made for Olin Mathieson, petroleum for Esso, soap and phosphate detergents for Lever Brothers, titanium dioxide for Du Pont, cadmolith colors for Glidden, silicofluorides for W. R. Grace, road oils and paving emulsions for American Bitumul, animal grease and tallow for Braun Rendering, and many more products that got one by the throat.

Now heading the roll call of heavy manufacturing was the thicket of conveyors, docks, and liver-red mills at the mouth of the harbor which once had inspired Colonel Agnus's story of the Scottish immigrants coming to America. "Is that the Goddess of Liberty?" asked a passenger who sighted the glow of light on the bay. "No," said the captain, "that is the Goddess of Industry—Sparrows Point."

The goddess flung her veil far and wide. Gaseous clouds of sulfur dioxide mingling with iron particulates traveled the wind currents of the bay, approaching distant shores like a sea fog. At Tilghman Island 35 miles away oystermen could sight the mill's eerie red glow on the horizon. At close range, a shroud of gritty orange smothered the landscape. On some mornings workmen reported seeing the sun over the blast furnaces, but on many more mornings one would never have even suspected it was there behind the angry-looking smoke that never ceased to stream from the stacks.

By night the peninsula burned bright from "mill reflections," the intense glare of light and heat that bounced off the clouds. Ed Gorman says because of the mill reflections he never saw the true sky until he was in the air force and stationed in Colorado. "Out there you could go out and count a million, billion stars. Well, that was astounding to a mill boy like me."[25]

*

HOMES STOOD HERE, said the caption of a photograph in the January 17, 1956, *Sun* showing bulldozers and bonfires "systematically eating into" the little cottages that still wore the prim roof moldings of their 1890 vintage. In building the world's biggest mill, the company tore down a third of the "Pullman of the East," paternalism being no longer necessary in a period of labor unions and easy mobility. About 200 houses were razed and 1,250 residents displaced. "It was a shock to us to have to move out," Mrs. George J. Nelson told the newspaper. She had raised her 12 children in a seven-room house on D Street, paying $28 a month in rent. The family moved to a converted barbershop in the city where they paid $75 a month. Her husband drove the eight miles to work at the wire mill with two of their sons.[26]

When the bulldozers were through, Rufus's model town was in shambles, sheared off on the west by the open hearths and blocked on the south by a crosshatching of hot-blast stoves and ore dumps. What remained of the central district along D Street—several rows of houses, a handful of retail stores, churches, and the high school—was pounded night and day by the mill. Big booms rumbled out of the No. 4 open hearth, shaking the earth and rattling Mary Gorman's cupboards.

"It broke my heart," she says. Although her house was not bulldozed to rubble, her sense of place was shattered. Having lived the whole of her life in Sparrows Point made the prospect of leaving it all the more terrifying. She disciplined herself not to "think depressing thoughts." The Gorman household underwent scourings like never before. Every day she swept away the red dust on the front stoop, wiping clean the changes that were taking place around her. "I'd been cleaning that dirt away for years. I figured I'd do it till I dropped dead."

Many of the residents were on their way out, persuaded by uncertainty over the town's future and by the company's offer to reimburse eight months' rent to those who moved voluntarily. For the daughters of John H. K. Shannahan, it was time to leave the town of their birth. They headed for the green pastures of Roland Park, an exclusive preserve in North Baltimore that was a preferred address for the superintendents and white-collar staff.

"Oh, it was hard giving up being a big wheel in a small town," Elizabeth McShane laments. "We really hated to leave, but the children were grown and we knew the end was coming."

"They needed more space all the time for the new mills," Katherine Roberts says. "But it wasn't like what we remembered."

"So much was gone—the B Street clubhouse, the store, the nice beach. And there was worry that the houses might be rented to people who were not appropriate. All the people we knew had dispersed."[27]

*

Ladles slung from overhead cranes caught the flood of boiling metal at the No. 4 shop. A crusty froth of slag slopped over, dripping thickly into the slag pot. Second and third helpers dashed to the lip of the platform and tossed in the bags of manganese and river coal. Then the ladle was swung over to the teeming pit for release into the narrow-gauge ingot trains, just as it had been done for 60-odd years.

The engineering superlatives of the No. 4 Shop?

"Phooey," said veteran Charlie Capp. "Those furnaces were only bigger, more dangerous, really." He cited the increased number of doors and the faster tap times as evidence that technology was only a fancy word for getting more sweat out of the working man. "They damn well wanted to work the helpers ragged with them new furnaces," says Capp. Bethlehem had always been run by tough guys, and working there was no pleasure. "We even had to fight for a decent cool-off room for break periods."

Joe Hlatki had to agree. The tap times and tonnage stats that excited the engineering department caused operational headaches for helpers like himself. "The company had ways of workin' them heats, your life was in your hands all the time. You got more tonnage in the furnace and the bottoms were more shallow, more spread out, than No. 3 shop. A fuel man told me, 'Nobody realizes it, Joe, but this No. 4 open hearth is just like an atomic bomb. You got gas under there, you got oil, you got tar, all in that cellar there, and it could blow sky high.' The only thing I liked about No. 4, I made more money. That was all. That was all we was in there for."

"If you're young, yeah, you can take it," Ben Womer said, "but it gets a little rough later on. Your body ain't equipped to withstand that any more. It's the heat and the heavy work." So Ben took the advice of his dad about getting old in a young man's job. "When he hit 62 he said, 'Put me on the labor gang, I'm getting off the floor.' And everyone of them old-timers raised Cain with him because he was cutting into his pension by taking a lesser job. And the old man said, 'I don't care. I'm not going to stay up on the floor 'cause I'll drop dead if I do.' And, don't you know, pretty near all them old-timers that stayed on the floor were dead and gone and he was still living."[28]

A mile and a half north, at the 56–68-inch hot-strip, a three-ton slab of metal was discharged every three minutes. Passing down a mechanical conveyor it entered the first of ten stands of rolls. Seconds later the slab was spit out a quarter mile down the line, a screeching, red-tongued strip of steel. The strip raced along a runway and disappeared—wound up into a spool by a coiler and ejected upon another conveyor that carried it to the storage room. A handful of men worked on the walkways, pushing dials and knocking roller pins with hammers. Chemicals foamed up over the grates. The stink of hot urine was strong.

But Bob Eney wasn't there any more to jot down the progress of the strips. "I couldn't stand it any more," he says. "I just had to be in a different environment." Quitting the suffocating atmosphere of the Point, Bob headed for New York. "I got a job with Lord & Taylor, starting in window display, and I made my way up to staff artist there."

For every native son who left, there were many others eager to get a job at the best-paying factory around. Mary Gorman's son, Ed, took a laborer's job after returning from the air force. After two years of scraping palm oil, he found that a scalesman's job was opening up. "That's what you looked for, a production job." Ed bid on the job and got it. "At the time I was damn pleased. I was still a kid and it gave me a future. I figured there wasn't much chance of being laid off because it was a go-go operation. I didn't know then what I know now."

Next door, the tin department was brimming with activity. The expansion program had placed two new tinning lines on stream, with enough capacity to make 10 billion cans a year. "What they wanted from us was as much overtime as possible," says Marian Wilson. "Every morning you'd just see all these stacks and stacks of plate to inspect."[29]

Down by the waterline Charlie Parrish heard the sound of steel from its opening notes as the freighters entered the harbor with their loads of ore. Nicknamed "redbellies," the ships approached the dock, their acres of heavy plating slicing the water until the coils of snubbed rope drew the behemoths in. There was always that sharp, focused moment when the dock finally grabbed hold of the boat and the groaning rope relaxed. So did the dock workers. Then more sounds. Coming up from the rear were the cranes whose innards Charlie kept in nice working order. The cranes rumbled into position, snapped up their metal-toothed buckets, and dropped them deep into the hatches.

The redbellies arrived at the rate of about ten a week. About half were assigned to the circuit between Sparrows Point and Palua, the Venezuelan port on the Orinoco, where over 6 million tons were shipped a year. The El Romeral mines in Chile supplied another 2 million tons. On January 5, 1957, Industria e Comercio de Minerios, S.A.–ICOMI, a joint venture between Bethlehem and Brazil, began shipping ore from a port established on the Amazon River. The material was transported to the river by railroad from remote Amapá near the equator.[30]

The company's proven supplies of iron ore in North and South America amounted to 500 million tons, a staggering total. And there was plenty of coal to smelt the ore. With the recent purchase of the High Power Mountain reserve in central West Virginia, Bethlehem had added 250 million tons of low-sulfur coal to its reserves. This boosted the steelmaker's coal holdings to over 1 billion tons, enough to continue and enlarge its operations for years to come.[31]

*

Nineteen fifty-seven was a superb year for the American steel industry. In the first six months, Bethlehem and U.S. Steel rolled up the highest output and earnings figures in history. Demand for canning plate, standard carbon structurals, cold-rolled strip, and sheet boomed. Even so, the industry raised prices by an average of $6 a ton, effective July 1. Spokesmen termed the rises necessary for higher wages won in collective bargaining and for higher capital costs. Bethlehem completed the year with the best operating ratio in the industry owing to a backlog of orders for capital goods. Net billings of $2.6 billion were at a record high, as was net income of nearly $200 million.[32]

The aging "boys" celebrated their good fortune in familiar style, awarding themselves the biggest bonuses since 1929. Grace reclaimed his title as the highest-compensated executive in America in 1956, earning $809,001 in salary and bonuses. This was $115,000 more than GM's Harlow Curtice, second best-paid executive. President Homer got $669,175. Overall, 11 of the 18 highest-paid corporate officers in the nation were Bethlehem officers. Vice presidents such as Stewart Cort were making more money than Henry Ford II.[33]

Trust busters like Senator Kefauver were disgusted. In hearings Kefauver excoriated the pay at Bethlehem as striking evidence of the steel industry's public-be-damned attitude. The same men who were always raising the price of the country's steel were feathering their own nests shamelessly. The Bethlehem board was as deaf to the public welfare as it was sensitive to the needs of its members, Kefauver complained.[34]

Internationally the year was one of pending upheaval. In Venezuela the dictatorship of General Marcos Pérez Jiménez was coming undone in 1957, threatening the Pax Americana that had reigned since 1948. There were riots on the streets of Caracas. Cuba was also in an explosive state. Suppression of the press and many murders by the secret police had followed Fulgencio Batista's illegal seizure of power in 1952. A small band of guerrillas led by Fidel Castro was organizing opposition. Castro had been declared dead by the government, but Herbert Matthews, a foreign correspondent for *The New York Times*, had found him alive and kicking in the same mountain range where Frederick Wood had discovered iron ore and where the tracks around mined-out Juragua now lay jungle-covered and rusting.[35]

Topping the news from Cuba was the launching of Sputnik by the USSR. The surprise development took place in October 1957, just days after the first heat of steel was tapped from the No. 4 open hearth at the Point. The first man-made satellite was small and light, only 23 inches in diameter and weighing under 200 pounds, but it foretold a new set of circumstances. The long pleasant ride of postwar hegemony was coming to a close as the world stood at the threshold of another age. The newspapers called it "the space age." Having broken through the space barrier, scientists soon would be generating temperatures far beyond the tolerances of mass-produced carbon steels, and industrial engineers soon had designs that called for materials which were of less weight and expense than steel. An era of technological sophistication and flexibility was arising, challenging the established order of industrial goliaths.

PART V

EPILOGUE

We're not in business to make steel, we're not in business to build ships, we're not in business to erect buildings. We're in business to make money.

—Frank Brugler, Bethlehem comptroller and director, speaking to *Fortune,* 1962.

CHAPTER 22
RUINS

In the months following Sputnik it would never have crossed the minds of Bethlehem's arrogant and overcompensated managers that their company, ninth-largest industrial of the *Fortune 500,* with 164,000 employees and record-shattering earnings, would rack up $1.47 billion in losses a quarter century later in the recession of 1982; nor that the projects Bethlehem excelled in—big ships, big buildings, and big mills—would be cited as relics of another age; nor that Sparrows Point, erstwhile largest steel mill on the globe, would hobble along at below half of its 1957 output, its pipe, wire, and nail mills shut and its daily employment rolls shrunken from 28,600 to under 10,000. The drama has been much the same in Pittsburgh, Gary, Youngstown and other once proud American steel towns.

The collapse of the nation's long-held superiority in steelmaking has its conventional economic explanations; prominent among them, high union wages, foreign competition, and environmental cleanup costs. They are elements of an oft-repeated argument that steel has been victimized by special circumstances and hence should be given special treatment by Washington. Government protection against imported steel has become the top priority of industry lobbyists; moreover, the trade has used the threat of more plant closings to extract wage rollbacks from the Steelworkers Union and environmental breaks from regulators. Yet none of these steps has stemmed the industry's deterioration significantly, for the simple reason that they have not addressed the root causes of the trade's blight. For these, one must look at the peculiar history and mindset of American steel.

In a competitive world economy, it is imperative for a business to link its future to new technology or new markets or substantial savings through improved productivity. America's steelmakers have seldom faced up to these realities. In a pattern that has not changed for 50 years, Bethlehem, U.S. Steel, and Republic (now part of bankrupt LTV) have endeavored to reduce all U.S. integrated mills to a common denominator of technology and markets—to freeze their hegemony—and to keep out efficient competitors overseas and at home. Since the days of Schwab and the monopolistic mergers of Morgan and Carnegie, the attitude of management has been that, because the trade supplied an essential commodity to the nation, it was *owed* a comfortable existence. After all, its leaders were in business to make money, not steel.

During the growth years from the Civil War through the 1950s, it was difficult not to make money, especially after the business had been organized into a few blocks of concentrated power, with fully owned raw-material supplies and exclusive market territories, and when management always acted as a monolith through the American Iron and Steel Institute. When economic challenges arose, the senior officers were unwilling to diverge from their time-honored modus operandi. They spurned product research. They choked off innovation. They relied on inflating the price of steel to keep up quarterly dividends, and they let the situation deteriorate to the point where their mills were out of date. In so doing, they created opportunities for their rivals—not only for the Japanese and other foreign steelmakers, but for domestic "mini mills" and makers of steel substitutes, such as aluminum and plastics.

A shrinking domestic market for traditional steel mill products, not overseas imports or high wages, has been the overriding source of steel's decline. In the face of rising prices, steel consumption in the United States grew sluggishly from 1957 to 1973, when it reached 122 million tons, then dropped off in cyclical swings. Today, Americans consume only about 90 million tons of finished steel a year from all sources. (Raw steel production to make the finished products is somewhat higher.)

Labor has borne the brunt of management's determination to pump up the bottom line by riding along the path of established practices and depreciated equipment. To cut costs, basic safety precautions were ignored or belittled. "If it ain't broke, don't fix it," was the ruling credo. Labor relations have remained perversely counterproductive, with the union blaming management for lack of investment, management blaming labor for inefficient work rules, and both blaming the government when demand dried up. Had the USWA and management talked more constructively outside the context of wages and benefits, they might have helped the industry. But then they would have hurt themselves in an atmosphere where the division between supervision and labor was absolute. Thus the union fell into the trap of opposing all change as "company talk," hastening its own decline.

The recession of 1958 flashed the first warning sign that steel had badly overexpanded. Production at Sparrows Point and other plants skidded to under 60 percent of capacity, the first significant drop in demand since 1938. An abrupt drop in consumer durables was deemed the immediate cause of steel's slump. Convinced that demand would soon rebound, as it always had in the past, the industry relied on another price hike to keep its earnings steady.

The $4.50-a-ton advance announced in the middle of the worst recession since the Depression caused yet another political tempest. While Eisenhower tried to defend the increase by citing high labor costs, Kefauver and other Democrats were not satisfied. To them the industry's action defied the laws of supply and demand. They were doubly outraged when it was disclosed later that Bethlehem's Arthur Homer had flattered himself with a paycheck of $511,249, thereby staying at the head of the salary and bonus parade of executive America. (Grace, bedridden from a stroke, received $150,000 as "honorary chairman of the board," although he was unable to attend any board meetings.)[1]

The price increase supplied the political ammunition which John Kennedy deployed in his famous face-off with the industry in 1962 over another price hike. But it did more: It opened the door to substitute products. The metal was no longer the cheap commodity that had built America. After including the 1958 increase, steel prices had advanced by 165 percent since 1945—2½ times as fast as concrete and plastics, double the rate of all industrial prices.[2] After seven price increases in eight years, the possibility that other materials could substitute for steel no longer seemed impractical to industrial buyers; indeed, it seemed essential to many of them.

*

Aluminum was the first major invader of steel's domain. For some time the leading companies (Alcoa, Reynolds Metals, and Kaiser Aluminum) had been foraging for new markets outside construction, engine parts, and transportation. The sector was seeking a mass market where, according to one official, "the product is consumed—not put in place to last forever." The 50-billion-can-a-year metal container business was one such market, worth $2 billion in sales. But to move into metal containers meant displacing steel. The first step was made during the 1958 recession. Esso Standard Oil switched from tin plate to aluminum plate for quart-sized oil cans made at its Bayonne, New Jersey, refinery. Reynolds had offered the oil company a creative deal to keep costs down: It agreed to buy back Esso's used aluminum cans from service stations for scrap. The payback from the resale of the cans offset the price differential between aluminum and tin plate. Esso was not only impressed with the better corrosion resistance of aluminum, but expressed concern with the runaway costs of tin plate supplied by Sparrows Point and the Fairless works of U.S. Steel, which had gone up by 25 percent since 1955.[3]

Esso's conversion to aluminum canning amounted to only 35 million cans a year, or under 0.05 percent of total yearly tin can production. But throughout the aluminum industry, engineers were at work to improve aluminum canmaking. An important advance was a can machine developed by Kaiser that punched out thin-walled cans from round blanks of aluminum sheet in a single press stroke. With this process Kaiser was able to make rigid containers with extremely thin walls at a cost competitive with tin cans—but these containers were suitable only for lightweight products such as grated cheese. Soon, however, a new type of "impact extrusion" enabled the manufacture of aluminum cans that could be filled with heavier products at a price competitive with tin plate. This process was pioneered by the small Aluminum & Chemical Corporation, which finished a pilot plant in Newport, Arkansas, in 1958.[4]

In January 1960, Reynolds invaded the frozen orange-juice business. The company placed the canmaking equipment from its research laboratory in Richmond on a truck and drove it to the Minute Maid Corporation plant in Florida City. Backed up to a receiving dock, the can line turned out 7 million cans in three weeks. Minute Maid was so impressed with the cost savings of aluminum cans—and the better corrosion resistance, which meant a longer shelf life for the frozen juice—that it leased the can line from Reynolds and ordered a second line. This development was followed by orders from Winter Garden Citrus Products and the Birds Eye division of General Foods at the end of the 1960 canning season.[5]

Steel's response was lethargic. First there was general disbelief that the market was threatened; then a belated move toward providing lighter canning stock. It had long been believed that conventional tin plate was unnecessarily heavy for many canning applications and could be reduced by 40 or 50 percent without sacrificing strength. By supplying more product per ton of sale, thin plate would have lessened the burden of higher tin-plate prices. Yet until challenged by aluminum, Bethlehem and U.S. Steel had been uninterested in making "skinny" tin plate because the new product required two passes in the rolling mills and thus slowed production.[6]

Only after aluminum got a foothold in the citrus market in 1960 did U.S. Steel announce that "skinny" plate would be made available to canners. *Would be available* was critical terminology. U.S. Steel was talking about something it didn't yet have in commercial quantity. It was not until mid-1961 that the company even got around to publishing a price for the new product. Bethlehem followed with identical prices for skinny plate produced at Sparrows Point. "We waited to see if the product would be accepted," Edmund Martin, Bethlehem's president, said later. For a brief period, tin plate was priced lower than aluminum on a per-can basis. Then Alcoa and Reynolds dropped the base price of aluminum to 24 cents a pound. Unwilling to go below their published price, U.S. Steel and Bethlehem announced that they would no longer supply tin plate to citrus packers.[7]

A 2-billion-can business was surrendered. While constituting only 4 percent of the metal packaging business, the defeat had demonstrated how tin-plate producers were reactive players in the marketplace. Despite their alleged muscle through long-term sales agreements, steelmakers were exposed as mired in inflexible pricing and marketing. The lesson was not lost on industrial customers. They now had convincing evidence that products improved when steel was placed under a little competition. Over the next 30 years, aluminum would wrest more and more business away from steel, winning even the coveted beer market which steel had claimed as its own during the Depression.

Today 99 percent of all beer cans are made of aluminum; the remainder are made of steel. (The major holdout is Pittsburgh Brewing, maker of Iron City Beer.) In soft drinks, 89.6 percent of all cans are aluminum. Overall aluminum packaging has increased over ten-fold since 1960, becoming the single largest market for the lightweight metal, while tin-plate tonnage has dwindled by one-third, in part because of the popularity of double-reduced plate, which lowered the demand for standard plate. Most of the market is now confined to canned fruits and vegetables, soups, paints, aerosols, and industrial products.[8]

In other markets, too, steel began to feel the pinch of alternative products. In construction, reinforced and prestressed concrete was moving into big-ticket construction from specialized applications. The first important inroad into standard steel construction was the use of prestressed concrete for short-length bridge spans in the interstate highway program. In 1960 Tishman Research Corporation sent shock waves through the building trades by erecting a prestressed-concrete parking garage at a Hempstead, Long Island, shopping center. Using new engineering techniques, the Hempstead "Tierpark" was built at $1,200 per parking space, compared to $1,700 for a conventional steel-girder structure. Over the next few years the construction axiom that only steel was strong enough for very large or tall buildings underwent a revision. Soon workers were pouring concrete in many important construction projects, among them the Americana Hotel and the CBS Building in Manhattan.[9]

The point was not that steel had been displaced from a market (concrete construction still used sizable amounts of steel reinforcing rods and bars, for example), but that the industry had to map out new directions for its growth. It had to develop ways to defend itself against market challenges as well as to find acceptable avenues of growth through specialized applications. Steel was confronted not only by new materials, but by what could be called the "Sputnik phenomenon." Engineers found that great economies could be achieved by reducing overall weight—for example, by using fewer structural girders in a large building. Reducing the weight of manufactured items became a priority to companies interested in lowering costs. Furthermore, there was a pronounced slowdown after 1958 in the growth of heavy users of steel. These included agricultural equipment producers, shipbuilders, and oil-drillers, as well as manufacturers of guns and other land-based ordnance for the military. Railroads in particular were stumbling.[10]

In a further challenge, years of research in plastics were paying big dividends to the U.S. chemical industry. Between 1955 and 1963 plastic sales had tripled to 8 billion pounds. Plastics were lightweight and could be "blow-molded" from master molds without machining or stamping. Reinforced with paperboard and other materials, they would not dent, corrode, melt, transmit heat, or conduct electricity. Extruded polypropylene, for example, made a clear film and sheeting with excellent protective characteristics for packaging dried fruits. Styrofoam, cellophane, and polyethylenes became the containers of choice for roadside fast-food chains like McDonald's.

Suggestive of the changeover from metals to plastics was the rise of the metal-plated plastics industry into a high-tech $100-million business. Once a cheap novelty item used as prizes in cereal boxes, plated plastics had become a big seller in Detroit. The reason again was economics. As steel became more expensive, automakers discovered that they could save money by switching to plastics coated with a thin trim of stainless steel or other metal. The Furniture City Plating Company of Grand Rapids, Michigan, was generally credited with developing a plating process that eliminated peeling and buckling.

Increasingly, plated plastics replaced metal in car instrument panels, dome lights, ventilation grills, and door locks. Several companies were developing plastics for such noncar uses as the trim on toasters and the knobs of radio and television sets. "I can count 120 potential applications inside and 150 outside for plated plastics on cars," James H. Fiser, president of EMC Plastics, said in 1965.[11]

*

In a period marked by the rise of substitute materials, steelmakers persisted in their longstanding habit of ranking at the bottom of research-and-development expenditures. Although Bethlehem had opened a $10 million research laboratory in 1961, no one took the facility seriously. Fewer than 0.5 percent of personnel worked in all aspects of R-and-D, and what they did was less than inspiring. Most of the work stressed the practical problems of daily production, improving the pellets of blast furnace feed, for example. The laboratory avoided serious research into sophisticated materials (such as plastics and fiber composites in steel-based products), which promised superior strength and better corrosion resistance because they might undercut sales in standard steel items. A 1966 report by the National Science Foundation found the leading steelmakers glaringly deficient in research, spending only 60 cents of every $100 in revenue for R-and-D, against $1.90 on average for all manufacturing sectors. The industries producing aluminum, concrete, plastics, and other steel substitutes invested more in research than steel—sometimes five and six times more.[12]

Low spending for research and development wasn't the only barrier to innovation. The self-satisfied motto of Andrew Carnegie, "Pioneering don't pay," pervaded the inner circle of Bethlehem's undistinguished management. Arthur Homer was named chairman of the board in July 1960 following the death of Eugene Grace a month short of his 84th birthday. The senior officers who surrounded Homer were, like he was, products of the elaborate management grooming system that Grace and Schwab had crafted. Few of the top 20 officers had less than 25 years of service at the company, and only one important department head had even worked for another steel company.[13] Trained in obedience, they followed the inner logic of the organization, drilling the daily cost figures across their desks and keeping quarterly dividends as high as possible since this accrued to their direct benefit.*

In a revealing interview with *Fortune* in 1962, Homer articulated the position that Bethlehem was rich enough to afford *not* to innovate. Despite sluggish sales, he reported that all of the plants and shipyards were making money and the company was comfortably profitable. "We have a nice business as it is," Homer gloated. Except for plans to build an integrated mill at Burns Harbor, Indiana (a carry-over of the old Grace-Schwab objective to move into the Chicago district through a merger with Youngstown Sheet & Tube), Homer said he had no plans to change the company's course, especially into new markets or technology that might prove "imprudent." Indeed, a major priority of his administration was to continue paying the high $2.40 dividend rate set in the 1950s. The long-term strategy of Chairman Homer, added Vice President James Slater, was to obtain all the tax credits available under federal laws and to hold back on capital expenditures until they paid out according to the honored "20 percent before taxes" profit rule. "Unlike some [companies] who can't wait, we don't have to add to capital unless we are sure it will pay out," Slater reported.[14]

Despite the lowest debt-to-capital ratio of the top 25 U.S. industrial corporations, including over $400 million in cash and marketable securities, Bethlehem had refrained from investing in the revolutionary basic oxygen furnace (BOF), invented in Austria in 1952 and introduced to North America in 1954 by a small company in Canada. The

*In 1960 a special dividend unit system was instituted by the officers to reduce the federal taxes they paid on their cash bonuses. Henceforth bonuses were spread out over the years through the crediting of annual "dividend units" which entitled an officer to payments equal to the value of the dividends of common stock during the life of his service and continuing through his retirement. In 1960 Arthur Homer received 3,634 dividend units on top of his cash salary of $300,000, and President Edmund Martin got 2,922 units on top of his $241,250 salary. Such compensation placed Bethlehem's management behind the senior officers of General Motors and Ford in terms of top business salaries for the year. (*Business Week*, May 21, 1961) two main reasons for the decision, Vice President John "Jake" Jacobs told *Fortune,* were that the BOFs were undergoing so many technical refinements that they might have to be scrapped too early for maximum value, and that they injured the value of the $250 million in open hearths built in the prior decade. Since the open hearths were relatively new, especially the "world's largest" shop at Sparrows Point, Jacobs figured that any replacements could wait until major repairs came due. "[W]e move only when improvements are so good we can no longer afford what we've got," he said smugly.[15]

A year later *Business Week* exposed the fallacy of Bethlehem's thinking. It disclosed that the company, together with U.S. Steel and other giants, had bought 40 million tons of "the wrong kind of capacity." Scaled up to handle high outputs by West German and Japanese steelmakers, the BOF was hopelessly outclassing the open hearths. It could make 200 tons of steel every 45 minutes, while the ponderous open hearths needed eight hours to tap a 425-ton heat. The equipment gave overseas producers a $5-a-ton operating advantage over the Americans. BOFs also saved on capital costs.[16]

Grudgingly, Bethlehem accepted the BOF, but only to a minimal extent. The real story was how limited management's commitment to the new technology was. Headquarters did not authorize construction of a BOF unit at Sparrows Point until 1964. At the end of 1965, two BOFs were opened inside rebuilt No. 1 shop, and No. 2 shop, where Mike Howard had worked and where Pop Weston had died, was shut down. This still left 65 percent of Sparrows Point's capacity stuck in old-fashioned open hearth steelmaking, and management wasn't willing to give up this capacity—at least not without a fight.[17] To keep No. 4 running, management opted for a low-cost fix. The shop was rigged with oxygen lances to speed up the heats. While this method did reduce heat times to four or five hours from the previous eight or nine, the measure was not commensurate with the improvements being made in BOF technology overseas. (Realizing this, Bethlehem masked the deficiencies of No. 4 to the press and public by calling the obsolete equipment "oxygen-blown steelmaking" in handouts.)[18]

While installing the new BOF, management failed to take advantage of another new production tool, the continuous caster. Continuous casting made it possible to eliminate the costly intermediate steps between steelmaking and primary rolling by directly making semi-solid billets or blooms from the steel furnaces. Continuous casting had been discussed in U.S. steel circles for more than 30 years, and for just as long had been rejected as impractical. In West Germany the method had been under active commercial development. In 1962 Roanoke Electric, a small independent company in Virginia, had introduced continuous casting to the United States. A year later, in 1963, Reynolds Metals began using rotary casting for making aluminum sheet.[19]

If continuous casting was so promising, especially in eliminating hundreds of mill jobs that Bethlehem complained were too expensive anyway, why did the continuous caster run into resistance? One reason, according to an insider, was that the device cut across the jurisdictions of three departments: steelmaking, transportation, and primary rolling. Jealously protective of their turfs, the superintendents and staff of these departments fought the encroachment of the caster.

The casting method also placed a premium on the purity of steel coming out of furnaces. Traditional steelmaking was wasteful, as experts had long noted, but it was fairly tolerant of variations in the composition of the steel. For example, "dirty steel" (metal containing impurities) could be mixed with batches of higher quality steel without causing serious production snags. Continuous casting demanded more precise metallurgical standards to succeed on a volume basis. The superintendents could argue that the continuous casting process was "impractical," reinforcing the prejudice of upper management to stick to familiar technology and, when threatened, to push harder in the direction they were going. Modernization schemes were designed to minimize change. New machinery such as the BOF was forced to work within the old structure.[20]

What was remarkable, the insider mused years later, was that nearly everyone was "oblivious" to Bethlehem's worsening competitiveness. Such was the hazard of a corporate culture that awarded conformity. Management's response to the inroads made by aluminum and plastics was not to talk about it. In this hermetically-sealed world, supervisors went about their daily routines, lulled by the acres of machinery that still made Sparrows Point the biggest mill in the middle 1960's. They were utterly convinced that steel was indispensable to the nation, no matter at what cost. "We listened to our propaganda for so long, we believe it," he noted ruefully.[21]

*

It was thus in markets and machinery where steel first stumbled and lost its competitive edge. The early triumph of Sparrows Point was the manufacture of high-quality, low-cost rails during the heyday of railroad building. New technology and cheap raw materials were the keys to this success. Charlie Schwab's important contribution was to move the Point forward into production of flat-rolled steels which made an excellent fit with consumer and automotive markets after World War I. His successors lacked his ability to look ahead and to adapt to changing economic realities.

What they did share with Schwab was an uncompromising hostility toward the USWA, which had legal rights to represent the third "M" of manufacturing, manpower. Strains between labor and management had led to several costly walkouts since 1946. The worst was the 116-day strike in 1959. Executives at Bethlehem and the other major steelmakers (who negotiated jointly with the union) entered into contract talks in 1959 determined to win a one-year wage freeze. Many observers believed that the union, led by the uninspired David McDonald, was prepared to accept a nominal increase, which would come mostly in supplemental benefits, not cash wages. At an average $3.10 an hour, unionized steel employees were better paid than any other blue-collar group in the nation except the coal miners. An opinion poll indicated that membership was strongly opposed to a strike.[22]

Heading the industry bargaining team for the first time was R. Conrad Cooper. A classic tough guy who had once considered professional boxing as a career, "Coop" did not hesitate to show his impatience at anything he considered a digression. "Oh, quit it," he said to questions he didn't like. Due to his background in industrial engineering, he was believed to be a major figure behind a second set of management demands issued in the 11th hour of negotations as a precondition to any offer of higher pay. They involved disciplining wildcat strikes and "management prerogatives" to change work assignments. Cooper demanded the elimination of paragraph 2B of the master contract, the past practices clause, which said that a local past practice could not be changed without a change in basic work conditions. Ironically, paragraph 2B had been added at the companies' request in previous negotiations as a way to put a cap on establishing work rules in local plants.

Even at the time there was talk by other companies that steel had made a serious tactical blunder by raising the "job reform" issue in national talks. For the most part, other companies dealt with plant manning problems on a local level, free from the emotions and time constraints of national bargaining. In steel there was a precedent for removing the issue: the agreement to resolve the issue of wage equalization through joint union-management committees in the '40's. Yet this was waved away by management negotiators.[23]

While many commentators were disposed to accept Cooper's charge that union work rules impeded production, they wanted examples. This became one of the conspicuous lapses of management's case. Saying that they knew the problem existed, Cooper and other spokesmen said they could not measure the extent of it. Was "featherbedding and loafing," as Cooper called it, worse in steel than in other industries, or worse on the shop floor than in management ranks? These questions were left unanswered in the noisy accusations that followed. In a visit that U.S. Steel Chairman Roger Blough and Cooper paid to Eisenhower they told the president that "runaway costs" were crippling the business and that the large companies were willing to risk a long strike, if necessary, to rectify the situation. Blough and Cooper urged the president not to interfere.

Given the prevailing distrust between the two sides, management's proposals struck a raw nerve at Sparrows Point and other mills. Veterans like John Duerbeck remembered what happened when supervision had the power to do what it wanted on jobs and automation in the hot mill. They abused it. For younger workers the issue threatened their livelihoods. Caught in the fear of job reduction and suffering from management's reluctance (or inability) to foster a cooperative atmosphere in the mills, they were keenly aware of what they and their fathers had done to gain seniority rights. They were ready to stand behind their union card regardless of the consequences.[24]

"We didn't even have to tell them," said a USWA staffer. "The companies did it for us." The walkout began on July 15. The shutdown dragged on through the summer and early fall, proving politically embarrassing during Nikita Khrushchev's trip to America in September. The AISI and Steel Companies Coordinating Committee spent over $1 million to try their case in newspaper advertisements. McDonald turned militant and emerged the hero of the rank and file.

As the strike turned into a siege, it seemed to feed upon the psychological needs of the combatants. Overweening pride and fear of appearing "weak" seemed to motivate the leaders of labor and management as each held the economy hostage to their self-interests.

Long aspiring to fill the shoes of Phil Murray without the work, McDonald seized upon management's overkill to solidify his shaky leadership. "I couldn't have written the script better myself," McDonald said of the work-rules issue doggedly pursued by Cooper.

With the exception of coal, no other U.S. industry so consistently thrust its internal affairs into the lap of the public, alternately damning and demanding government intervention. The lament of Frances Perkins back in 1933 that the men entrusted with the stewardship of this important industry coped with conflict "like 11-year-old boys at their first party" was still regretably true. When shortages began causing layoffs in auto plants (GM had built up a three-month stockpile before the stoppage), Eisenhower reluctantly invoked his emergency powers under the Taft-Hartley Act. The men returned to work, but nothing had been resolved.[25]

A different style of labor relations prevailed in the aluminum industry. Company negotiators were savvy enough to realize that confronting labor by blunt force was a foolish way to conduct business. In 1959 the USWA contract with Alcoa, Reynolds, and Kaiser had expired, but production was not disrupted as bargaining continued. Late in the year, a new contract was signed between McDonald and aluminum which called for a cash and benefit package totalling 28 cents an hour over three years.[26]

In December 1959 Eisenhower left the country on a three-continent peace mission, leaving Vice President Richard Nixon to handle the steel crisis. Nixon was credited with getting management to offer an improved wage package, while Secretary of Labor James Mitchell got McDonald to soften his demands. The price of settlement: no changes in work rules and a hefty 40-cents-an-hour increase in wages and benefits over three years.[27]

*

It was an article of faith among steel executives, frequently echoed in the press, that the 1959 strike opened the door to imports. But steel imports had been on the rise since the beginning of 1958. They nearly doubled between January 1958 and June 1958. Then, following the industry's mid-recession price hike in July, imports surged. By December 1958 imports exceeded exports by a significant margin for the first time, excluding strikes, in the 20th century. The gap widened during the first six months of 1959 and increased by another 25 percent during the 116-day strike. Ever since, the United States has been the world's largest importer of steel products.[28]

Labor *was* cheaper overseas. But wages alone did not determine the cost of steel; labor productivity and capital expenditures were critical variables. In the past, superior technology and output had compensated for higher American wages. In fact, according to Donald Barnett, former chief economist for the American Iron and Steel Institute, the overseas wage advantage began to diminish in the 1960's as America's imports of steel were rising. "[T]he extent to which U.S. wage rates exceeded those of its international competitors," he wrote in 1983, "was relatively greater in the first fifteen years after World War II than at any time before or since."[29]

The same management decisions that placed Sparrows Point and other steel plants in a position of inferiority compared to producers of substitute materials also proved costly in meeting foreign competition. The Americans found themselves in the 1960s progressively outclassed by steelmakers in West Germany, Canada, Belgium, and even Brazil. But the chief competition was from Japan. The advance of the Japanese was an object lesson in strategic planning by business and government. Japan had low wages, but it had serious cost barriers nevertheless. Unlike resource-rich North America, the nation had no workable iron ore deposits, and coal was expensive to mine. To overcome these handicaps, the Japanese went to South Africa for ore and pig iron and used the latest technology to reduce the importance of raw materials in overall costs. BOF furnaces were found to be especially useful in reducing the import of high-priced scrap from the United States.[30]

By 1962, Yoshihiro Inayama, president of Yawata Steel and later chairman of Nippon Steel, could boast of a tidewater mill at Tobata that incorporated the latest in BOFs and casting methodology. His steel was soon considered superior in terms of chemical purity and corrosion resistance to the basic-carbon products supplied by American mills.[31]

Another company, Fuji Iron & Steel, began exporting a new product to the United States: chrome-plated strip steel. Developed in response to the dwindling supply of tin, this extremely thin yet flexible product cost less than either aluminum or tin plate. It was to find great favor among American canners. (Eventually Bethlehem signed a licensing agreement with Fuji to produce the chrome plate at Sparrows Point, which was then marketed under its own name.)[32]

Japanese imports of steel to the United States increased exponentially from a negligible 31,466 tons in 1957 to 4.5 million tons in 1967. Steel became the island's top export and contributed to the nation's trade surplus that was the engine of its remarkable growth. The economics that drove up imports were simple: Wire rod shipped 10,000 miles from Japan to New York was of comparable, if not superior, quality to rod from Sparrows Point 200 miles away—and about 15 percent cheaper. The domestic companies refused to "meet the competition" with price cuts, even on items where their costs were comparable to those of the Japanese. In 1967, when imports were running at record levels, U.S. Steel, Bethlehem, and Republic chose this time, of all times, to raise their prices.[33]

Faced with the Asian challenge, the centralized trade association, the American Iron and Steel Institute, embarked on a multimillion-dollar lobbying campaign to persuade Congress to place temporary levies and other cost barriers on Japanese and other steel. In state legislatures, the AISI pushed unabashedly for "Buy American" measures which required the purchase of domestic steel. Regular alarums were sounded from high quarters on the evils of foreign steel. "Bethlehem and other U.S. producers can justify continued investment in steel manufacturing properties *only* if they have the opportunity to participate in market growth," warned Bethlehem's Edmund Martin in a statement to stockholders and the press.[34]

Playing on the fears of plant closings and job losses, of course, was not totally out of character. In the worst days of the Depression, Charlie Schwab had tried to stampede the hapless Herbert Hoover into a new round of tariff barriers on finished steel products; this on top of the notorious Smoot-Hawley tariff, signed by Hoover, which was commonly viewed as a contributing factor to the deepening of the crisis. Now the AISI wanted to create the impression that foreign steel was being "dumped" at below-cost prices. Unless the government "leveled the playing field" with tariffs or quotas, American companies could not be expected to compete with overseas producers often aided by international agencies.

This argument, while plausible, was not substantiated in various studies. Japanese steel companies, for example, received around $176 million in aid between 1957 and 1960 from the Export-Import Bank, the World Bank, the International Development Bank, and others.[35] This was some $25 million *less* than Bethlehem Steel had expended in its open hearth binge at Sparrows Point in 1956–57. Outspending American companies was not the strategy that distinguished the Japanese producers; it was their ability to use investment capital wisely to erase longstanding liabilities.

The import controversy further presupposed that domestic supply and demand were in perfect harmony. They were not. With steel capacity concentrated in a narrow band between the Middle Atlantic states and Illinois, two-thirds of the country west of the Mississippi had to import their steel anyway. Customers in California and Texas took one out of four tons of steel, but had less than 4 percent of the nation's capacity. The domestic industry had made little effort to invest in the West and Southwest. Moreover, international trade was not a one-way street. While complaining of imports, Bethlehem continued to import inexpensive iron ore from its South American mines. And Kaiser Steel earned extra revenue by shipping coal and ore to Japan—often carried away in the same boats that were bringing in Japanese steel.[36]

Despite the flimsiness of their arguments, steel companies had the political leverage to win partial trade protection through the joint action of the outgoing Johnson administration and incoming Nixon administration. Early in 1969, the State Department persuaded Japan and the European Coal & Steel Community to sign "voluntary restraint agreements" on steel. Imports were reduced to 14 million tons, about one-quarter below the current levels. Imports were allowed to increase gradually over the next three years. This program was replaced later by trigger price mechanisms aimed at keeping foreign steel at roughly 20 percent of total domestic consumption.[37]

*

The effect of the anti-import campaign was to enable the most entrenched and least efficient companies to consolidate their political base and form a powerful coalition against structural change. Bethlehem and U.S. Steel had better raw material sources and access to large customer accounts through the "sales reciprocity" system, while Armco and Inland had better production and quality. The latter were willing to defer to the majors not only out of habit, but because they figured they could operate at higher profit under the inflated price umbrella of U.S. Steel and Bethlehem.

For a short time the strategy worked. Spurred by the escalation of the Vietnam War, which stimulated production in all sectors of the economy, steel experienced a surge in demand. Management used the tight market and partial protection from imports to resume their "stair-step" price behavior. Steel prices rose steeply after 1969, even after the Nixon administration instituted price controls in August 1971. The steel price index advanced 63 points between 1969 and 1974 compared to a 48-point advance for industrial commodities in general and 38 points for the consumer price index.[38]

Steel justified the price increase as necessary to pay for higher labor costs—they had gone up at two-thirds of the price rate—and to raise capital for plant modernization. Massive rehabilitation programs (coupled with higher import restrictions) were hailed by the industry as the solution to its problems, even though it was the inflexibility of the gigantic mills to which the industry was historically committed that contributed to its failure to compete under the price sensitive rules of free enterprise.

Meanwhile, a new kind of domestic competition was growing. "Mini mills" departed from the dogma that steel had to be big and integrated to survive. Pioneering minis like Roanoke Electric and Florida Steel were very small and very efficient. They used down-scaled equipment to pare costs. Their raw material, scrap melted down in electric furnaces, was cheap. With start-up costs as low as $4 million, the independents could turn out simple iron and steel products that were competitive with the majors. Their marketing strategy was simple. They punctured the majors' price umbrella by offering steel at prices below the standard list price and got all the business they wanted. Mini-mill capacity grew from insignificance in 1960 to 6 million tons in 1972. Altogether about 20 companies were producing wire, reinforcing bars, and rods, many of them located in the South and Southwest where there was a chronic shortage of steel.[39]

Compared with mini mills, the integrated goliaths ate up vast amounts of capital. As long as blast furnaces, coke ovens, and open hearths remained the workhorses of the trade, adding to or replacing capacity was increasingly costly. According to U.S. Steel, a fully integrated new plant would start at $1.5 billion, or over $500 per ton of installed capacity. One way to reduce costs would have been to utilize new techniques of "direct reduction" of iron ore, which bypassed the blast-furnace stage. However, in 1972, Bethlehem's board of directors rejected this course (a compromise between mini-mill methods and standard practices). Instead, buoyed by internal forecasts of a 25-million-ton increase in domestic steel demand by 1980, the board decided to "modernize" Sparrows Point with a new blast furnace.[40]

Convinced that they were building the next-generation blast furnace, not a white elephant, the directors allocated $175 million for "Big L," to be sited on what remained of the company town. Substantial funds also were committed to upgrade and expand Bethlehem's iron-ore properties overseas, build a new ore-handling pier, new ore conveyors, new oil tanks, additional coal yards, and the "hemisphere's largest" sinter strand. Total projected tab: about $275 million.

It wound up costing much more. Plagued by technical bugs, excruciating delays, and "cost overruns that left you gasping," said one engineer, Big L was finally opened late in 1978. The company had itself a showcase, a gleaming triangular edifice that towered 298 feet —higher than most office buildings in downtown Baltimore. It was rated to produce 8,000 tons of pig iron a day, enough to keep a dozen ore carriers busy. "'L' Furnace—Symbol of Progress" was the company's assessment in a gala opening ceremony for local officeholders and businessmen.[41]

But in the meantime a series of rude jolts had upended the premises upon which such massive structures were based. The expansionistic fever of 1972–74 had vanished by 1975. With sales diving, Bethlehem soon found itself awash in red ink. A net loss of $448.2 million was reported for 1977, the heaviest single-year loss in U.S. corporate history until eclipsed by Chrysler in 1979. Major parts of Johnstown and Lackawanna were closed. Seven thousand mill jobs were eliminated. While Chairman Lewis Foy blamed imports for the disaster, the fact was that weak management had failed to come to grips with a bloated capital structure until events overcame them. Johnstown and Lackawanna were aging plants in poor locations which duplicated other facilities and should have been closed years before. In a speech to financial analysts, Foy admitted that steelmaking costs at the mills were 30 to 40 percent higher than at more modern mills.[42]

Big L at Sparrows Point did little to break the company's fall. What Bethlehem needed was new markets, not a $200-million blast furnace. Aluminum and plastics continued to bite into the tin-plate and sheet markets, while domestic mini mills, even more than foreign imports, were taking away business. Competing head-on with Bethlehem, Nucor Steel of North Carolina sold reinforced bars, rod, and wire items at about 15 percent below the cost of Sparrows Point's products.

What kept the 50-year-old rod and wire mill open was that minis didn't have the capacity to fill all customer orders.[43]

"Cobbles," or accidents, in the rodmaking line happened on average once a shift, according to a foreman. It became routine for employees to see red-hot strands of rod flapping wildly around the mill before the operator shut down the equipment. Every breakdown cost money, but funds were not allocated to overhaul the machinery. "We were running junker equipment, all right, and it showed," the foreman recalled. "Customers were rejecting our stuff. And we were always going into the line and bashing out the bad rod to get the damn mill running again."[44]

Aside from poor quality, deferred maintenance contributed to a spate of worker injuries and deaths. Always hazardous, steel work got downright dangerous. A total of 10,226 injuries among 17,000 production workers was recorded in the 1978–79 period, according to Bethlehem's own records. This was the worst accident record in about 18 years. Thirteen of those injuries resulted in death. Investigations by the Maryland Occupational Safety and Health Administration concluded that faulty equipment and/or inadequate safety procedures contributed to 7 of the 13 deaths. The state agency slapped Bethlehem with $62,800 in fines for "serious" and "willful" violations of the federal Occupational Safety and Health Act.[45]

Three of the deaths were in the aging No. 4 open hearth. Two workers had collapsed from heat stroke on an extremely hot summer evening, one dying within hours and the other lingering on for months even though brain-dead. Later it was found that the "cool-off" booths where men rested between operations were not properly maintained and had not been functioning for weeks. Another employee died from a backbreaking fall into a slag hole. The lid covering the hole had not fit right and was loose. In another accident, a 29-year-old wire drawer became entangled in the wire running past his station at 1,200 feet per minute. He was hurled headfirst into a wire-spooling machine. The equipment did not have safety guards. Two other workmen died from toxic leaks of carbon monoxide in the blast-furnace department. Deteriorated pipelines were the cause.[46]

In discussing the company's maintenance of equipment, Russell R. Jones, general manager, stuck to the dollars-and-cents approach, telling superintendents at a country club speech in 1978, "Both maintenance forces and costs must follow operational levels. In other words, when we are in periods of high production and we are making money, we can spend it. The converse is also true. When production levels are down and we are not making money, we cannot spend it." Jones, following company procedures, refused to be interviewed by the press on the accidents. Except for a brief, unilluminating press release, silence was the response of management.[47]

Schwab and Grace always had preached of the virtues of retaining the best lawyers. In days past, it was Paul D. Cravath, dean of Wall Street attorneys; in the '70s it was the prominent Baltimore management attorney Earle Shawe. Shawe was hired to get the company out of its jam over the "willfull" safety violations. A feisty and shrewd man who once worked for the National Labor Relations Board, Shawe did his work well. He outgunned the state lawyers and won an important reversal when he appealed the heat-stroke citation in state court.[48]

The company threw additional legal reserves into headlong battles with the Environmental Protection Agency and the Maryland Department of Natural Resources. The heavy pollution of the Chesapeake Bay by Sparrows Point, now entering its fourth decade, was catching up to the company. In 1977 the state fined the mill $500,000 for missing pollution control deadlines—and then extended those deadlines for three years. According to an independent report by the Council on Economic Priorities, Sparrows Point was the worst polluting steel plant in the nation. The impact on local water quality of the three major pollutants from cokemaking—ammonia, cyanide, phenol—was labeled "horrendous" by the report. Phenol discharges alone, averaging 3,335 pounds per day, were ten times higher than at the steel mill with the second-worst record. The continuous heavy dumping of sulfuric acid wastes had so polluted public beaches that many had been closed for swimming for years. Furthermore, the mill had problems with PCBs. Eighteen "minor" leaks of the poisonous chemical were discovered in a canvass of electrical equipment.[49]

It was not coincidental that the worst polluting equipment was also the most outmoded. Contrary to industry representations, pollution standards did not for the most part take money away from needed plant modernization; they directed attention to where modernization was most urgent. Moreover, pollution control equipment opened up a badly needed new market for the industry. The biggest regional customer of Sparrows Point plate in 1975 was the Koppers Company, which used the steel to manufacture industrial pollution abatement machinery.

The fact that Chairman Foy didn't see fit to target such a steel-intensive market for diversification demonstrated the kind of opportunities lost when an organization was stuck in a defensive mode. In an interview with *The Wall Street Journal*, Foy said the company had no intention of altering its product line or attempting to acquire a major nonsteel subsidiary. He said he was holding to internal forecasts of robust demand in the near future. As further evidence, he pointed to a study by the U.S. Department of Commerce. The government agency had projected that steelmakers would need to expand capacity by 12 percent during the 1980s to meet increased domestic demand. If a tight market returned, Foy pointed out, even the most marginal of the corporation's facilities could be moneymakers.[50]

No doubt Foy was too busy to focus on new markets. A handsome fellow who spent most of his spare time on the Bethlehem golf course, and who dispatched his business with a great reliance on a long line of assistant vice presidents, Foy enjoyed the perks of his position. The final months before his retirement in 1980 were typically hectic. Celebration of the 75th anniversary of Schwab's purchase of Bethlehem, involving a heavy round of receptions in Bethlehem, New York, and Washington, took up a lot of his time. In April, 1980 Foy transported about 250 plant officials and their wives to Boca Raton, Florida, for a party at the company's expense. The guests fished in the Florida waters and enjoyed the nourishment of a never-closed free bar.

Foy and his wife Marjorie then held a retirement party at the Saucon Valley Club where Donald Trautlein, his successor, and his wife Mary were presented to several hundred guests. The Foys and Trautleins boarded the corporate jet for a globe-girdling tour, touching down at such important steel centers as Singapore, Cairo, and London.[51]

*

The bottom fell out 18 months after Trautlein returned from his junket and assumed the chairman's post in mid-1980. After a moderately successful 1981, demand dropped in early 1982 under the destructive wave of a recession engineered by Paul Volcker and the Federal Reserve Board to squeeze out Carter-era inflation. Mill production dropped to 55 percent of capacity and Bethlehem reported a quarterly deficit of $67 million. Trautlein responded with massive layoffs. On May 7, 1982, the number of workmen scratched from the mill boards increased from 3,150 to 5,030 at Sparrows Point, the highest number of layoffs since 1933. The increase was due largely to the shutdown of No. 4 open hearth. Operating for 19 years after the company's engineering staff had concluded that BOF furnaces were of superior cost savings, the shop had epitomized the "Bethlehem way" of hugging the bottom end of mill equipment. In retiring No. 4, the company now relied on the BOFs that were getting old by world standards.[52]

A $750-million modernization effort to pull the company out of its doldrums, announced in 1981 by Trautlein, was put on hold pending action by the government to provide further protection against foreign steel. On May 13, 1982, a company official warned in a speech to the Maryland Chamber of Commerce that Sparrows Point's coastal location made it very vulnerable to "unfair" steel imports. Unless local business and government officials backed industry petitions for antidumping duties on foreign steel in Washington, the company would ax unspecified portions of the capital program, the official indicated.[53]

In the second quarter Bethlehem and the seven other large integrated companies sustained pretax losses of nearly $700 million. Output was a sickening 42.4 percent of capacity. Sparrows Point was running at 40 percent. Eastern Baltimore County went into shock as the pipe mill, the rod and wire mill, a coke oven battery, and a sinter unit were shut. Many laid-off workers hoped that the departments would reopen in a few months, but as the layoff rolls grew longer instead of shorter, it became obvious that the situation was desperate.

Meeting with the membership on September 30, 1982, Ed Gorman, recording secretary of Local 2609, gave the alarming news. Out of 8,500 members, only 3,500 still had jobs. One thousand workers were on indefinite furlough and 3,500 were laid off on a week-to-week basis. Puffing furiously on his cigar, Gorman exclaimed, "All my life, I've never seen anything like this. Horrible! Horrible! And no light at the end of the tunnel."

He waved a copy of *U.S.A. Today.* It carried a story that quoted Treasury Secretary Donald Regan's reaction to the downturn. STEELWORKERS: LOOK FOR A NEW JOB was the headline. Turning to the business page of the *Baltimore Evening Sun* Ed read out loud how the shipyard had lost a $104-million contract to build two T-5 support tankers for the navy. Company officials said the work would have provided 2,000 jobs in the plate mill and shipyard. The contract was lost not to foreign competitors, but to a shipyard owned by George M. Steinbrenner, owner of the New York Yankees.

Ed stared at the sea of unemployed workers. "We're part of the mess," he spat out. "On election day two years ago some of us sat on our duffs and said, 'Well, there's no difference between Jimmy Carter and that cowboy.' And now that cowboy is going to get the labor unions on their knees, beggin', and they're going to get us back on the minimum wage."[54]

"Politics is economics," 2609 President Ed Bartee exclaimed. Blaming "Reaganomics" for the mess, he denounced calls by the Steel Companies Coordinating Committee to reopen the master contract for renegotiation to reduce wage costs. "Concessions don't create jobs," Bartee said. "The autoworkers are learning that right now."*

There was an undertow of resentment among the younger workers. Some were black; nearly all of them expected to be laid off in the next week or two. One asked Dave Wilson, District 8 director, where it would end. "The old guys, they got their jobs. But us young guys, 'cause of low seniority, ain't got much of nothin'."

After an awkward pause Wilson said that the matter was "in the hands of the company and the politicians." If Bethlehem could get competitive again, he said, maybe the jobs would come back.

Another young worker prefaced his remarks by saying that he had never spoken at a union meeting or any public forum. "I was wondering if these concessions ain't that bad an idea after all. I mean maybe we should be working for $8 rather than $12 an hour, so the work can be spread around."

Several men jumped up. "Now who says if you give up your money, Bethlehem is gonna keep your job?" asked one of them. Said another, "The company don't give you nothin'. They take."

Marian Wilson whispered sadly, "Grown men are crying when they see the pink slips. They're sayin' they're not working men any more." She still had a job owing to her 30 years of seniority, but those with 10

*Cash wages paid to production workers in steel averaged $13.96 an hour in 1982, according to the Bureau of Labor Statistics's *Supplement to Employment and Earnings*. This compared to $13.01 an hour for automobile workers, $13.71 for workers in primary aluminum, $12.66 for coal miners, $12.13 for metal can workers, $11.85 for aircraft assembly, and $11.50 for railroaders. With fringe benefits added in, the average cost of an employed steelworker was between $17 and $18 per hour. However, due to the rapid increase of unemployed workers, who were paid supplemental unemployment benefits by the companies, total hourly employment costs rose to $23 and higher.

and 15 years were threatened. In some cases the shock was so great the company ambulance was called in. "We've had heart seizures, that's how bad it is. The mill is like a graveyard."

A food and clothing drive by Local 2609 raised more than $5,000 in cash from working steelworkers over the next few months. Volunteers like Marian collected cans of food, piles of clothes, and other essentials, and helped distribute the goods to the unemployed. As winter approached some workers moved. Many didn't have the resources to leave. They spent their time learning to do other things—drive a truck, install drywall, do repairs for neighbors. Some went back to school. Some did nothing. "Experienced layoffs we call them," said district director Wilson.[55]

The bad news piled up. Bethlehem reported a yearly loss of $1.47 billion, contributing nearly half of the $3.2 billion in red ink reported in 1982 by the top 15 integrated producers. Foreign producers captured one-fourth of the market, aided by the strong dollar, which reduced the price of imported goods. Another 20 percent was going to domestic mini mills. Collectively, U.S. production dropped far behind that of the USSR and Japan.[56]

The effects of the paralysis of Sparrows Point were felt throughout the state. A report by Chase Econometrics disclosed that the Baltimore metropolitan area ranked at the bottom of Northeast regions in terms of economic growth. Since 1980 the area had a net loss of 20,000 jobs, while Boston added 27,000, metropolitan New York 47,000, and Washington 15,000. It ranked a dismal 45th out of 52 metropolitan areas, slightly under steel-dependent Pittsburgh. "Baltimore is a heavy-goods manufacturing center with an aged capital plant, which makes it highly responsive to national economic downturns," the report commented.[57]

The backbone of blue-collar Baltimore was bending under the strain. Now overdependence on smokestack industries was faulted by the same establishment that had once embraced the factory chimney as a symbol of progress. Editorial writers scolded unions for being too greedy; *givebacks* and *cutbacks* crept into the vocabulary of collective bargaining. Jobs were in service industries, financial institutions, and data processing, not in factory and mechanical trades, said Patrick R. Arnold, director of research for the Maryland Department of Human Resources. ENDANGERED SPECIES: THE AMERICAN WORKER, replied the bumper stickers on cars in the embattled neighborhoods around the Point.[58]

Once again, managerial myopia contributed to the rout. Panicked by the losses, Chairman Trautlein and President Walter F. Williams wiped out whole departments without crafting a coherent strategy for renewal, hastening the very decline they were charged with stemming. While Nucor specialized and economized, Bethlehem cut off its production fingers by eliminating all finishing departments that were not projected to yield a long-term profit. The company, for example, abandoned the bolt-fastening business in 1985, saying it had no future, and took $35 million in write-offs. At the same time Nucor broke ground for the first domestic bolt plant in decades. Nucor has since established a $45-million business in bolts (paying its top workmen $30,000 to $35,000 a year), while Bethlehem's Lebanon plant lies empty and padlocked.[59]

Equally shortsighted, Trautlein Williams ordered cutbacks in already inadequate R-and-D spending as soon as the company began incurring losses in 1982. This was followed by further cuts in 1983 and 1984. Finally, in the name of raising cash, the company's research laboratory outside of Bethlehem was put up for sale. By contrast, Japanese steelmakers have stepped up R-and-D when demand started to sag in the mid-'80s.[60]

Adding to this sorry story of incompetence was the recurring theme of greed. What heavily damaged employee morale was the disclosure in 1986 that Trautlein and a few cronies had handed themselves generous salary increases as well as a "golden parachute" severance package in case of a takeover. USWA officials, who had accepted wage concessions in 1983 to help the company, felt betrayed. They balked at a second wage cut. Negotiations were resumed only when the golden parachutes were rescinded. Trautlein's pay had gone up 11 percent, to $542,060. Advisors had warned him of a probable backlash over the boost, but he had waved away the warnings, complaining that he was not getting as much compensation as the chairman of U.S. Steel, David M. Roderick. "I was below market," *The Wall Street Journal* quoted him as saying.[61]

"The architect of any real disaster," economist John Kenneth Galbraith has written, "achieves a notoriety that all but invariably assures more remunerative employment. It is our compassionate system of upward failure." Trautlein retired early in March 1986, after closing the books on $2.03 billion in losses. Thirteen vice presidents had departed the stricken steelmaker with severance packages exceeding $1 million each, according to John Strohmeyer. Trautlein was replaced by company president Williams. A 35-year veteran and member of the board of directors since January 1, 1978, he had been promoted to president and chief operating officer of the company under Trautlein in 1980. Williams seems determined to adhere to the well-worn grooves of the Bethlehem system. As *The New York Times* noted in a profile after his appointment, "Mr. Williams has made it clear that he would apply no new philosophies in his efforts to restore profitability. In an interview in Bethlemen's 21st-floor executive suite overlooking the Lehigh Valley, Mr. Williams discussed his commitment to the course put into place by his predecessor, Donald H. Trautlein."[62]

Williams succeeded in restoring modest profitability in 1987, chiefly through onetime asset sales and adroit use of federal investment tax credits. His regime also has benefited from the 1984 decision by the Reagan administration to place quotas on imports of semifinished slabs. With demand for such material rebounding in 1987, Bethlehem has found itself with a largely captive market. Prices for semifinished slabs have increased sufficiently for the refiring of several antiquated open hearths at Sparrows Point. Even obsolete equipment can operate profitably under the import restrictions (administered by the Commerce Department headed by ex-Armco head William Verity), and Bethlehem is expected to report a profit in 1988.[63]

Despite the current upturn in earnings, the long-term outlook for Bethlehem—and for Big Steel in America—is not bright. Its prosperity is based on protectionism and high prices. While Bethlehem and other companies have cut costs sharply by closing down mills and laying off workers, the industry still has not dealt with the fundamentals: finding new markets and catching up with foreign technology.

Extension of the quotas into the 1990s is foremost on the agenda of the steelmakers. Lobbyists for Bethlehem, U.S. Steel (now USX), and the Iron and Steel Institute have been pressing Congress for an early commitment on an extension of the quotas, due to expire in September 1989.[64] "Piecemeal protectionism" has perhaps slowed but not arrested the decline, nor have other public benefits, ranging from tax credits to Buy American programs, spurred significant redirection or renewal.

*

Today there are 10,000 fewer jobs at the Point than in 1980; 21,000 less than in 1957. Those who linger on, a dispirited group of only 7,900 men and women, pass through the guard booths into a world of empty railroad tracks and partly used buildings. The plant gets along by supplying sheet and canning coils for such customers as Campbell's Soup and General Motors, and semifinished slabs and galvanized products for industrial buyers. A continuous caster, so long delayed, has been installed recently, increasing quality and lowering costs. Ironically, the machinery was imported from Austria, a nation whose "unfair" steel imports Bethlehem strives so mightily to curtail.[65]

Labor relations in 1988 are at an all-time low, according to USWA district director Wilson. Union and management remain locked in the mutual hostility that has separated the two sides since Charlie Schwab's time. "We are re-living the past," says district director Wilson, commenting on the current quarrel over work rule revisions. So far, the new plant manager has declined to meet with union officers to iron out the difficulties. Employee grievances at the Point have shot up to a record 2,600 unresolved cases, yet Wilson can see no future for the plant other than that defined by a discredited management and a resentful union. "They'll never sell it," Wilson says, dismissing the possibility of a collective purchase of the property through an employee stock ownership plan (ESOP), the imaginative solution that was used by non-USWA workers to save Weirton Steel.[66]

The oldest open hearth sheds, too expensive to demolish, dominate the west side of the peninsula, the cavernous shells dead and partially dismantled. To the south, the furnaces where Charlie Parrish once struggled with the pain of racism have been dynamited and carted away by truck. Acres of abandoned coke ovens cast a sepulchral shadow over the waterfront. Amid the rusting Erector-set towers, it is hard to imagine what had been here, the tens of thousands of strong men who came to the Point, sweating, shouting, muscling mountains of foreign ore into rivers of American steel.

Little remains of the town where generations of families purchased their groceries at the company store, paid their rent to the company collector, and were straightened out by "Boots" Campbell if they stepped out of line. Only a few rows of sycamores give any hint of what was the most complete and comprehensive steel town in the East.

Still it is difficult to say good-bye. Across Bear Creek from the mill, Ben Womer has created his own memorial to the peninsula's glorious epoch. He has been collecting artifacts of plant and town since his retirement from the open hearth. They surround him in a room at the Dundalk YMCA Building. Spilling out of one box are 11 wooden molds used to form furnace castings. Poking around the room, one can find a rafter from the Sparrows Point Methodist Church, spikes from the streetcar tracks than ran down D Street, a grave marker from the cemetery, and many American flags. To represent the boom days of World War I, Ben has come up with a swagger stick purportedly used by General Pershing in France, and, in a sly commentary on the company town, there's an old 78 titled, "Don't Sell Daddy Any More Whiskey."

Looking through a pile of plant photographs a while back, he exclaimed happily, "Now this could quite possibly be me. That's what we looked like—a hat, a coat, pants. You can see how dirty you got. And that's old No. 1 shop, with them 50-ton furnaces.

"Now I got three pictures of the open hearth—they're in color. Beautiful pictures. Just beautiful of the steel. Orangish-white when she's runnin' into the ladle."

He slapped the pictures down and fetched an article from the *Dundalk Times.* While not much for putting pen to paper, having dropped out of Sparrows Point High, he had wanted to get his memories down for his grown children, none of whom do steel work any more, and for others. "We made many a heat of steel for skyscrapers, boats, tin cans and Lord only knows what else," he wrote. "The men who worked in the blast furnaces, coke ovens, or open hearths had to work the holidays most of the time because those three departments were not able to shut down like the rest of the mills. Many a man who worked daylight on Christmas never saw his children on Christmas morning unless they were gotten up at five o'clock to see their toys when he was ready to go to work. The same with Thanksgiving and Christmas dinners. This kind of thing could only mean that your devotion to your family, to your town and to the company you worked for was your trademark."

The article ended, "When the pages of history are written about the Patapsco Neck section of Baltimore County, the steel company, the men who ran the mills along with the workers, will certainly head the list."

"You know," commented Ben, "I got many calls about what I wrote. People stopped me on the street and thanked me. They said it made them feel better."

ACKNOWLEDGMENTS

Robert Schrank helped to get this book started and Michael Lacey helped to get it done. Then a project specialist at the Ford Foundation, Bob took a chance and provided me with a grant to examine the Frederick Wood Papers. Mike Lacey's Program on American Society and Politics, at the Woodrow Wilson Center for Scholars in Washington, was the refuge where the book began to take shape.

Along the way I was fortunate to have the guidance of dozens of people. Peggy J. Miller played an active and continuing role in conceptualizing the book. She kept me from running aground many times. Warren I. Susman, a colleague at the Wilson Center, gave generously to the early chapters despite his worsening health. His death in 1985 was a personal loss. My interest in American history and literature was sparked by Kenneth S. Lynn's incandescent seminars at Johns Hopkins University in 1971–72. His suggestions and encouragement during the early stages of this project were invaluable.

The book profited from the wise counsel of Steven M. Luxenberg, my old partner on the *Baltimore Sun,* who listened to my ideas and read (and reread) parts of the manuscript. Mary Jo Kirschman brought her imaginative involvement to several sections of the book. She assisted in the interviews of the women who had worked in the tin mill and conducted research into the organizing campaigns at Sparrows Point. To James D. Dilts I owe special thanks for his knowledge of 19th-century railroading and enthusiasm for 20th-century track walking.

I am indebted as well to the late Herbert G. Gutman, who offered steadfast support; Michael Nash and the staff of the Hagley Museum; and, more recently, to the Joseph Regenstein and John Crerar Libraries at the University of Chicago. Technical assistance was supplied by Therese Chappell, Ralph Clayton, and David Shayt. Also sustaining me over the long haul: Ronnie Dugger, Meg Gallagher, Art Levine, Harry and Eugenia Miller, Kathy Miller, Kurt and Kathleen Reutter, Robert Ruby, and the Sip & Bite Restaurant in East Baltimore.

It is impossible to thank individually the many people whom I interviewed. There were 117 interviews conducted among workers and retirees at Sparrows Point alone; additional information came from interviews at government offices and at other mills. Constructing the major portraits in the book required lengthy questioning and fact checking; the people who appear on these pages accepted my intrusions and gave freely of their time and knowledge. Unfortunately, some of them—Elizabeth Alexander, Mary Gorman, and Mike Howard included—will not be able to see the results.

I am grateful to a number of groups and individuals who helped me get to know the steel communities around Baltimore. Dave Wilson permitted me to attend meetings at Local 2609 and made available a spare room in the union hall to talk to workers. I benefitted from the help of Charles Capp, the Dundalk-Patapsco Neck Historical Society, Alan Fisher, Lawrence (Reds) Forbes, Bill Harvey, Sirkka Lee Holm, the Rev. Everett Miller, the Senior Steelworkers Associations of Locals 2609 and 2610, and Linda Zeidman.

Several members of management who agreed to talk to me cannot be thanked by name. Anonymity was the condition they requested in exchange for their candor.

Arthur H. Samuelson, my original editor at Summit, believed in this book. Reading all of the early drafts, he asked penetrating questions that helped to draw out the social and public-policy implications of the story. His resolve was indispensable. His recent successor, Dominick V. Anfuso, has handled the editing and production chores with dispatch and good cheer. Others who have contributed to the final product are James Silberman, publisher; Michael Caine, copy editor; Edward J. Acton, literary agent; and Steven Lagerfeld, senior editor of *The Wilson Quarterly*.

The book is dedicated to my parents, Bettie Lytle Reutter and E. Edmund Reutter, Jr., and to my wife, Peggy J. Miller. My deepest thanks are to them.

Chicago, Illinois
July, 1988

NOTES

Chapter 1: Goddess of Industry

1. *Baltimore American*, May 31, 1890.
2. List of participants and schedule of events were recorded in the Maryland Steel Company Papers of Frederick W. Wood (henceforth FWW Papers). Accession 884, Hagley Museum and Library, Greenville, Del.
3. From *Baltimore American, Baltimore Sun, New York Times*, and *Baltimore Herald*, May 31, 1890. In some early accounts the peninsula was spelled "Sparrow's Point."
4. *Baltimore American*, May 31, 1890.
5. *Baltimore American*, May 31, 1890.
6. From FWW Papers.
7. *Baltimore American*, May 31, 1890.
8. FWW Papers and *Baltimore American*.
9. FWW Papers. The sequence of speakers varied a bit in the *American*, which gave precedence to Gen. Agnus's oratory.
10. Metallurgy's Middle Eastern origins are described in Douglas Fisher's *The Epic of Steel* (New York: Harper & Row, 1963) and the chemistry of iron is rendered understandable in Norman F. Smith's *The Inside Story of Metals* (New York: Julian Messner, 1977).
11. Fisher, *Epic of Steel*, 52–54.
12. "Cherry-red" heat is about 1200–1300° F.
13. Elting E. Morison, *Men, Machines and Modern Times* (Cambridge, Mass.: MIT Press, 1966), 133.
14. "History of Pennsylvania Steel Company," FWW Papers, c. 1890; *Dictionary of American Biography* (New York: Charles Scribner's Sons, 1931), vol. 6. 318–19; Jeanne McHugh, *Alexander Holley and the Makers of Steel* (Baltimore: Johns Hopkins Univ. Press, 1980), 212. Prior to Pennsylvania Steel, two experimental plants had been established; neither, however, produced commercial quantities of steel before 1867.
15. William M. Sellew, *Steel Rails: Their History, Properties, Strength and Manufacture with Notes on the Principles of Rolling Stock and Track Design* (New York and London: D. Van Nostrand, 1913), 1–5. While a 1-inch-square bar of iron fractures under a weight in excess of 20,000 pounds, the strength of plain carbon steel is 100,000 pounds per square inch or more.
16. George H. Burgess and Miles C. Kennedy, *Centennial History of the Pennsylvania Railroad Company* (Philadelphia: Pennsylvania Railroad Co., 1949), 293

17. James M. Swank, *History of the Manufacture of Iron in All Ages* (Philadelphia: American Iron and Steel Association, 1892), 414–15; Joseph M. Levering, *A History of Bethlehem, Pa.* (Bethlehem, Pa.: Bethlehem Times Publishing, 1903), 72–75; Col. Frederick L. Hitchcock, *History of Scranton and Its People* (New York: Lewis Historical Publishing, 1914), 23–31.
18. Swank, *History*, 411–12; Bessie Louise Pierce, *A History of Chicago* (New York: Knopf, 1957), vol. 3, 156–57; James Howard Bridge, *The Inside History of the Carnegie Steel Company.* (New York: Aldine Book Co., 1903), 71–93; "List of Tonnage of Steel Makers, East and West," FWW Papers.
19. FWW Papers; William T. Hogan, *Economic History of the Iron and Steel Industry in the United States* (Lexington, Mass.: Lexington Books, 1971), 19–23.
20. Bent noted that 1,000 miles of water transport was equal in cost to 100 miles on land by railroad.
21. Cuba Papers, FWW Papers.
22. *National Cyclopaedia of American Biography* (New York: James T. White & Co., 1947), vol. 33, 238–39; "Class Notes," *MIT Review*, March 1944; Robert Hallowell Richards, *R. H. R.: His Mark* (Boston: Little, Brown & Co., 1936), 118.
23. Cuba Papers; William J. Clark, *Commercial Cuba: A Book for Business Men* (New York: Charles Scribner's Sons, 1898), 61; I. A. Wright, *The Early History of Cuba, 1492–1586* (New York: Macmillan, 1916), 14–16; Robert P. Porter, *Industrial Cuba* (New York: G. P. Putnam's Sons, 1899), 318–20.
24. Cuba Papers, 1882 pocket notebook.
25. Cuba Papers.
26. Porter, *Industrial Cuba*, 319–20.
27. *Iron Age*, March 8 and April 5, 1883; Cuba Papers; "Cuban Ore Production," *Statistics of the American Iron Trade*, (New York: American Iron and Steel Institute, 1913), 31–33.
28. Incorporation papers, Juragua Iron Co., FWW Papers; *Iron Age*, July 17, 1884.
29. FWW Papers, log books, 1887.
30. Baltimore County property records, Towson, Md.; Wood notebooks.
31. Notebooks and sketches, 1887, FWW Papers.
32. Letters between Rufus and Frederick, 1887, FWW Papers.
33. Letters from Rufus, FWW Papers.
34. *Baltimore American*, May 30, 1890; FWW Papers.
35. From construction accounts, FWW Papers.
36. As a Maryland state senator in 1886 Jackson pushed through the legislature a 30-year property-tax exemption for the right of way of his railroad line. *Laws of Maryland, 1886*, State Hall of Records, Annapolis, Md. Also see Frank R. Kent, *The Story of Maryland Politics* (Baltimore: Thomas and Evans, 1911).
37. From Bent correspondence, FWW Papers.

38. FWW Papers. For background see Kent, *Maryland Politics.*
39. Letter, Preston to Bent, FWW Papers, Dec. 22, 1887.
40. Letter, Preston to Bent, FWW Papers, Jan. 28, 1888.
41. Correspondence and tax files, FWW Papers.

Chapter 2: Creating the Works

1. FWW Papers and incorporation papers filed at Baltimore County Courthouse, Towson, Md.
2. Letter, Rufus to Frederick, FWW Papers, June 23, 1889.
3. My complaint is with an all-purpose, value-free "progress" that attempts to make managerial decisions appear immutable.
4. E. C. Potter, "Rails, Past and Present," *Iron Age,* Feb. 17, 1898.
5. Sir James George Frazer, *Myths of the Origin of Fire* (London: Macmillan, 1930), 132–33; Cyril Stanley Smith, *Sources for the History of the Science of Steel, 1532–1786* (Cambridge, Mass.: MIT Press, 1968), 35. Della Porta's book was first published in 1589 and printed in English in 1658.
6. Quoted in Henry Nash Smith, ed., *Popular Culture and Industrialism, 1865–1890* (New York: New York University Press, 1967).
7. Day Allen Willy, "A Concentrated Industry," *Cassier's,* March 1906; "The Maryland Steel Company: The Works at Sparrows Point," *Iron Age,* Dec. 9, 1897.
8. FWW Papers; *Baltimore American,* Aug. 2, 1891; *Iron Age,* Dec. 9, 1897; "Biography of a Rail," *Baltimore Sun,* March 12, 1911. The two converters were rated at 20 tons, but could each handle up to 25 tons per blow.
9. Wood's internal notes, FWW Papers, and *Iron Age,* Dec. 9, 1897.
10. *Iron Age,* Dec. 9, 1897.
11. The mill was the first in the nation to roll rails in 180-foot lenghts. Aside from the savings in scrap, the method permitted finishing at a lower temperature and made for tougher track. The rail was cut then into standard 30-foot lengths.
12. Morison, *Men, Machines;* Peter Temin, *Iron and Steel in 19th Century America: An Economic Inquiry* (Cambridge, Mass.: MIT Press, 1964); Alfred D. Chandler, Jr., *The Visible Hand: The Managerial Revolution in American Business* (Cambridge, Mass.: Belknap Press, 1977); and *Cassier's,* March 1906.
13. See Carl Condit's excellent *American Building Art* (New York: Oxford University Press, 1960); also John R. Stilgoe, *Metropolitan Corridor: Railroads and the American Scene* (New Haven: Yale Univ. Press, 1983); Christopher Pick, *The Railway Route Book* (London: Willow Books, 1986).
14. *Baltimore Sun,* Dec. 7, 1891. Stilgoe, *Metropolitan Corridor,* discusses the works in the context of railroading, 77–103.

15. *Baltimore American,* April 22, 1893; Richard T. Ely, *The Labor Movement in America* (New York: Thomas Y. Crowell & Co., 1886), 56.
16. Here were the hours of the Rail Straightening Department in 1892:

DAY TURN (ONE WEEK)

		Hrs.
Sun.	off	—
Mon.	7 A.M.–6 P.M.	11
Tues.	same	11
Wed.	same	11
Thurs.	7 A.M.–5 P.M.	10
Fri.	7 A.M.–6 P.M.	11
Sat.	7A.M.–5 P.M.	10

NIGHT TURN (ONE WEEK)

Sun.	7 A.M.–7 A.M. (Mon.)	24
Mon.	6 P.M.–7 A.M.	13
Tues.	same	13
Wed.	same	13
Thurs.	5 P.M.–7 A.M.	14
Fri.	6 P.M.–7 A.M.	13
Sat.	5 P.M.–7 A.M.	14

Hours in other countries were described in *American Industrial Conditions and Competition* (or *Jeans Report*) (London: British Iron Trade Assoc., 1902).
17. *Sunday Herald,* April 9, 1893.
18. Pocket notebooks, FWW Papers, 1890–95.
19. *Sunday Herald,* April 9, 1893.
20. From 1895 Correspondence, FWW to Bent, FWW Papers.
21. When Steelton was started, "men were getting from $4 to $10 a day for skilled labor, but such times are past. Now the average per diem for skilled and unskilled labor is $1.75." Maryland Bureau of Industrial Statistics, *Fourth Biennial Report* (Baltimore: State Printer, 1892), 125.
22. Maryland Bureau of Industrial Statistics, *1896 Report,* on factory wages. David Brody's *Steelworkers in America: The Nonunion Era* (Cambridge, Mass.: Harvard Univ. Press, 1960) describes similar cuts in labor costs achieved by other railrollers.
23. For "American standard of living," see E. R. L. Gould, *The Social Condition of Labor,* Johns Hopkins University Studies in Historical and Political Science (11th series), (Baltimore: Johns Hopkins Univ. Press, 1893).
24. From Thomas B. Weeks, "Introduction of Commissioner," Maryland Statistics of Labor, *Second Biennial Report* (Baltimore: State Printer, 1887), 6, 17.
25. From Brody, *Steelworkers.*

26. FWW Papers; *Critic,* Aug. 1 and 8, 1891; U.S. Immigration Commission, *Immigrants in Industries, Part 2: Iron and Steel Manufacture* (Washington, D.C.: U.S. Govt. Printing Office, 1911), vol. 8, 645–46.
27. *Critic,* Aug. 8, 1891.
28. Axel Sahlin, "Direction, Management and Labour," in *American Industrial Conditions* (London: British Iron Trade Assoc., 1902). 503.
29. Wolfgang Schivelbusch, *The Railway Journey: Trains and Travel in the 19th Century* (New York: Urizen Books, 1979), 133.
30. Schivelbusch, 133.
31. *Baltimore American* and *Baltimore Sun,* Dec. 8, 1891.
32. *Baltimore American* and *Baltimore Sun,* Dec. 8, 1891.
33. *Baltimore American,* Dec. 8, 1891.
34. Wood's statement was published in the above articles.
35. Jury report was published in *Baltimore Sun,* Dec. 8, 1891.
36. This case seems to have been lost in the judicial cracks. A docket entry says the suit was "settled, 1901," but the actual case file shows no record of such a settlement.
37. In a typical case a $5,000 negligence suit filed in the Court of Common Pleas was dropped when claimant Patrick Naughton accepted a $75 cash payment and signed the following: "I, Patrick Naughton, do hereby release and forever discharge the said Maryland Steel Company and their successors from all claims and demands of every kind and description whatsoever, which I have, or can, or may have against the said Maryland Steel Company for or by reason of any and all injuries sustained by me." *Naughton* v. *Maryland Steel,* Baltimore Court of Common Pleas, 1901.
38. FWW Papers.
39. See E. H. Downey, *Workmen's Compensation* (New York: Macmillan, 1924).
40. *Wood* v. *Heiges,* in Maryland Hall of Records, Annapolis, Md.
41. Herbert Casson, *The Romance of Steel* (New York: A. S. Barnes & Co., 1907), 363, 365.

Chapter 3: Designing the Town

1. Budgett Meakin, *Model Factories and Villages: Ideal Conditions of Labour and Housing* (London: T. Fisher Unwin, 1905); U.S. Bureau of Labor, "Housing of the Working People in the United State," *Bulletin of Bureau of Labor,* no. 54 (Sept. 1904); correspondence, Rufus Wood to Frederick Wood, FWW Papers.
2. U.S. Bureau of Labor, *Report on Conditions of Employment in the Iron and Steel Industry* (Washington, D.C.: U.S. Govt. Printing Office, 1913), vol. 3, 420. Sparrows Point was identified in the report as "Community A. Eastern District."
3. Charles Hirschfeld, *Baltimore, 1870–1900: Studies in Social History,* JHU Studies in Historical and Political Science (59th series) (Baltimore: Johns Hopkins Univ. Press, 1941), 96–98; Amy C. Crewe, *No Backward Step*

Was Taken (Baltimore: Teachers Assoc., 1949), 209; Baltimore County School Commissioners, *Reports of Board, 1888–1901.*

4. Typescript statement of company and a speech by Rufus in 1907, FWW Papers; Baltimore County School Commissioners.
5. Company statement, FWW Papers; School Commissioners.
6. Company statement, FWW Papers.
7. *Baltimore Sun,* Jan. 30, 1893; Rufus letter, FWW Papers. Feb. 24, 1889; Dundalk-Patapsco Neck Historical Society booklet, *Reflections, Sparrows Point, 1887–1975,* 1976.
8. Letter, FWW Papers, June 23, 1889.
9. *Sunday Herald,* April 9, 1893, and *Baltimore News,* April 15, 1903 (on saloon district); *Critic,* Dec. 12, 1891; Rufus letter to Circuit Court, FWW Papers, April 15, 1903.
10. For Lowell's place in U.S. industrial history, see Arthur L. Eno, Jr., ed., *Cotton Was King: A History of Lowell, Massachusetts* (Lowell, Mass.: New Hampshire Publishing Co., 1976); Dickens quote in Roy Lubove, ed., *Pittsburgh,* (New York: New Viewpoints, 1976).
11. Based on blueprints and street maps of Sparrows Point, 1888—93; street addresses of residents; and letters, Rufus to Frederick, FWW Papers.
12. Dwelling Costs Ledger and house photographs, FWW Papers.
13. Rufus's letters are filled with references to Boston.
14. Family data in FWW Papers and 1900 U.S. Census report of the village.
15. *Sunday Herald,* April 9, 1893.
16. Interview with Elizabeth McShane and Katherine Roberts, Feb. 1982.
17. Author's analysis of the door-to-door census material (known as Manuscript Census) taken of the village for the 1900 U.S. Census, found at National Archives, Washington, D.C.
18. Recollections of Margaret Lindemon and company rent books.
19. Manuscript Census.
20. Sahlin, "Direction, Management and Labour," 504; Manuscript Census.
21. Manuscript Census; company blueprints.
22. Interview with Florence Parks, March 1982.
23. Parks interview; census data.
24. U.S. Bureau of Labor, *Report on Conditions of Employment,* vol. 3, 424, 426.
25. Manuscript Census.
26. Sahlin, 503.
27. Sahlin, 503–04.
28. The Rev. Henry A. Miles, *Lowell, As It Was, and As It Is* (Lowell, Mass.: Powers & Bagley, 1845), 128, 142. The Lowell foremen were responsible for the "moral care" of their employees and were "one of the most essential parts of the moral machinery of the mills," wrote Miles.
29. Rufus Wood correspondence, FWW Papers; recollections of Florence and Nathaniel Parks, March 1982.
30. House lease, FWW Papers.

31. "Summary of Town Accounts, 1894–97," FWW Papers.
32. Rufus Wood correspondence, FWW Papers; Hirschfeld, *Baltimore.*
33. Rufus Wood correspondence, FWW Papers.
34. Company Store Statements, FWW Papers. Very little was transacted in cash. An 1892 letter reported that cash sales were 20 percent of charged sales in 1891.
35. "Stockholders of S. Pt. Store Co., of Balto. Co., Md—Corrected" (an initial filing was inaccurate), FWW Papers; author's review of store's financial statements.
36. Quoted in *Federal Writers' Project Guide to 1930s Illinois,* (Chicago: A.C. McClurg & Co., 1939), 108.
37. *Baltimore Sun,* Oct. 21, 1906. Mencken joined the paper as Sunday editor in 1906 after six years on the *Herald.*
38. *Baltimore Sun,* Oct. 21, 1906.

Chapter 4: Foreign Affairs

1. Cuba Papers.
2. Russell H. Fitzgibbon, *Cuba and the United States* (Menasha, Wis.: George Banta Publishing, 1935), 20–25.
3. Fitzgibbon, *Cuba and the U.S.* Sherry Olson reported that schoolchildren in Baltimore were assembled at Ford's Theater for their public war chanting. In Olson, *Baltimore* (Baltimore: Johns Hopkins Univ. Press, 1980), 243.
4. *New York Times* dispatches, July 1 and 2, 1898; Henry Watterson's *History of the Spanish-American War* (Cincinnati: Martin-Degarmo Co., 1898), 100–102.
5. Cuba Papers; deeds of Spanish-American Iron Company, FWW Papers. Also see Porter, *Industrial Cuba,* 320–21.
6. Porter, 322.
7. Fitzgibbon, 28–31.
8. *National Cyclopaedia of American Biography,* vol. 28, 107; David F. Healy, *The United States and Cuba, 1898–1902* (Madison, Wis.: Univ. of Wisconsin Press, 1963), 189.
9. Healy, *United States and Cuba,* 84.
10. Fitzgibbon, 54–56; Healy, 10.
11. Civil Order No. 53, published in *Report of the Military Governor of Cuba,* Annual Reports of the War Department (Washington, D.C.: U.S. Govt. Printing Office, 1901), part 1, 283; on royalties and property taxes, *(Supplemental) Report of the Military Governor,* part 3, 8.
12. Harry F. Guggenheim, *The United States and Cuba: A Study in International Relations* (New York: Macmillan, 1934), 48–49. Guggenheim, U.S. ambassador to Cuba between 1929 and 1933, had a deep love of the island and its people and wrote a sorrowful book about the "unsatisfactory relationship" that had evolved between the two countries since 1898.
13. Personal letters, correspondence files, FWW Papers.

14. Personal letters, FWW Papers.
15. Cuba Papers; Incorporation Papers, Pennsylvania Steel Company (of New Jersey); and Underwriters' Agreement, 1901.
16. Incorporation Papers.
17. U.S. Commissioner of Corporations, *Summary of Report on the Steel Industry* (Washington, D.C.: U.S. Govt. Printing Office, 1911), 1–12, and chart of companies whose stocks were acquired by U.S. Steel Corporation at or shortly after its organization in 1901; Abraham Berglund, *The U.S. Steel Corporation* (New York: Columbia Univ. Press, 1907), 5–15.
18. *United States* v. *United States Steel Corporation et al.*, 251 U.S. 417 (1920).
19. U.S. Commissioner of Corporations, *Summary of Report;* also see Horace L. Wilgus, *A Study of U.S. Steel Corporation in its Industrial and Legal Aspects* (Chicago: Callaghan & Co., 1901), 2–4.
20. U.S. Commissioner of Corporations, *Summary of Report,* 38.
21. Testimony of Jacob Schonfarber in U.S. Industrial Commission, *Capital and Labor Employed in Manufactures and General Business,* (Washington, D.C.: U.S. Govt. Printing Office, 1901), vol. 7, 449. Regarding the millions gained by Carnegie and his partners—altogether the Morgan syndicate paid $447,416,640 in U.S. Steel stocks and bonds for the Carnegie company—see Bridge, *The Inside History,* 363–64; Stewart H. Holbrook, *Iron Brew* (New York: Macmillan, 1939), 269–71; and Melvin I. Urofsky, *Big Steel and the Wilson Administration* (Columbus, Ohio: Ohio State Univ. Press, 1969), prologue.
22. Correspondence file, FWW Papers; *Iron Age,* April 9, 1908. The Spanish-American Iron Co. became an operating subsidiary of the Pennsylvania Steel holding company.
23. FWW Papers; Berglund, *U.S. Steel,* 168.
24. From pooling percentages, FWW Papers. From 1887 to 1901 the railmakers agreed on market allocations almost as frequently as they broke away from them. Wood's papers include references to quarterly meetings of steel rail presidents in 1892 and 1893 to try to strike agreements on market shares. In the 1897–99 period especially, the industry was in great flux as Andrew Carnegie conducted territorial raids on other railmakers, including Pennsylvania-Maryland Steel. Stability returned after 1901.
25. *Baltimore Sun,* June 13, 1901.
26. "Iron Mining in Cuba," *Iron Age,* April 9, 1908.
27. Quoted in Hogan, *Economic History,* 367; FWW Papers.
28. From Annual Reports, Spanish-American Iron Co., in Cuba Papers.
29. From cost and production figures, FWW Papers, 1901–06.
30. From Annual Report, 1907, Spanish-American Iron Co., and FWW Papers. The Palma government also gave the mine subsidiary title to Cagimaya Island, 450 acres, on Nipe Bay to establish an ore port as well as a railroad right of way between the port and the Mayari deposit. A fee of 50 cents per hectare (2.471 acres) was charged for this and other property.

31. Fitzgibbon, 112–125. The troops' patrol functions were incorporated into various reports such as *Road Notes, Cuba,* by the U.S. War Department (Washington, D.C.: U.S. Govt. Printing Office, 1909).
32. "The Mayari Iron Ore District of Cuba," *Iron Age,* Aug. 15, 1907.
33. James E. Little, "Sintering of Mayari Iron Ores in Cuba," *Iron Age,* Sept. 14, 1911; Annual Reports, 1908 and 1909, Spanish-American Company.
34. *Iron Age,* Aug. 15, 1907.
35. W. T. Stead, *The Americanization of the World; or the Trend of the Twentieth Century* (London and New York: H. Markley, 1902); John A. Hobson, *Imperialism. A Study* (New York: James Pott, 1902), 11. Pennsylvania Station in New York was opened in 1910 and is used by Amtrak today.
36. Victor S. Clark, "Labor Conditions in Cuba," *Bulletin of the U.S. Department of Labor,* July 1902, 666–67; Guggenheim, *U.S. and Cuba,* xii–xiv; and Porter, *Industrial Cuba,* 100–105.
37. Clark, 685; Cuba Papers. Clark's account of labor policies helps decipher the meaning of some of the correspondence in the FWW Papers.
38. Correspondence, 1899, FWW Papers.
39. Clark, 704; "Iron Mining in Cuba," *Iron Age,* April 9, 1908.
40. Clark, 685.
41. Clark, 686.
42. Charles E. Woodruff, *The Effects of Tropical Light on White Men* (New York and London: Rebman Co., 1905), 129–141. Sabbatical leaves were granted for the wet season of June through September.
43. According to Clark (702–05), the daily pay for other industrial jobs in Cuba was: lumbering, quarrying, and bricklaying, 87 cents to $1.05; manganese mining, 85 cents; sawmill planers and helpers, $1.50; and carpenters, $2. *La tarea* was credited with reducing the average loading time of ore boats from 25 hours to 17 hours, 27 minutes.
44. "Iron Mining in Cuba," *Iron Age,* April 9, 1908.

Chapter 5: Secret Submarines

1. *Moody's Analyses of Investments* (New York: Moody's Investors Service, 1913), 755; "Sudden Death of R. K. Wood," *Baltimore American,* May 17, 1909; *National Cyclopaedia of American Biography,* vol. 45, 36.
2. Data from *Iron Age,* March 26, 1914 and March 11, 1915, and Sellew, *Steel Rails.*
3. "Biography of a Rail," *Baltimore Sun,* March 12, 1911. Remember, two rails are needed for a length of track.
4. FWW Papers on rail sales; and Pennsylvania-Maryland Steel's 263-page catalog, *Rails, splice bars and accessory rolled sections for track* (Philadelphia, Pa. Steel Co., 1912).
5. *Iron Age,* Jan. 30, 1913 (testimony of W. E. Corey); plus earlier articles on the pool in *Iron Age:* "An International Rail Pool," Nov. 3, 1904, and "The International Rail Pool," July 6, 1905.

6. Correspondence files, FWW Papers. This letter was dated Dec. 1901.
7. *Iron Age,* Jan. 30, 1913; FWW Papers on rail sales.
8. Correspondence files, FWW Papers.
9. Speech by Quincy Bent describing the properties of steel and armor plate in *Baltimore Engineer,* Sept. 1934; Arundel Cotter, *The Story of Bethlehem Steel* (New York: Moody Magazine and Book Co., 1916), 4–5; Helmuth C. Engelbrecht and F. C. Hanighen, *Merchants of Death: A Study of the International Armament Industry* (New York: Dodd, Mead, 1934), 53.
10. Bethlehem Steel Co., *Mobile Artillery Material* (South Bethlehem, Pa.: Bethlehem Steel Co., 1916), 4; *Acieries de la Compagnie de Bethlehem, Exposition Universelle de 1900 Paris* (Philadelphia: Bethlehem Steel Co., 1900). Courtesy of U.S. Army Military History Institute. For a detailed description of high-speed steel and the role of Taylor in its development, see Frank Barkley Copley, *Frederick W. Taylor: Father of Scientific Management* (New York: Harper & Brothers, 1923), vol. 2, 79–90.
11. Clippings printed in Robert Hessen, *Steel Titan: The Life of Charles M. Schwab* (New York: Oxford Univ. Press, 1975), 165.
12. Good sources on the European arms industry include J. D. Scott, *Vickers* (London: Weidenfeld & Nicolson, 1962); Engelbrecht and Hanighen, *Merchants of Death* (well researched if polemical); Richard Lewinsohn, *The Mystery Man of Europe: Sir Basil Zaharoff* (Philadelphia: J. B. Lippincott, 1929); Philip Noel-Baker, *The Private Manufacture of Armaments* (London: V. Gollancz, 1936).
13. Scott, *Vickers,* 86–87.
14. Hessen, *Titan,* 179.
15. The government orders were reported in Bethlehem's catalogue *Mobile Artillery Material;* Schwab's peripatetic ways were described in, among other places, Merle Crowell, "Schwab's Own Story," *American Magazine,* Oct. 1916.
16. Bethlehem Steel Annual Reports, 1909 and 1910.
17. Letter, G. Petriccione to Charles Rand and Frederick Wood, Oct. 5, 1910, FWW Papers. Petriccione was the company's confidential agent in Havana.
18. *New York Times,* Oct. 10 and 11, 1911.
19. Cotter, *Story of Bethlehem Steel,* 43.
20. "Bethlehem Purchase of Fore River Shipyard," *Iron Age,* May 1, 1913; Cotter, 45. For an illuminating history of the submarine invented by American John P. Holland, see Herbert C. Fyfe, *Submarine Warfare Past and Present* (London: E. G. Richards, 1907).
21. Barbara W. Tuchman, *The Guns of August* (New York: Macmillan, 1962), 17–27.
22. Tuchman, 22.
23. *New York Times,* Oct. 22, 1914.
24. Scott, *Vickers,* 98–103; Cotter, *Story of Bethlehem Steel,* 33.
25. *New York Times,* Nov. 26, 1914.

26. Gaddis Smith, *Britain's Clandestine Submarines 1914–1915* (New Haven: Yale University Press, 1964), 30–31, 37–38.
27. Smith, 38–39.
28. Arthur J. Marder, ed., *Fear God and Dread Nought: The Correspondence of Admiral of the Fleet Lord Fisher of Kilverstone* (London: Jonathan Cape, 1959), vol. 3, 66. The deal was made with "the great Schwab of the American Steel Corporation [sic]. He is more English than the English! He had two million sterling more orders than Krupps last year! and is far bigger than Elswick and Vickers put together." (Fisher to George Lambert, civil lord of the admiralty, Nov. 3, 1914, published by Marder in 1959).
29. Hessen, 213. This biography—the only written on Schwab so far—is disappointedly superficial about his wartime activities and follows the company's patriotic version of his career slavishly.
30. *New York Times*, Nov. 21, 1914.
31. *New York Times*, Nov. 26, 1914. As early as Nov. 10, the paper disclosed aspects of the sub contract, although the reporting was more gossip than news. In a Nov. 11 dispatch, for example, the mogul's rental of a lavish suite of rooms at the Savoy was of greater interest to the reporter than information suggesting that he was conferring with the War Office.
32. Munitions salesman Johnston blamed the leak on a German spy in the British Admiralty (Hessen, 331); *New York Times*, Aug. 19, 1914, on Wilson's call for strict neutrality; "Public Circular Issued by the Secretary of State, October 15, 1914, Regarding Neutrality and Trade in Contraband," *Papers Relating to the Foreign Relations of the U.S., 1914 Supplement* (Washington, D.C.: U.S. Govt. Printing Office, 1928), 574.
33. Ray Stannard Baker, *Woodrow Wilson: Life and Letters* (New York: Doubleday, 1935), vol. 5, 188.
34. *New York Times*, Dec. 3, 1914; Samuel Flagg Bemis, *A Diplomatic History of the United States*, (New York: H. Holt & Co., 1936), 97–99.
35. Bemis, *Diplomatic*, 378–79, 412–13. A second Confederate cruiser, the *Florida*, was also involved.
36. Baker, *Woodrow Wilson*, 188–89.
37. Bryan's press release was issued on Dec. 7, 1914. *Papers Relating to Foreign Relations, 1914 Suppl.*, 578.
38. *New York Times*, Dec. 8, 1914.
39. *New York Times*, Dec. 5, 1914; Smith, *Clandestine Submarines*, 78–80.
40. Eugene G. Grace, *Charles M. Schwab* (South Bethlehem, Pa: Bethlehem Steel Co., 1947), 42; also see Urofsky, *Big Steel*, 98–103.
41. Smith, 81.
42. Smith, 81.
43. *Wall Street Journal*, Dec. 24, 1914, quoted in Smith. *New York Times*, Dec. 24, reported the same substance ("I went to Europe and canceled the contracts").
44. Smith, 83; Hessen, 215.
45. Bernstorff to Bryan, Jan. 27, 1915, *Papers Relating to Foreign Relations*,

1915 Supplement (Washington, D.C.: U.S. Govt. Printing Office, 1928), 781.

46. Smith, 89. Smith's sleuthing took him to the records of the U.S. State Department, British Admiralty and archives of Vickers, Bethlehem Steel, and Electric Boat, which gave him a much different version of the truth than the Bryan press release and a slew of history books based on the assumption that the sub contract was canceled.
47. Bryan to Bernstorff, Feb. 17, 1915, *Papers Relating to Foreign Relations, 1915 Suppl.*, 782.
48. After the war Schwab told Josephus Daniels, U.S. secretary of the navy, that he made $4 million off the submarine deal, which seems reasonable given the inflated base price of the vessel over peacetime contracts. In E. David Cronon, ed., *The Cabinet Diaries of Josephus Daniels, 1913–1921* (Lincoln, Neb.: Univ. of Neb. Press, 1963), 576.
49. *Moody's Analyses of Investments, 1912,* 666–67 and 755–56; and Bethlehem Annual Reports, 1912, of two companies.
50. Maryland Steel Co., *Annual Stockholders' Report,* typescript, 1914; annual reports, 1914 and 1915; "Rail Sales," FWW Papers; *Iron Age,* June 25, 1914. Net loss on rail and marine operations for the year was $390,827.
51. Frederick to Caroline Wood, July 29, 1914, FWW Papers. Caroline was vacationing at the family farm in Townsend, Mass.
52. Letter, Mrs. Liebig to Wood, Nov. 27, 1914, FWW Papers. Information on Liebig family in 1910 Manuscript Census.
53. FWW Papers; *Iron Age,* Feb. 18 and 25, 1915.
54. Cotter, *Story of Bethlehem Steel,* 34–35; *New York Times,* Jan. 21, 1916.
55. *New York Times,* Oct. 6 and 16, 1915. Donner bio in *National Cyclopaedia of American Biography,* 44, 247; Stotesbury in Lucius Beebe, *The Big Spenders* (Garden City, N.Y.: Doubleday, 1966), 360, 379–80, and *National Cyclopaedia of American Biography,* vol. B, 104.
56. Pennsylvania Steel Co. (of New Jersey), "To the Holders of Preferred and Common Stock," April 3, 1916; Pennsylvania Steel Co., "Suggested Apportionment," typescript, March 6, 1916; *New York Times,* Feb. 20, 1916.
57. FWW Papers; "To the Holders"; and *Iron Age,* Feb. 24, 1916.
58. "Bethlehem Acquires Pennsylvania Steel Co.," *Iron Age,* Feb. 24, 1916; "To the Holders;" FWW Correspondence with Schwab, Feb. 28, 1916, and meetings with W. F. Roberts and H. S. Snyder, March 11 and 14, 1916, FWW Papers.
59. *Iron Age,* Feb. 24, 1916; *New York Times,* Oct. 6, 1915.
60. Cotter, *Story of Bethlehem Steel,* 46.
61. Cotter, 40–41.
62. *Baltimore American,* Nov. 23, 1916.

Chapter 6: The Munitions Machine

1. A transcript of the mayor's presentation was published in *Baltimore Municipal Journal,* Dec. 22, 1916. Also see J. H. Preston Papers in Baltimore Archives.
2. Schwab's subordinates had worked out the savings in a report: "Due to the less freight on ores and coal from the mines to the works, Sparrows Point should, with equal labor conditions and equivalent plants, produce pig iron for $1.50 per ton cheaper than Bethlehem [plant].

$.60 on ore or	$1.00 per ton of pig
.25 on coke or	.35 " " " "
.25 on coke oven operation	.35 " " " "
	1.70
Penalty on limestone	.20
Net	1.50

"$1.50 per ton of pig, would be the equivalent of roughly 2.00 per ton of plates." The report went on to indicate that with a $2-per-ton cost advantage, Sparrows Point would save the company $3 million a year in manufacturing commercial steel products (1.5 million tons × $2 per ton).
3. *Baltimore American,* Nov. 22, 1916.
4. *Baltimore Evening Sun,* Nov. 22, 1916.
5. Baltimore budget from *Baltimore Magazine,* Dec. 1916; *Baltimore Evening Sun,* Nov. 22, 1916.
6. Wilson speech of Aug. 18, 1914.
7. Bethlehem acquired the machine shops through the purchase of American Iron and Steel Manufacturing Company in 1916; the May's Landing Proving Ground was owned by a subsidiary, Bethlehem Loading Co.
8. "Bethlehem is the only ordnance works in America that makes projectiles complete, both plain steel and armor piercing, and ammunition in complete rounds. It is greater than Germany's Essen," said *American Magazine* in "Schwab's Own Story," Oct. 1916. Bethlehem's advertising catalog, confirmed its world-class status as a furnisher of "ordnance materials of all kinds."
9. *New York Herald,* Sept. 28, 1915. Also see, "Charles M. Schwab—The American Krupp," *Forum,* Aug. 1916.
10. From "Cost and Profit in Shells Approximated," *Iron Age,* April 27, 1916; Bethlehem Steel Co., *Ordnance Material* (South Bethlehem, Pa.: Bethlehem Steel Co., 1914), published in English, French, and Spanish.
11. Bethlehem Steel Co., *Ordnance Material,* "Machining High Explosive Shells," *Iron Age,* Oct. 7, 1915.
12. Bethlehem Steel Co., *Mobile Artillery Material,* 20–21.

13. Tuchman, *Guns of August,* 166–67.
14. Bethlehem Steel Co., *Mobile Artillery Material,* 48–50.
15. Records of J. P. Morgan & Co. published in *Munitions Industry: Hearings before the Special Committee Investigating the Munitions Industry,* Part 26, 74th Congress, 2d, 1936, Exhibits 2156–58.
16. *Munitions* hearings, testimony of Thomas W. Lamont, Jan. 7, 1936, 7507–08.
17. *New York Times,* Jan. 24, 1917 and Bethlehem Steel Corp. Annual Report, 1916. Prior to the stock distribution, Schwab holdings of 60,000 shares of Bethlehem common and 90,000 shares of preferred constituted 50.4 percent of the shares outstanding. The new class B shares were nonvoting stock since Schwab did not want to dilute his control of the company.
18. The inventor of the modern submarine, John Holland, made this observation. In Fyfe, *Submarine Warfare,* 122. To fight U-boats an array of new devices was developed, including depth-bombs, antisubmarine walls, and mines.
19. Gaddis Smith, *Clandestine Submarines,* 130. Two of the Mediterranean submarines managed to sink another submarine. Lamentably one of the victims belonged to Italy, an ally.
20. Smith, 132–35.
21. *Munitions* hearings, Exhibits 2156 and 2157. Six-inch, 90-pound shells were quoted at $19.85 apiece in *Iron Age* on April 27, 1916. Fuses went for under $1 each.
22. Allied and German preparations at the Somme were detailed by Sir Douglas Haig, commander in chief of the British armies in France. See J. H. Boraston, ed., *Sir Douglas Haig's Despatches* (London: J. M. Dent & Sons, 1920), 19–24.
23. Boraston, *Haig's Despatches;* Winston S. Churchill, *The World Crisis, 1916–1918* (New York: Charles Scribner's Sons, 1927), vol. 3, 177.
24. Boraston, *Haig's Despatches.* Haig and his minions claimed "success" on that first awful day.
25. Churchill, 182–83.
26. Churchill, 183.
27. Somme losses in *Encyclopedia Americana* (New York: Grolier, 1979), 29, 723, and *Munitions* hearings. There were three payments from Britain totaling $14,142,642.24 and ten payments from France for $9,124,492.94
28. *Munitions* hearings. Total payments to Bethlehem contracted through Morgan were $242,640,154. Not included was Schwab's $40-million munitions deal with Kitchener and $15 million for the subs, making the grand total in the text.
29. "Trade Supremacy Large Factor in the War—Address of President E. H. Gary," *Iron Age,* Nov. 5, 1914.
30. *Iron Age,* Oct. 28, 1915.
31. Quoted in Burton J. Hendrick, *The Life of Andrew Carnegie* (New York: Doubleday, 1932), vol. 2, 282.

32. *Munitions* hearings. U.S. Steel shipments were made through subsidiary U.S. Steel Products Company.
33. *Munitions* hearings, exhibits and testimony of George Whitney, partner, J. P. Morgan & Co., and others, Jan. 7, 1936, 7643–54.
34. *Iron Age,* Jan. 6 and Oct. 5, 1916.
35. See Arthur S. Link, *Wilson: Campaigns for Progressivism and Peace, 1916–1917* (Princeton, N.J.: Princeton Univ. Press, 1965), 160–64.
36. Olson, *Baltimore,* 292; *Baltimore Sun,* Jan. 24, 1917.
37. *New York Times,* April 3, 1917.
38. New Bulkhead Line, Sparrows Point Plant, March 10, 1917. Blueprint of Plant Extensions, 1917; *Baltimore Sun,* July 10, 1917.
39. Blueprint, 1917; J. H. K. Shannahan, "History of Sparrows Point," 1926, FWW Papers.
40. Shannahan, 20.
41. "Bethlehem's Tin-Plate Plant," *Iron Age,* April 5, 1917.
42. *Iron Age,* April 5, 1917; J. H. K. Shannahan, "The Romance of the Tin Can," typescript, c. 1927, Dundalk-Patapsco Neck Historical Society.
43. Shannahan, "History," 23–24; FWW Papers on plate mill construction.
44. Shannahan, "History"; Bethlehem Annual Report, 1917.
45. FWW Papers; Grosvenor B. Clarkson, *Industrial America in the World War: The Strategy Behind the Line, 1917–1918* (Boston: Houghton Mifflin, 1923), 330. Sparrows Point exported a total of 189,580 gross tons of rail to France between 1916 and 1918.
46. Arthur Strawn, "A Man of Heart," *American Mercury,* Oct. 1927.
47. FWW Papers; "Class Notes—Obituaries," *MIT Technology Review,* March 1944; *Baltimore Sun,* Dec. 24, 1943.
48. "New Commander at Sparrows Point," *Baltimore Sun,* April 27, 1918.
49. *Baltimore Sun,* April 27, 1918.
50. *Baltimore Sun,* April 27, 1918; *Iron Age,* Jan. 3, 1918; *New York Times,* May 22, 1918.
51. W. C. Mattox, *Building the Emergency Fleet* (Cleveland: Penton, 1920), 17–80.
52. Quoted in Hessen, *Titan,* 237–38.
53. Conditions spelled out in *U.S.* v. *Bethlehem Steel,* 1925. *(See* Chapter 7.)
54. *New York Times,* April 17, 1918.
55. *New York Times,* June 18, 1918; *Emergency Fleet News,* June 24, 1918.
56. Hessen, *Titan,* 242.
57. *New York Times,* July 5, 1918.
58. *New York Times,* July 24, Sept. 2, and Sept. 21, 1918.
59. From Bethlehem Annual Reports, 1914–18; U.S. Department of Commerce, Historical Statistics, "Consumer Price Index" (Washington, D.C.: U.S. Govt. Printing Office, 1975), 211.
60. *New York Times,* Dec. 8, 1918; *Iron Age,* Dec. 12, 1918.

Chapter 7: Chasing Dollars

1. *New York Times,* March 21, 1920.
2. Urofsky, *Big Steel,* 228–32; *AISI Annual Statistical Reports,* 1914–18.
3. Bethlehem Annual Reports, 1914–19; B. C. Forbes, *American Magazine,* July 1920.
4. Frederick Lewis Allen, *Only Yesterday* (New York: Harper & Brothers, 1931), 160.
5. *New York Times,* Jan. 4 and April 6, 1924.
6. Preston William Slosson, *The Great Crusade and After, 1914–28* (New York: Macmillan, 1930), 167–68.
7. Memo to Wood at Hog Island, 1918, FWW Papers.
8. Slosson, 168.
9. See James Warren Prothro, *The Dollar Decade: Business Ideas in the 1920s* (Baton Rouge: Louisiana State Univ. Press, 1954).
10. *New York Times,* March 21, 1920; also Ivy Lee Papers, Princeton Univ. Library, Princeton, N.J.
11. Speech at Princeton; also see Charles M. Schwab, *Succeeding with What You Have* (New York: The Century Co., 1917, 1919).
12. This example was given by Schwab in a conversation with *Wall Street Journal* publisher Clarence Barron. Quoted in Arthur Pound and Samuel Taylor Moore, eds., *They Told Barron: Conversations and Revelations of an American Pepys in Wall Street* (New York: Harper & Brothers, 1930), 81.
13. Schwab, *Succeeding,* 42–44; *Iron Age,* April 26, 1917.
14. FWW Papers.
15. Schwab, *Succeeding,* 47.
16. Copley, *Frederick W. Taylor,* vol. 11, 150–64. Copley quotes Taylor in a letter to General William Crozier, U.S. chief of ordnance: "I think I told you that the moment Schwab took charge of the Bethlehem works...he ordered our whole system thrown out. He saw no use whatever in paying premiums for fast work; much less in having time study and slide rule men, 'supernumeraries,' as he called them, in the works at all."
17. Schwab, *Succeeding,* 13–14.
18. Schwab, *Succeeding,* 12, 14–15.
19. U.S. Department of Commerce, *Historical Statistics,* "Consumer Price Index" (Washington, D.C.: U.S. Govt. Printing Office, 1975), 211. Yearly earnings from interviews with workers at Sparrows Point (for partial list *see* Chapter 9 notes). The steelmaker did not disclose employment figures except as a corporate-wide hourly average that *included* mill supervision. This average was 62.2 cents per hour in 1925.
20. Interviews *(see* Chapter 9 notes).
21. Interviews with Bill Forbes, John Duerbeck, and Anthony Maggitti, June 1981; advertisement in *Dundalk Community Press,* 1925.
22. Grace, *Charles H. Schwab,* 31.
23. Schwab, *Succeeding,* 34; Cotter, *Story of Bethlehem Steel,* 55–59.

24. From Bethlehem Annual Reports 1919–1925. The minority board members of the corporation were business friends of Schwab. They were Wall Street stockbroker Grayson M.-P. Murphy, stock speculator/banker Eugene V. P. Thayer, and Allan A. Ryan, son of Thomas Fortune Ryan and central figure in the infamous Stutz Corner of 1920.
25. Cotter, *Story of Bethlehem Steel,* 59.
26. Grace, *Schwab,* 23; "Bethlehem Steel," *Fortune,* April 1941; Pound and Moore, *They Told Barron,* 82.
27. *Bethlehem Bonus System* (New York: Bethlehem Steel Co., 1931), appendix.
28. *Bethlehem Bonus System.*
29. B. C. Forbes, "'Gene' Grace—Whose Story Reads Like a Fairy Tale," *American Magazine,* July 1920. Forbes was the founder of the business magazine that bears his name.
30. Forbes, "'Gene' Grace"; *Life,* Jan. 26, 1942.
31. Grace, *Schwab,* 30, 35.
32. Grace biography in *National Cyclopaedia of American Biography,* vol. 50, 1.
33. Forbes "'Gene' Grace"; and Cotter, *Story of Bethlehem Steel,* 59.
34. *New York Times,* April 18 and 19, 1925; "Report of Special Master and Referee, Feb. 7, 1936," *U.S.* v. *Bethlehem Steel Corporation, et. al.,* U.S. District Court of Eastern Pennsylvania.
35. "Report of Special Master" was published in the *Munitions* hearings, Part 36, Feb. 1936; the Bowles-Radford letter is on p. 12371.
36. *New York Times,* April 19, 1936.
37. *New York Times,* April 18, 1936.
38. "Report of Special Master," 12373.
39. "Report of Special Master," 12374–75.
40. *They Told Barron,* 80.
41. *They Told Barron,* 81–82.
42. *New York Times,* March 22, 1925.
43. *Report and Award of National War Labor Board, Bethlehem Steel Corporation,* Dept. of Labor Library, Washington, D.C., Aug. 4, 1918.
44. *Report and Award, NWLB,* Aug. 4, 1918.
45. *New York Times,* Sept. 15, 1918; Urofsky, *Big Steel,* 268. Grace paid time-and-a-half overtime and double time on Sunday until the armistice, then reverted to the bonus-without-overtime system.
46. Press release, Bethlehem Steel, Bethlehem, Pa., Oct. 1, 1918; *Iron Age,* Oct. 24, 1918; also, correspondence published in National Labor Relations Board, *In the Matter of Bethlehem Steel Corporation [et. al.] and Steel Workers Organizing Committee,* Cases C-170 and R-177, Aug. 14, 1939, in *Decisions and Orders,* vol. 14, Aug. 1–31, 1939 (Washington, D.C.: U.S. Govt. Printing Office, 1940).
47. "His Master's Voice," in Henry F. Pringle, *Big Frogs* (New York: Macy-Masius, 1928), 104–05.

48. NLRB, *In Matter of Bethlehem Steel,* 548–50; Stuart D. Brandes, *American Welfare Capitalism* (Chicago: Univ. of Chicago Press, 1976), 124–25; "Bethlehem Plan of Employee Representation," *Iron Age,* Oct. 24, 1918.
49. *Iron Age,* Oct. 3, 1918; *Report and Award, NWLB,* Aug. 4, 1918.
50. At first Taft and the Labor Board insisted that Bethlehem hold shop elections under NWLB supervision and follow other procedures to try to keep the shop committees free of company domination. By February 1919, though, they withdrew all inspectors from the South Bethlehem mill and let the company handle the ERP the way it wanted to. See *Iron Age,* Feb. 13, 1919, and *New York Times,* Feb. 5, 1919.
51. The bylaws and procedures were published in *Iron Age,* Oct. 24, 1918.
52. *Iron Age,* Oct. 24, 1918; and long feature of plan and participants by John Calder, "Five Years of Employee Representation Under 'The Bethlehem Plan,'" *Iron Age,* June 14, 1923.
53. Bylaws of ERP, *Iron Age,* Oct. 24, 1918.
54. Bylaws of ERP, *Iron Age,* Oct. 24, 1918.
55. Grace declared the strike over on Sept. 30, 1919. On Sept. 29, 1919, the *Times* reported that the priest of St. Lukes Church, the Rev. John Gaynor, had taken to the pulpit and vigorously implored the men to "remain steadfast" with their employer. His words were said to have had a big impact in convincing the men not to strike. Concerning the lack of union strength at the Point during the 1919 strike, see William Z. Foster, *The Great Steel Strike and Its Lessons* (New York: B. W. Huebsch, 1920), 181. Also see "Strikes," *Report of Maryland State Board of Labor and Statistics, 1919* (Annapolis, Md.), 177–78.
56. Interview with Frank Amann, May 1981.

Chapter 8: Tin Cans and Takeovers

1. Forbes, "'Gene' Grace," *American Magazine,* July 1920.
2. Henry A. White, *James D. Dole, Industrial Pioneer of the Pacific: Founder of Hawaii's Pineapple Industry* (New York: Newcomen Society, 1957), 10–22.
3. J. H. K. Shannahan, "Romance"; interview with Elizabeth Alexander, chief inspection forelady, Oct. 1981.
4. White, *James Dole,* 18; "Pineapples in Paradise," *Fortune,* Nov. 1930.
5. "Campbell's Soups," *Fortune,* Nov. 1935; Shannahan, "Romance"; interview with Elizabeth Alexander.
6. William E. Leuchtenburg, *The Perils of Prosperity, 1914–32* (Chicago: Univ. of Chicago Press, 1958), 180–81; White, *James Dole,* 23.
7. "Profits in Cans," *Fortune,* April 1934; Leuchtenburg, *Perils,* 192; Shannahan, "Romance."
8. William Hogan, *Economic History,* 756; "Profits in Cans," *Fortune.*
9. Shannahan, "Romance" and "History." Sparrows Point had the capacity

to manufacture 115,000 gallons of benzol-benzene per month. Aspirin, nylon stockings, industrial solvents, and early plastics also used this "miracle" by-product in their manufacture.

10. *Iron Age,* Feb. 10, March 2, and April 6, 1916.
11. Shannahan, "History of Sparrows Point," (tinplate department); *Fortune,* April 1934. *Moody's* "Price Range of Commodities" showed the price of a base box of tinplate ranged from $4.75–$5.50 in the 1921–23 period.
12. Fisher, *Epic of Steel,* 138–39.
13. Shannahan, "History"; "Report of the Baltimore Industrial Commission" (typescript, 1939); *New York Times,* March 14 and 30, 1922 and Dec. 27, 1924.
14. Bethlehem Annual Report, 1918, and *Recollections* (Public Affairs Dept. Bethlehem Steel Co., 1979).
15. The sale of steel rails to railroads stabilized at about 2.7 million tons a year in the 1920s, or about 500,000 tons below the pre-war boom years. *AISI Annual Statistical Reports* for 1920–1929.
16. Bethlehem Annual Reports, 1917–1922; Western Maryland Railway, Annual Reports, 1916–1920.
17. Baltimore Harbor Board, *Summary Report* (Baltimore, 1923).
18. "The Bethlehem Steel Company's Tofo Mines," *Iron Age,* June 11, 1914; "Chilean Iron Ore for Bethlehem," *Iron Age,* Jan. 23, 1913; "An 'Iron Age' Interview with Mr. Schwab," *Iron Age,* Jan. 1, 1926; Bethlehem Annual Reports, 1920–22.
19. *Iron Age* articles above.
20. "Production of Iron Ore by Leading Countries," *AISI Annual Statistical Report* for 1930, 114. For practical purposes all Chilean ore was exported to Sparrows Point. Bethlehem equipped several of its ore carriers with tanker vaults so bulk oil could be shipped to South America on return trips.
21. *United States* v. *United States Steel Corp.,* filed Oct. 26, 1911, in the U.S. District Court of New Jersey; U.S. Steel Corp. Annual Reports.
22. "Proposed Merger of Seven Steel Companies," *Iron Age,* Dec. 8, 1921; Annual Reports of respective companies, 1921; *Moody's Analysis of Investments,* 1922.
23. *New York Times,* Dec. 2, 1921; *Iron Age,* Dec. 8, 1921; Chadbourne biography in *National Cyclopaedia of American Biography,* vol. 30, 308.
24. *New York Times,* May 12 and 13, 1922; *Iron Age,* May 18, 1922.
25. *New York Times,* May 17, 20, 23, 1922.
26. *New York Times,* May 31, 1922.
27. *New York Times,* June 6, 1922.
28. *New York Times,* July 22, 1922.
29. *Iron Age,* Aug. 24, 1922; Hogan, *Economic History,* 906.
30. *New York Times,* Sept. 30, 1922; Bethlehem Annual Report, 1922; Hogan, 908–09; *Iron Age Review,* Jan. 4, 1923.
31. *New York Times,* Sept. 29, 1922

32. *Moody's Analysis of Investments,* 1923.
33. Grace, *Schwab,* 36.
34. Grace, 37.
35. *Iron Age,* Nov. 30, 1922; Bethlehem Annual Report, 1923.
36. *Moody's Analysis of Investments.*
37. "Keeping Business Out of Jail, an authorized interview with the Attorney General," *Nation's Business,* Nov. 1922.
38. *New York Times,* Jan. 28, 1922.
39. *New York Times,* June 1, 1927; Gertrude G. Schroeder, *The Growth of the Major Steel Companies, 1900–1950,* JHU Studies in Historical and Political Science (70th Series) (Baltimore: Johns Hopkins Univ. Press, 1952), 102

Chapter 9: Charlie and His Workmen

1. *They Told Barron,* 232.
2. "Residence for Charles M. Schwab," *Architectural Record,* Oct. 1902; "Charles M. Schwab's New Mansion in New York City," *Harper's Weekly,* Aug. 2, 1902; Casson, *The Romance of Steel,* 159; Hessen, *Titan,* 132–33; Andrew Tully, *Era of Elegance* (New York: Funk & Wagnalls, 1947), 147. When the mansion was completed in 1904 the U.S. consumer price index was 27. The CPI was 338.7 in May 1987.
3. Hessen, *Titan,* 132–33; Beebe, *Spenders,* 376–78; Casson, *Romance,* 159; *New York Times,* May 13, 1917.
4. Beebe, *Spenders,* 376, 283–87; *Architectural Record,* Oct. 1902.
5. *See* Chapter 6.
6. Hessen, xv.
7. *Bicentennial Commemorative History of Loretto, Pa.* (Ebensburg, Pa.: Damin Printing, 1976); author's visit to Loretto, Aug. 1985. My thanks to The Rev. Demetrius Schank for allowing me to tour the estate and to Joseph Bentivegna for sharing his research on the village and mansion. Schwab was born in nearby Williamsburg, Pa., on Feb. 18, 1862.
8. Frank Seymour, "Immergrun: The Schwab Estate," in *Bicentennial History of Loretto.*
9. Robert Imlay, "The Gardens of Charles M. Schwab, Esq., Loretto, Pa.," *Architectural Record,* May 1920; "The Gardens on the Estate of Mr. Charles M. Schwab at Loretto, Pa.," *Country Life,* June 1920; Seymour, 206–07; Hessen, 250; and author's visit.
10. Seymour, 209–11; author's visit.
11. Dedication ceremonies were held on August 20, 1919.
12. Imlay; Seymour, 208; author's visit.
13. Imlay; *Country Life,* June 1920.
14. Seymour, 209; interview with Father Demetrius Schank, Mt. Assisi Monastery, Loretto, Pa., Aug. 1985.

15. Father Demetrius.
16. "Charles M. Schwab: The World's Greatest Salesman and Business Personality," *National Magazine*, July 1924; *Colliers*, Sept. 3, 1927.
17. Ray Eldon Hiebert, *Courtier to the Crowd: The Story of Ivy Lee and the Development of Public Relations* (Ames, Iowa: Iowa State Univ. Press, 1966), 165. In fact, as noted in Chapter 5, Schwab had vowed to make Bethlehem the world's greatest gun factory.
18. Arthur Strawn, "A Man of Heart," *American Mercury*, Oct. 1927. The story was inspired by Schwab's 65th birthday,.
19. Bethlehem Annual Report, 1928; Shannahan, "History."
20. Annual Reports, 1925–28; Shannahan, "History."
21. Compiled from Shannahan, "History," and *AISI Directory of Iron and Steel Works*, 1928. In several departments where tonnage figures differed between the two reports, I have favored Shannahan.
22. "Schwab Dreams Again: That Sparrows Point Will Be Greatest Steel Works in the World," *Baltimore*, Dec. 1926.
23. Author's estimate from "Summary of Employment" figures issued by plant.
24. Interviews with James Allen, Louis Brandner, Nathaniel Parks, Joe and Catherine Hlatki, Senja Jarvis, Joe Lewis, Katherine Roberts, George White, Tass Stathem, Albert Wilhelm, Ben Womer, and Sirkka Lee Holm.
25. Interview with Senja Jarvis and Sirkka Lee Holm, March 1981.
26. Interview with Louis Brandner, June 1981.
27. U.S. Census Reports, "Total Negro Laborers in Blast Furnaces and Steel Rolling Mills," in Horace R. Cayton and George S. Mitchell, *Black Workers and the New Unions* (Chapel Hill, N.C.: Univ. of North Carolina Press, 1939), 14.; interview with Odell Payne, daughter of Edward Watkins, Aug. 1980.
28. Interviews with Tass Stathem, March 1981, and James Allen, Feb. 1981.
29. *New York Times*, Jan. 1, 1924. For background see Interchurch World Movement, Committee of Inquiry, *Report on the Steel Strike of 1919* (New York: Harcourt, Brace, 1920).
30. Interviews with workmen. Also see E. M. Hartl and E. G. Ernst, "The Steel Mills Today: The Twelve-Hour Day and the Seven-Day Week Go On," *New Republic*, Feb. 19, 1930.
31. "Report on Industrial Situation." This typescript report was found in the BRISC Files in Baltimore; Hartl and Ernst.
32. This sketch was derived from tape-recorded interviews with Ben R. Womer in 1982.
33. Interviews with Charlie Edward Parrish, Jan. and July 1983.
34. Accident reports, FWW Papers; Bethlehem Annual Report, 1956, on lost-time injury rate, 1916–56; *Baltimore Sun* vertical file, "Sparrows Point, Accidents," 1920–29.
35. Horace B. Davis, *The Condition of Labor in the American Iron and Steel Industry* (New York: International Publishers, 1933), 39–40.

36. Interview with Joe Lewis, June 1981.
37. *Baltimore Sun,* Dec. 1, 1927 and Jan. 1, 1928.
38. Dundalk-Patapsco Neck Historical Society, *Reflections* (pamphlet), 1976; Womer interview.

Chapter 10: Smash-up

1. "Clash of Steel," *Fortune,* June 1930; Bethlehem Annual Report, 1929.
2. *Chicago Tribune,* Dec. 11, 1929.
3. *Iron Age Review,* Jan. 4, 1923; "Youngstown Purchase of Steel & Tube Company," *Iron Age,* Jan. 11, 1923.
4. *See* "National Steel: A Phenomenon," *Fortune,* June 1932.
5. From *Moody's Manual of Investments, 1929.* Ranked immediately below Bethlehem were Anaconda Copper ($764 million in assets), Ford ($761 million) and Standard Oil of Indiana ($697 million). Translated into tonnage, a Bethlehem-Youngstown union would have owned 11 million ingot tons of capacity, or almost as much as Germany, the number-2 steelmaking nation in 1930.
6. *New York Times,* March 10, 1930; *Iron Age,* March 20, 1930; and testimony of acquisition, Court of Common Pleas, Youngstown, July–Sept. 1930.
7. "Eaton of the Eaton Group," *Business Week,* Dec. 18, 1929; *National Cyclopaedia of American Biography* vol. C, 88; "Irate Youngstown Tries to Stop a Wedding," *Business Week,* March 19, 1930.
8. *Fortune,* June 1930.
9. *Fortune,* June 1930.
10. "Clash of Steel"; *New York Times,* March 12, 1930.
11. *Youngstown Vindicator,* March 17, 1930; *New York Times,* March 20, 1930; *Iron Age,* March 20, 1930; "Ohio Versus Wall Street," *World's Work,* June 1930.
12. *Fortune,* June 1930.
13. *New York Times,* March 27, 1930; *Fortune,* June 1930.
14. *New York Times,* March 27, 1930.
15. *New York Times,* March 27, 1930.
16. *New York Times,* March 28, 1930.
17. *New York Times,* March 28, 1930.
18. *New York Times,* April 12, 1930; *Iron Age,* May 8, 1930.
19. *New York Times,* April 12, 1930.
20. *New York Times,* May 15, 1930.
21. *New York Times,* May 15, 1930.
22. *Iron Age,* May 15, 1930.
23. *Business Week,* July 9, 1930; *New York Times,* June 10, 1930.
24. *New York Times* and *Baltimore Sun,* July 18, 1930.

25. *New York Times* and *Baltimore Sun,* July 18, 1930; *Iron Age,* July 24, 1930.
26. *New York Times,* July 19, 1930.
27. *New York Times,* July 22, 1930; *Iron Age,* July 24, 1930; supplemental material in *Bethlehem Bonus System* (letter, Schwab to Bethlehem stockholders, March 2, 1931).
28. *New York Times,* July 27, 1930; "Beth's and Other Salaries Probed," *Fortune,* Sept. 1930; "Stars of Bethlehem," *Nation,* Aug. 6, 1930. The only major media figure who defended the bonus system publicly was B. C. Forbes.
29. *New York Times,* Aug. 1, 1930; Hessen, *Titan,* 276.
30. *New York Times,* July 11 and 27, Aug. 25 and 26, 1930.
31. *New York Times,* Aug. 26 and 27, 1930.
32. *New York Times,* June 28 and Sept. 21, 1930. The legal fees were $1.25 million for the Eaton forces and $1 million for the Schwab forces, according to a 1933 suit.
33. *Iron Age,* Jan. 1, 1931; *New York Times,* Dec. 30, 1930. "The prestige of Cyrus S. Eaton of Cleveland is greatly enhanced by victory in the suit to stop the merger of Youngstown with Bethlehem," opined *Business Week.*
34. *Iron Age,* Jan. 8, 1931.
35. *New York Times,* Jan. 14 and 15, 1931.
36. *New York Times,* Jan. 15, 1931.
37. Hessen, 273.
38. Hessen, 273.
39. *Bethlehem Bonus System;* "Mr. Schwab Forearms Against a Bonus Attack," *Business Week,* March 11, 1931.
40. *New York Times* and *Baltimore Sun,* April 15, 1931.
41. Hessen, 277.
42. *New York Times,* July 3, 1931; "Bethlehem Stockholders Make Bonus Peace," *Business Week,* July 15, 1931.
43. "Pay or Plunder?" *Nation,* June 24, 1931.
44. "Bethlehem Bonuses," *New Republic,* July 15, 1931.
45. *Iron Age,* Jan. 1, 1931.
46. "Annual Financial Report, 1931," *New York Times,* Dec. 31, 1931.
47. *New York Times,* Dec. 31, 1931; Bethlehem Annual Report, 1931; *Iron Age,* Jan 7, 1932.
48. *Iron Age,* Jan. 7, 1932; *Baltimore Sun,* June 8, 1931.
49. *Iron Age,* Jan. 7, 1932; White, *James Dole,* 25.
50. *Business Week,* Aug. 12, 1931; *Iron Age,* Oct. 23, 1931; *New York Times,* Dec. 31, 1931.
51. Bethlehem Annual Report, 1931; *New York Times,* Oct. 2, 1931.
52. Quoted in Richard A. Lauderbaugh, "Business, Labor, and Foreign Policy: U.S. Steel, the International Steel Cartel, and Recognition of the Steel Workers Organizing Committee," *Politics and Society,* VI: 4 (1976), 438.
53. *New York Times,* October 16, 1932.

Chapter 11: Depression Snapshots

1. "Industry Experiences a Calamitous Year," *Iron Age*, Jan. 7, 1932; *AISI Annual Statistical Report* for 1931.
2. Even *Iron Age* conceded that the excesses of the '20s had contributed to steel's collapse in 1931. See Lewis Haney, "Confidence in Stability of Money Is One Great Need," *Iron Age*, Jan. 7, 1932.
3. Interviews with workers, *see* below.
4. *Baltimore Sun*, Dec. 19 and 20, 1931; *Dundalk Community Press*, Jan. 15, 1932.
5. "Production of Iron Ore by Countries," *AISI Annual Statistical Reports*.
6. Guggenheim, *U.S. and Cuba*, 131–32.
7. "How Bethlehem Steel Has Effected Employment Stabilization," *Iron Age*, Dec. 22, 1932; Frances Perkins, *People at Work* (New York: John Day, 1934), 120; Bethlehem Annual Reports, 1930 and 1931. The daily workforce at Bethlehem declined from 64,316 in 1929 to 45,258 in 1931. At Sparrows Point about 6,000 men were laid off.
8. Interviews with Charlie Parrish and wife, Feb. 1983.
9. Interview with Ben Womer, Sept. 1982.
10. Interview with John Duerbeck, June 1982.
11. Interview with Nathaniel Parks, Sept. 1981.
12. Interview with Joe Lewis, June 1981.
13. From articles and advertisements in *Dundalk Community Press*, Jan. 15 and 22, 1932.
14. Al Richmond, *A Long View from the Left: Memoirs of an American Revolutionary* (Boston: Houghton Mifflin, 1973), 104–05.
15. Richmond, 101; Davis, *Condition of Labor*, 223.
16. Foster, *Great Steel Strike*.
17. From Richmond and Cayton and Mitchell's excellent *Black Workers*, 82–83.
18. Louis Adamic, "The Steel Strike Collapses," *Nation*, July 4, 1934.
19. Cayton and Mitchell, 84–85; *Baltimore Sun*, March 2, 1930.
20. Reprinted in Davis, 26, 34, 87.
21. Richmond, 101, 104.
22. *Iron Age*, Jan. 5, 1933; "Production of Steel by Countries," *AISI Annual Statistical Reports*; Richard A. Lauderbaugh, *American Steel Makers and the Coming of the Second World War* (Ann Arbor, Mich.: Univ. of Michigan Research, 1980), 145.
23. *AISI Annual Statistical Reports*.
24. *AISI Annual Statistical Reports*.
25. "Arctic Weather Fails to Retard Siberian Steel Program," *Iron Age*, June 25, 1931. The Kuznetsk site was one of the few places in the world where workable deposits of ore, coal, and limestone were in close proximity. Its drawback was severe weather. One of the technical achievements of the Russians was the pouring of many thousands of yards of concrete at temperatures as low as 40° F below zero.

26. *New York Times,* Feb. 19, 1932.
27. *New York Times,* Feb. 19, 1932.
28. Wages were reduced effective May 16, 1932.
29. Irving Bernstein, *The Lean Years* (Boston: Houghton Mifflin, 1960), 502–03.
30. Reconstruction Finance Corporation, H.R. 7360, 72d Congress, 1st Session.
31. From *Iron Age,* Jan. 5, 1933, summary of 1932.
32. *Iron Age* summary of 1932.
33. Schwab speech published in "Steel Institute to Broaden Scope as Leaders Assert Faith in Future," *Iron Age,* May 26, 1932; and *Iron Age,* Jan. 5, 1933.
34. From Schwab speech.
35. Bethlehem Annual Report, 1931, published in April 1932.
36. Bernstein, *Lean,* 508; *Baltimore Sun,* Nov. 8–9, 1932.
37. *New York Times,* Oct. 19, 1932 and March 1, 1935. Schwab reportedly had made $100,000 in campaign contributions to Hughes in 1916.
38. *Iron Age,* Jan. 5, 1933.
39. *Baltimore Sun,* Dec. 12, 1932; interviews with workmen.
40. *Dundalk Community Press,* Nov. 24 and Dec. 3, 1932.

Chapter 12: The New Deal

1. *New York Times,* March 5, 1933.
2. *New York Times,* March 3 and 4, 1933.
3. "Text of the Inaugural Address," *New York Times,* March 5, 1933.
4. *New York Times,* March 5, 1933.
5. *New York Times,* Feb. 23 and March 3, 1933.
6. George Martin, *Madam Secretary: Frances Perkins* (Boston: Houghton Mifflin, 1976), 238–41; Frances Perkins, *People,* 128–37.
7. Perkins, *People,* 163.
8. *National Cyclopaedia of American Biography,* vol. F, 28–29.
9. Russell Lord, "Madame Secretary," *New Yorker,* Sept. 2–9, 1933.
10. Martin, 50; Lord.
11. Martin, 56–64.
12. Paul U. Kellogg, ed., *The Pittsburgh Survey* (6 vols.) (New York: Charities Publication, 1910).
13. Wrote Perkins: "I had already had a conviction, a 'concern' as the Quakers say, about social injustice; it was clear in my own mind that the promotion of social justice could be made to work practically" *(The Roosevelt I Knew* [New York: Viking, 1946], 10.
14. Perkins, *Roosevelt,* 13.
15. Perkins, *Roosevelt,* 5.
16. Martin, 103.
17. Perkins, *People,* 125–31. Also see Rexford G. Tugwell, *The Battle for Democracy* (New York: Columbia Univ. Press, 1935), 44–46.

18. Perkins, *Roosevelt,* 214; *People,* 135.
19. Tugwell, *Battle,* 14; Perkins, *People,* 169; Irving Bernstein, *The Turbulent Years* (Boston: Houghton Mifflin, 1970) 28–32.
20. William E. Leuchtenberg, *Franklin D. Roosevelt and the New Deal* (New York: Harper & Row, 1963), 41–62.
21. National Industrial Recovery Act, H.R. 5755, 73d Congress, 1st Session; *New York Times,* June 17, 1933.
22. Recovery Act; *New York Times,* June 17, 1933.
23. *New York Times,* July 18, 1933. Also see Hugh Johnson, *The Blue Eagle from Egg to Earth* (Garden City, N.Y: Doubleday, 1935).
24. Perkins, *Roosevelt,* 215.
25. *Proposed Code of Fair Competition for the Iron and Steel Industry,* AISI, New York, July 15, 1933.
26. Perkins, *Roosevelt,* 216, and "The Reminiscences of Frances Perkins," Oral History Research Office, Columbia University, vol. 5, part 5, 321–323. Roosevelt agreed she should visit Sparrows Point and Homestead. "But don't get yourself arrested."
27. *New York Times,* July 29, 1933, and Perkins, Oral History, 338–41.
28. *Evening Sun,* July 29, 1933.
29. Perkins, Oral History, 334–45; *Evening Sun,* July 29, 1933.
30. Perkins, Oral History, 344–45; *Evening Sun,* July 29, 1933; interview with John Duerbeck, June 1982.
31. "Miss Perkins's Criticism of the Steel Code," *New York Times,* Aug. 1, 1933.
32. *New York Times,* Aug. 1, 1933.
33. *New York Times,* Aug. 1, 1933.
34. Perkins, *Roosevelt,* 223; and Oral History, vol. 5, 315–16; *New York Times,* Aug. 16, 1933.
35. *New York Times,* Aug. 16–17, 1933; Arthur M. Schlesinger, *The Coming of the New Deal* (Boston: Houghton Mifflin, 1959), 117.
36. *New York Times,* Oct. 20, 1926, and Aug. 17, 1933; Schlesinger, 117. A slightly different version of the exchange was recorded by journalist Ernest K. Lindley in *The Roosevelt Revolution: First Phase* (London: V. Gollancz, 1934), 212.
37. *Code of Fair Competition* (Schedule D); "Steel Code as Accepted," *Iron Age,* Aug. 24, 1933; Cayton and Mitchell, *Black Workers,* 97–99; Perkins, *People,* 163–64.
38. "Advances in Pig Iron," *Iron Age,* Aug. 31, 1933; "New Basing Points and Delivered Prices of Pig Iron," *Steel* Sept. 4, 1933; *Code of Fair Competition.*
39. *Steel,* Sept. 4, 1933; *New York Times,* Nov. 26, 1933.
40. "Iron and Steel Industry Undertakes Self-Government," *Iron Age,* Jan. 4, 1934; *New York Times,* Nov. 26, 1933.
41. "A Chronological Record of Recovery Legislation and Administration," *Iron Age,* Jan. 4, 1934; *New York Times,* Sept. 26 and Oct. 7, 1933.

42. "The Steel Rail," *Fortune,* Dec. 1933.
43. *Fortune,* Dec. 1933.
44. *Steel,* Sept. 4, 1933; *Baltimore Sun,* Feb. 20, 1934.
45. Bethlehem Annual Report, 1933.
46. Interviews with workmen.
47. *Baltimore Sun,* Oct. 16, 1933.
48. *Steel,* Oct. 2, 1933; *Iron Age,* May 31, 1934.
49. *New York Times,* Aug. 19 and Sept. 30, 1934; "The Steel Strike Collapses," *Nation,* July 4, 1934.
50. *New York Times,* Oct. 6, 1934; Perkins, Oral History, vol. 7, 503–04.
51. *New York Times,* Dec. 19, 1934; Cayton and Mitchell, 151–52.

Chapter 13: The Coming of the CIO

1. *A.L.A. Schechter Poultry Corp. et al* v. *United States,* 295 U.S. 495–555. The ruling did not affect the public works program.
2. *New York Times,* June 1, 1935; Leuchtenberg, *Roosevelt,* 144–47.
3. National Labor Relations Act, S. 1958, 74th Congress.
4. Perkins, Oral History, vol. 7, 135–47. Perkins initially expressed reservations about the broad scope of the Wagner Act.
5. "The American Federation of Labor," *Fortune,* Dec. 1933.
6. Cayton and Mitchell, 154–57; interview with Charlie Barranco, April 1981.
7. *New York Times,* Oct. 18, 1935. For a full account, see Bernstein, *Turbulent,* 352–96.
8. Bernstein, *Turbulent,* 400.
9. The captive mine dispute flared during the code period when the Frick Coke Company, a U.S. Steel subsidiary, refused to recognize the UMW.
10. Vincent D. Sweeney, *The United Steelworkers of America* (Pittsburgh: USWA, 1956), 11–14; "The Great Labor Upheaval," *Fortune,* Oct. 1936; Robert R. R. Brooks, *As Steel Goes...* (New Haven: Yale Univ. Press, 1940), 153–55.
11. Bernstein, *Turbulent,* 452; *Fortune,* Oct. 1936.
12. "John Llewellyn Lewis," *Fortune,* Oct. 1936; Bernstein, *Lean,* 117–20. Recent scholarship has unmasked a man whose education was quite at odds with his image. See Melvyn Dubofsky and Warren Van Tine, *John L. Lewis* (New York: Quadrangle, 1977).
13. Perkins, Oral History, vol. 8, 389; *Fortune,* Oct. 1936.
14. *New York Times,* July 3 and 6, 1936.
15. Reprinted in *Vital Speeches of the Day,* Aug. 1, 1936.
16. *Vital Speeches.*
17. Walter Galenson, *The CIO Challenge to the AFL* (Cambridge, Mass.: Harvard Univ. Press, 1960), 86.
18. Review of issues of *Steel Labor,* courtesy of U.S. Department of Labor Library, Washington, D.C.
19. *New York Times,* July 24, 1936; *Steel Labor.*

20. Review of issues of *Labor Herald*, Pikesville, Md.
21. *Labor Herald*, 1936; interview with Mike Howard, April 1982.
22. *Labor Herald*, Aug. 16, 1936; also *Steel Labor*, Aug. 28, 1936.
23. *Labor Herald*, Jan. 8 and 22, 1937; interviews with Mike Howard, Charles Barranco, and Ellen Pinter, secretary for SWOC district office.
24. Mike Howard interview.
25. Bernstein, *Turbulent*, 455–57; Brooks, *Steel*, 90, 93–97.
26. *New York Times*, March 3, 1937. The Steel settlement capped a giant rise in membership in the CIO. *Fortune* conceded the point in a special story, "The Great Labor Upheaval" (Oct. 1936).
27. Both Taylor and Lewis were mum on the details of their talks, causing a great many unfounded rumors regarding the "real" reasons for their pact. See, for example, "It Happened in Steel," *Fortune*, May 1937.
28. *Fortune*, May 1937.
29. *New York Times*, Nov. 14, 1936, and March 2, 1937; Galenson, *CIO*, 95.
30. *National Cyclopaedia of American Biography*, vol. D, 68–69, and vol. G, 75–76; "The Corporation," *Fortune*, March 1936; "It Happened in Steel," *Fortune*, May 1937.
31. Perkins, Oral History, vol. 6, 153–56; also see vol. 7, 503–05.
32. Brooks, *Steel*, 134–35; Galenson, *CIO*, 96–97; Bethlehem Annual Report, 1937; *Moody's Industrials, 1937*.
33. Brooks, 134; Schwab, *Succeeding*, 25.
34. "Restlessness in Steel," *Fortune*, Sept. 1933; Tom M. Girdler, in collaboration with Boyden Sparkes, *Boot Straps* (New York: Charles Scribner's Sons, 1943), 14, 72, 189–97.
35. *Dundalk Community Press*, March 5, 1937; *Baltimore Sun*, March 19, 1937; Bernstein, *Turbulent*, 480.
36. NLRB, *In the Matter of Bethlehem Steel Corporation and Steel Workers Organizing Committee*, Cases C-170 and R-177, Aug. 14, 1939, 574–75.
37. Jerold S. Auerbach, *Labor and Liberty: The La Follette Committee and the New Deal* (Indianapolis: Bobbs-Merrill, 1966), 101.
38. NLRB, 626.
39. Auerbach, 101; Bernstein, *Turbulent*, 482–83; Brooks, 135–37.
40. Auerbach, 124–28.
41. Auerbach, 127.
42. Sweeney, *United Steelworkers*, 38; "The Industrial War," *Fortune*, Nov. 1937; Auerbach, 124; Brooks, 137–44.
43. *Baltimore Sun*, June 14, 1937.
44. *New York Times*, June 18, 1937; Bernstein, 491. Also see Keith Sward, "The Johnstown Strike of 1937: A Case Study of Large-Scale Conflict," in *Industrial Conflict: A Psychological Interpretation* (New York: Cordon, 1939), 74–102.
45. Speech published in NLRB case, 617–19.
46. NLRB, 619–22.
47. NLRB, 622; Bernstein, 493; *New York Times*, June 20 and 21, 1937.

48. *New York Times,* June 24, 25, and 26, 1937. "We cannot but believe that the bitterness and suspicion which separate the two sides would be allayed by a man-to-man discussion around the conference table," remarked the Taft Board. Inland Steel struck a compromise with SWOC. It recognized the union as a bargaining agent for its members and the SWOC strike at its Indiana Harbor plant ended.
49. Perkins, Oral History, vol. 7, 503–06.
50. *New York Times,* June 25, 26, 27, and 28, 1937; Bernstein, 494.
51. *New York Times,* June 30, 1937.
52. *Baltimore Sun,* June 21, 1937; NLRB case.
53. *Baltimore Sun,* June 21, 1937; interview with Howard.

Chapter 14: Technology's Wrench

1. "The Steel Industry: What of its Future?" *Steel,* Jan. 2, 1933; Langdon White and Edwin J. Foscue, "The Iron and Steel Industry of Sparrows Point," *Geographical Review,* April 1931; "Restlessness in Steel," *Fortune,* Sept. 1933.
2. Bethlehem won the contract to supply all steel for the towers and superstructure of the Golden Gate—100,000 tons in all. Most of the steel plate and angles were manufactured at Sparrows Point, shipped by boat to San Francisco, and erected by subsidiary McClintic-Marshall. John A. Roebling's Sons did the cable work. When opened in 1937, the Golden Gate was the longest suspension bridge in the world—its main span 4,200 feet long and each tower 746 feet above mean low water level, or nearly 200 feet taller than the Washington Monument. From *The Golden Gate Bridge* (Bethlehem Steel, 1937) and other sources.
3. "Bethlehem Steel," *Fortune,* April 1941.
4. Bridges, *Inside Story,* 268; interviews with John Duerbeck, Anthony Maggitti, Charles Ripkin, and other hot-mill workers.
5. Shannahan, "Romance"; D. E. Dunbar, *The Tin-Plate Industry* (New York: Houghton Mifflin, 1915), 5–7.
6. Dunbar, 7; interviews.
7. Interviews.
8. Fisher, *Epic of Steel,* 143; Hogan, *Economic History,* 845–47.
9. "Armco's Great Contribution," *Fortune,* Sept. 1931; Fisher, 144.
10. Hogan, 847–851.
11. For example, *Iron Age* wrote, "This is regarded as the first radical development that has been successful in an industry that is over 200 years old, and is the result of long experience of the inventor, J. B. Tytus, and years of experiments of the staff of the American Rolling Mill Co."
12. In "Schwab Dreams Again," *Baltimore,* Dec. 1926; also see Shannahan, "History," and Hogan, 852–54.
13. "National Steel: A Phenomenon," *Fortune,* June 1932; Hogan, 855.
14. Hogan, 852–57; Fisher, 145–47.
15. Prices from *Iron Age.*

16. George W. Stocking, *Basing Point Pricing and Regional Development* (Chapel Hill, N.C.: Univ. of N.C. Press, 1954), 95; *AISI Steel Works Directory* (New York: AISI, 1935); also see *Economic Concentration and World War II,* Report of the Smaller War Plants Corp. (Washington, D.C.: U.S. Govt. Printing Office, 1943), 90–95.
17. *Moody's Industrials* profile of American and Continental Can, 1935; "Profits in Cans," *Fortune,* April 1934.
18. *Fortune,* April 1934.
19. *Fortune,* Nov. 1930 and April 1934.
20. William C. Stolk. *American Can Company: Revolution in Containers* (New York: Newcomen, 1960), 18–19; Hogan, 1154.
21. Hogan, 1385; "Beer Into Cans," *Fortune,* Jan. 1936.
22. *Fortune,* Jan. 1936.
23. Stolk, 19; *Fortune,* Jan. 1936.
24. *Fortune,* Jan. 1936.
25. Bethlehem Annual Reports, 1936–38; *Iron Age,* July 15, 1937; *New York Times,* April 30, 1938.
26. "The Production of BethColite at Sparrows Point," (South Bethlehem, Pa.: Bethlehem Steel, 1938); *Iron Age,* Oct. 14, 1937.
27. Interview with engineer, March 1980.
28. Interview with Duerbeck, June 1982.
29. Interviews, 1982–84.
30. Harold J. Ruttenberg, research director of SWOC, publicized the displacement issue in "The 85,000 Victims of Progress," *New Republic,* Feb. 16, 1938, and "The Big Morgue," *Survey Graphic,* April 1939. Reports of displacement were first made to the Wage and Policy Convention of SWOC, Dec. 13–14, 1937.
31. *Steel Labor,* Oct. 28, 1938; interviews with Charles Barranco and Brendan Sexton; SWOC leaflets courtesy of Local 2609.

Chapter 15: End of an Era

1. NLRB, *In the Matter of Bethlehem Steel Corporation and Steel Workers Organizing Committee,* Cases C-170 and R-177.
2. From record of NLRB case; *Steel Labor,* Sept. 3, 1937, and Jan. 21, 1938.
3. NLRB case.
4. NLRB case.
5. NLRB case; *Baltimore Sun,* Nov. 8, 1938.
6. NLRB decision, 595.
7. NLRB decision, 629.
8. *New York Times,* Aug. 16, 1939; *Baltimore Sun,* Aug. 15, 16, 17, 1939.
9. The most important court test of the Wagner Act was *NLRB* v. *Jones & Laughlin Steel Corp.,* decided in 1937.
10. The company lobbied for the "Big Navy" and for higher import quotas on finished iron and steel, but not on imported iron ore, through the '30s.
11. Stettinius was elevated to the top slot in April 1938.
12. *Hearings before the Special Committee Investigating the Munitions In-*

dustry, Part 26, 74th Congress, 2d, 1936; *New York Times*, Feb. 19 and 23, 1936; B. C. Forbes quoted in Hessen, 299.

13. *New York Times*, Feb. 19, 1936. The Shearer incident first emerged in 1929 in hearings before the Senate Subcommittee on Naval Affairs.
14. Hiebert, *Courtier to the Crowd*, 309–11.
15. *New York Times*, May 20, 1938.
16. Bethlehem Annual Reports, 1937–38; *New York Times*, May 20, 1938.
17. *New York Times*, April 10, 1935.
18. *New York Times*, April 10, 1935, and April 14, 1937.
19. *New York Times*, April 13, 1938.
20. Hessen, 290; *Baltimore Sun*, Sept. 21, 1929.
21. Hessen, 142, 251, 285; Pound and Moore, eds., *They Told Barron*, 37; also see Beebe, *The Big Spenders*, 283–87.
22. Pieced together from court records filed in New York and Ebensburg, Pa., after Schwab's death.
23. *New York Times*, Dec. 19, 1936; Hessen, 300.
24. Hessen, 291–92.
25. *New York Times*, Jan. 13 and 20, March 10, and Sept. 1 and 19, 1939.
26. *New York Times*, Sept. 20–22, 1939; March 7 and May 14, 1941; "Bankrupt Millionaire," *Time*, May 26, 1941.
27. "Bankrupt Millionaire"; Lewis D. Gilbert, *Dividends and Democracy* (New Rochelle, N.Y.: Amer. Research Council, 1954), 38–39.
28. His will was filed in New York on Sept. 30, 1939.
29. Interview with Father Demetrius, T.O.R., at Mt. Assisi Monastery; "Castle and College," *Time*, Nov. 2, 1942; *New York Times*, April 17 and Oct. 4, 1942.
30. H. J. Silliman, "Schwab to Cramer to Dempsey," *Saturday Evening Post*, Nov. 8, 1941. After suffering considerable neglect over the years, the Schwab car has been stored by rail enthusiasts at Altoona, Pa.
31. Sweeney, *Steelworkers*, 47–51; Galenson, *CIO*, 116.
32. Perkins, Oral History, vol. 6, 252.
33. Bernstein, *Turbulent*, 721–26.
34. Sweeney, *Steelworkers*, 48–50; interview with Oscar Durandetto, April 1981; also with SWOC organizers Brendan Sexton and Mike Harris.
35. From *Steel Labor* and *Baltimore Sun*, Nov. 8, 1940. Also see "Bethlehem Background," *New Republic*, April 7, 1941.
36. Interviews with Durandetto and with Prof. Roderick N. Ryon at Towson State University.
37. Pamphlets courtesy of Locals 2609 and 2610.
38. Interview with the Parrishes, July 1983; *Baltimore Afro-American*, Feb. 22, 1941.
39. Interview with Ellen Pinter.
40. Interviews with Howard, Durandetto, Lewis, and Barranco.
41. Interviews with McShane and Mary Barnhart.
42. *New York Times*, April 8, 1941.
43. *New York Times*, May 13, 1941.

44. Sweeney, *Steelworkers,* 50.
45. *Baltimore Evening Sun,* Aug. 19, 1941; Durandetto interview. There was a second walkout by crane operators four days later that halted production for two hours at the plate mills.
46. *Evening Sun,* Sept. 25, 1941; *Sun,* Sept. 26, 1941. Proclaimed the union's leaflets prepared for vote day: UNION MEN ARE PATRIOTIC MEN. 100% UNION MEANS 100% CONTRACT. WE WILL NEVER GO BACK TO INDUSTRIAL SLAVERY. They were in red, white, and blue.
47. *Evening Sun,* Sept. 25, 1941.
48. Interviews with workmen quoted.
49. *Baltimore Sun,* Sept. 26, 1941; *Maryland Labor Herald,* Sept. 26, 1941.
50. *Baltimore Afro-American,* Sept. 27, 1941.
51. *Maryland Labor Herald,* Oct. 2, 1941; Sweeney, *Steelworkers,* 50.

Chapter 16: In the Combat Zone

1. *Baltimore Sun,* Feb. 22 and April 18, 1937; "Production of Steel by Countries," *AISI Annual Statistical Reports.*
2. Bethlehem Annual Reports, 1939–40; *Baltimore Afro-American,* Sept. 30, 1939.
3. *New York Times,* June 5, 1941.
4. *New York Times,* Aug. 1, 1945.
5. *New York Times,* Nov. 17, 1941, and *Iron Age,* Sept. 27, 1945. The original contract called for the allotment of $55.77 million to Bethlehem, the added amount going for a wide-plate mill at Sparrows Point. The mill was canceled when it was determined that the tonnage could be rolled on the hot-strip.
6. "The War Goes to Mr. Jesse Jones," *Fortune,* Dec. 1941; *Iron Age,* Sept. 27, 1945.
7. "Bethlehem Ship," *Fortune,* Aug. 1945.
8. *Baltimore Evening Sun,* Feb. 4, March 18, and Sept. 27, 1941; "The No. 1 Bottleneck Now Is Lack of Ships," *Fortune,* May 1942. Lionel Corp., the toy train maker, produced the binnacles and compasses for the Liberty Ships, Congoleum Rug made the universal gear joints, and Hershey Chocolate Company supplied the 10-inch mooring bits on the upper deck.
9. "Eugene Grace: Bethlehem Steel Corporation's Seasoned President Runs a Large Section of the U.S. War Effort," *Life,* Jan. 26, 1942; Bethlehem Annual Report, 1941; "West Coast Yards, Navy Repairmen," *Fortune,* Aug. 1945.
10. Senator Harry S. Truman gained fame for exposing lapses of efficiency among defense contractors as chairman of the Senate Special Committee to Investigate the National Defense Program.
11. "What's Itching Labor?" *Fortune,* Nov. 1942, and "No. 1 Bottleneck," *Fortune,* May 1942.
12. *New York Times,* Feb. 7, Feb. 11, and July 17, 1942; Galenson, *CIO,* 117–18.

13. *New York Times,* May 5, 1942.
14. *Life,* Jan. 26, 1942.
15. Bethlehem Annual Reports, 1930–45; also "Historical Operating Data" in Annual Report, 1952.
16. A survey of the war's changing technology was in *New York Times,* Jan. 3, 1943, section 9. In 1939 the dollar value of U.S. airplane output was $225 million; by 1942 it was $4.5 billion as 49,000 aircraft were produced.
17. "RADAR: The Technique" and "RADAR: The Industry," *Fortune,* Oct. 1945.
18. *New York Times,* Jan. 3, 1943; "Radar," *Fortune,* Oct. 1945; The Aluminum Association, *Aluminum Statistical Review, 1940–45* (New York).
19. *AISI Annual Statistical Reports,* 1914–18 and 1941–45.
20. *AISI Statistical Reports.*
21. "How War Affected Steel Producing Areas," *Business Week,* Oct. 20, 1945; *Evening Sun,* May 26, 1945.
22. Data from Bethlehem Steel and *AISI Works Directory.*
23. From workmen interviews and *First Helper's Manual,* Sparrows Point Open Hearth Department, 1959. A survey by the Maryland Health Department found that 30 percent of the mill steel force was exposed to carbon dioxide, 32.8 percent to metallic dusts, 19 percent to oils, 15.4 percent to nonmetallic dusts, 10.2 percent to lead, 11.3 percent to sulfur dioxide, and 7 percent to silica dust.
24. Interviews with Howard, May–June 1981 and April 1982.
25. Interviews with Womer, June and Sept. 1982.
26. *Baltimore Evening Sun,* March 9, 1942; "Bethlehem Ship," *Fortune,* Aug. 1945.
27. "Steel," *Fortune,* May 1945.
28. "Glenn L. Martin Co.," *Fortune,* Dec. 1939; *Iron Age,* Feb. 8, 1945; "The Battle of Baltimore," *Life,* Feb. 14, 1944; *Baltimore Sun,* March 4, 1943; "Shot, Shell, and Bombs," *Fortune,* Sept. 1945.
29. *New York Times,* Jan. 3, 1943; pamphlet from Bethlehem Steel, c. 1943.
30. *Baltimore Sun,* Nov. 7, 1942; "The Changing Scene at Sparrows Point," *Management Conference Agenda,* July 1976.
31. *Management Agenda,* July 1976; Ben Womer interviews.
32. Interviews with Joe Hlatki, Sept. 1982, and Elizabeth McShane.
33. Lt. Col. Whitman S. Bartley, *Iwo Jima: Amphibious Epic* (Washington, D.C.: U.S. Marine Corps, 1954).
34. *Evening Sun,* May 26, 1945.
35. Bethlehem Annual Reports for 1935 and 1945; "Bethlehem Ship," *Fortune,* Aug. 1945; *Iron and Steel Works of the World* (London: Quin Press, 1958), 779–82.

Chapter 17: Brave New World

1. "Coal Smoke: Vital in War, Magic in Peace," *Steel for Victory,* (New York: AISI, August 1944); "Steel: Report of the War Years," *Fortune,* May 1945.

2. Average weekly wages were computed by the Bureau of Labor Statistics.
3. *New York Times,* Sept. 12, 1945, and *Historical Statistics,* CPI (Washington, D.C.: U.S. Dept. of Commerce), 209.
4. "This Is Why They Strike," *Nation,* Feb. 2, 1946. Also radio address of Phil Murray, Jan. 20, 1946.
5. *New York Times,* Oct. 30, 1945; Bethlehem Annual Reports, 1941–44.
6. *Nation,* Feb. 2, 1946; "Ring Out the Old, Ring in the New," *Sparrows Point Steelworker,* Local 2609, Dec. 1945; Securities and Exchange Commission reports on Grace compensation in *New York Times,* May 29, 1942, and April 30, 1943; Grace quote in *Fortune,* May 1945.
7. Joe Hlatki kept his pay stubs.
8. *National Cyclopaedia of American Biography,* vol. G, 71; "Steel," *Fortune,* May 1945.
9. The entente between U.S. Steel and Bethlehem became obvious when Fairless rejected the USWA wage demand as AISI spokesman on November 13, 1945.
10. *New York Times* summary, Jan. 20, 1946.
11. *New York Times,* Nov. 29 and Dec. 5, 1945.
12. *New York Times,* Jan. 13 and Jan. 17, 1946.
13. *New York Times* and *Baltimore Sun,* Jan. 18, 1946.
14. *New York Times,* Jan. 19, 1946.
15. *New York Times,* Jan. 20, 1946. Henry Kaiser had built a popular following through his success in West Coast shipbuilding during the war. His clout in steel was limited because his Fontana Works was small, 600,000 ingot tons.
16. "The Great Steel Strike Begins," *Life,* Feb. 4, 1946.
17. *Baltimore Sun,* Jan. 21 and 22, 1946.
18. Interviews with Elizabeth McShane and Mike Howard; *Baltimore Sun,* Jan. 25, 1946.
19. "Mr. Fairless Is Wrong," *Life,* Feb. 4, 1946; *Business Week,* Jan. 26, 1946, on *Daily News* and other reaction; *New York Times,* Feb. 1, 1946, on Grace.
20. *New York Times,* Feb. 16, 1946.
21. *New York Times,* Feb. 16, 17, and 18, 1946.
22. "The Boom," *Fortune,* June 1946.
23. *New York Times,* Jan. 20, 1946; John Gunther, *Inside U.S.A.* (New York: Harper & Bros., 1947), 615; material from Motor Vehicle Manufacturers' Association, Detroit, Mich.
24. *New York Times,* Aug. 29 and 30, 1946; Bethlehem Annual Report, 1946; *AISI Annual Statistical Report,* 1946; "Boom," *Fortune,* June 1946.
25. Charles H. Hession, "The Tin Can Industry" in Walter Adams, *The Structure of American Industry* (New York: Macmillan, 1954), 404; Fisher, *Epic of Steel,* 244–45; "Beer Can Evolution" and other material from Can Manufacturers Institute, Washington, D.C.

26. *Bethlehem Review,* 1946, 12–13; "Post-War Growth and Prosperity," *Sparrows Point Agenda,* Nov. 1976.
27. "Steel," *Fortune,* 1945; *New York Times,* Nov. 13, 1946.
28. Morris L. Cooke and Philip Murray, *Organized Labor and Production: New Steps in Industrial Democracy* (New York: Harper & Bros., 1940). Also see, "Guaranteed Annual Wage for Labor?" *New York Times Magazine,* April 8, 1945.
29. Clinton S. Golden and Harold J. Ruttenberg, *The Dynamics of Industrial Democracy* (New York, 1942); "U.S.A., United Steelworkers of America," *Fortune,* Nov. 1946; Thomas J. Peters and Robert H. Waterman, Jr., *In Search of Excellence* (New York: Harper & Row, 1982).
30. "Steel," *Fortune,* May 1945.
31. "Bethlehem Ship," *Fortune,* Aug. 1945; Grace speech was reprinted in booklet form: E. G. Grace, *Charles M. Schwab* (South Bethlehem, Pa.: Bethlehem Steel, 1947).
32. Grace, 14.
33. "Fortune Faces," *Fortune,* March 1946; *National Cyclopaedia of American Biography,* vol. 57, 559–560.
34. *Baltimore Sun,* Nov. 30 and Dec. 5, 1947; *National Cyclopaedia of American Biography,* vol. 47, 313.
35. Grace, 52–53.
36. Shannahan, "History," and other company documents.
37. *AISI Works Directory,* 1946; Bethlehem Annual Reports, 1945–1946.
38. See, for example, S. Kip Farrington, Jr., *Railroading from the Head End* (New York: Doubleday, 1943).
39. *Bethlehem Review,* 1946, 13; *Baltimore Sun,* Jan. 11, 1948; "The Industrial Scene," *Fortune,* Oct. 1948.
40. Company documents and hot-mill data.
41. Robert R. Bennett, *Physical Characteristics of Sparrows Point,* U.S. Dept. of Interior, Geological Survey, 1945.
42. Most of the following material was taken from "Papers of Abel Wolman," Library of Congress Manuscript Room, Washington, D.C., Folders 54–56, Bethlehem Steel, and interviews with Wolman in 1980.
43. Wolman Papers, Library of Congress.
44. Interviews with Wolman.
45. Ordnance No. 907, Baltimore Board of Estimates; "Mayor Howard Jackson Papers" at Baltimore City Archives; Abel Wolman, "Industrial Water Supply from Processed Sewage Treatment Plant Effluent at Baltimore, Md.," *Sewage Works Journal,* Jan. 1947, 19.
46. Wolman was retained as Bethlehem's water consultant in 1940 and remained in that capacity until 1964. It was he who requested the $250-a-day consulting fee "for my personal services, plus necessary costs and expenses for associates and travel" in letters between himself and Cort (Nov. 23 and 25, 1940). Regarding the ASCE salary survey, it was pub-

lished in the February 1940 issue of *Civil Engineering*. In 1944 the ASCE adopted a compensation plan for consultants that basically followed the $50-a-day maximum formula.

47. Ezra Whitman personally approved the sewage project despite objections by Wilson T. Ballard, the chief roads engineer, that the pipeline might interfere with vehicle egress on North Point Highway. His engineering firm, Whitman, Requardt, designed and was general engineer of the pipeline.
48. Interviews with Forbes and other employees, March 1983.
49. Wolman Papers.
50. Wolman interviews. Blast-furnace emissions based on James S. Cannon and Frederick Armentrout, *Environmental Steel Update: Pollution in the Iron and Steel Industry* (New York: Council on Economic Priorities, 1977).
51. Interview with Bob Eney, April 1983.
52. Interviews wtih McShane and Roberts.

Chapter 18: Union Business

1. "Sparrows Point Plant" (typescript), Bethlehem Steel, April 1950; Bethlehem Annual Reports, 1946–50.
2. "Sparrows Point Plant"; *Baltimore Sun*, July 13, 1947.
3. "Bringing Out Jungle Ore for U.S. Mills," *Business Week*, June 18, 1949; *Baltimore Sun*, Jan. 11, 1948, and March 22, 1951; "Iron Ore in the Americas," *Fortune*, Dec. 1945.
4. *Fortune*, May 1949.
5. "Bethlehem Steel," *Baltimore*, August 1958. The scrap yard was owned by subsidiary Patapsco Scrap.
6. Bethlehem Annual Reports, 1947–48, and memorabilia collected by Dundalk-Patapsco Neck Historical Society.
7. *Proceedings of the Fourth Constitutional Convention*, United Steelworkers of America, 1948, "Reports of Officers"; Sweeney, *Steelworkers*.
8. Gunther, *Inside U.S.A.*, 622; "U.S.A.: United Steelworkers of America," *Fortune*, Nov. 1946; material from District 8 office, USWA.
9. From District 8 office.
10. From District 8; interview with David Wilson, former president of Local 2609, Jan. 1981.
11. Interview with Joe Hlatki, Sept. 1981.
12. Grievance No. 61-101, filed at District 8 office.
13. Grievance file.
14. Grievance file.
15. Arbitration Case No. 2099-3860, filed at District 8. Thanks to Bernie Parrish.
16. Based on taped interviews with Charlie Parrish, 1983.
17. From arbitration transcript, slightly condensed for clarity, 76–78.
18. Arbitration Case, 97–98.

19. Aribitration Case, 99–100.
20. Arbitration Case, 105.
21. *Opinion and Award of the Impartial Umpire,* March 25, 1949.
22. *Baltimore Evening Sun,* April 15, 1964.
23. Based on taped interviews with Mike Howard, May 1981 and April 1982.
24. *Proceedings of Fourth Constitutional Convention,* USWA, 131; Sweeney, *Steelworkers,* 75–76.
25. *Communism in the District of Columbia–Maryland Area.* Hearings before the Committee on Un-American Activities, 82d Congress, 1st, 1951, 4467–4469; *Baltimore Sun,* July 12, 1951.
26. *Communism,* HUAC, 4503.
27. *Baltimore Sun,* June 21, 1951.
28. *Baltimore Sun,* June 21, 1951.
29. *Baltimore Sun,* June 21, 1951.
30. *Baltimore Sun,* June 21, 1951. Editorially the paper heaped praise on HUAC, saying its investigation countered "the world-wide Communist conspiracy against free governments." According to the paper, people who took the Fifth should be fired from their jobs. (June 22, 1951).
31. Sweeney, *Steelworkers,* 74–75, 87–95.
32. Sweeney, *Steelworkers,* 143; also see John Herling, *Right to Challenge* (New York: Harper & Row, 1972).
33. Sweeney, *Steelworkers,* 139, 143.

Chapter 19: Women's Work

1. "Women in Important Job at Bethlehem's Tin Mill," *Baltimore American and News-Post,* photocopy, c. 1959; *AISI Works Directory, 1954;* "Production of BethColite at Sparrows Point."
2. From interviews with tin division personnel. Thanks to Ed Gorman and Walter Webb.
3. Remarks of tin sorters reminiscing at the Senior Steelworkers luncheon in 1981. "If only she were alive to talk to you, what a story she could tell!" was a common exclamation. Losing track of the boss in retirement, the women has consigned her to the grave. But in the telephone book for the Saint Helena section of Baltimore County there was a listing for a Mrs. Alexander, and she turned out to be the legendary head forelady. She was contacted, an appointment was arranged, and, as the sorters had promised, quite a story unfolded.
4. What follows was distilled from six interviews with Mrs. Alexander; also from an Oct. 15, 1948 profile of her appearing in the *Dundalk Community Press,* "40 Years in Steel Work Earns Mrs. Alexander Title of 'First Lady.'"
5. See annual surveys of wages of *Maryland State Board of Labor and Statistics;* also U.S. Women's Bureau, *Women in Maryland Industries,* Bulletin 24, 1922.
6. Interviews with Mary Gorman, June–Aug. 1981.

7. From programs, "Banquet and Dance by Ladies of the B. H. (Sorting) Department," also photographs, courtesy of Mrs. Alexander.
8. Interviews with Marian Wilson, Sept. 1982.

Chapter 20: Complacent Giant

1. Interview with Ed Gorman, Jan. 1983.
2. Interviews with Bob Eney, March–April, 1983.
3. "Production of Steel by Countries," *AISI Annual Statistical Report*, 1953.
4. *The Encyclopedia Americana Annual: Events of 1951* (New York: Grolier, 1952), 364.
5. Bethlehem Steel Annual Reports; *AISI Annual Statistical Reports*; "Bethlehem Steel and the Intruder," *Fortune*, March 1953.
6. *New York Times*, April 18, 1950; "A Spark in Steel," *Fortune*, Dec. 1948.
7. Steel mill price index and unit labor costs published in *AISI Annual Statistical Reports;* also see, *Administered Prices, Report of the Senate Subcommittee on Antitrust and Monopoly*, 85th Congress, 2d, 1958, 35–40.
8. Walter Adams, "The Steel Industry," *Structure of American Industry*, 196.
9. *Study of Monopoly Power, Hearings before the Subcommittee on Study of Monopoly Power of the House Committee on the Judiciary*, Part 4A, *Steel*, 81st Congress, 2d, 1950, 117–136; *New York Times*, April 18, 1950.
10. *Study of Monopoly Power*, 76.
11. *Study of Monopoly Power*, 186–192; *New York Times*, April 20, 1950.
12. *Study of Monopoly Power*, 326.
13. *Study of Monopoly Power*, 931 and 934.
14. *Study of Monopoly Power*, 357. (*See* Chapter 5 on the workings of the old International Rail Pool.)
15. *Study of Monopoly Power*, 752–765.
16. *Study of Monopoly Power*, 752–765.
17. *New York Times*, April 28, 1950.
18. *New York Times*, Aug. 26, 1956.
19. "Bethlehem Steel," *Fortune*, March 1953.
20. John Strohmeyer, *Crisis in Bethlehem: Big Steel's Struggle to Survive* (Bethesda, Md.: Adler & Adler, 1986), 31–32.
21. Strohmeyer, 47; interviews with management personnel.
22. *New York Times*, Nov. 10, 1951; *Fortune*, March 1953.
23. Interviews with Joe Hlatki and Ben Womer.
24. *Fortune*, March 1953.
25. "The Industry's Leaders Agree: Steel Is Ready to Grow," *Business Week*, June 4, 1955; *AISI Annual Statistical Report*, 1955.
26. *New York Times*, May 27, 1955.
27. *Business Week*, June 4, 1955.
28. *AISI Annual Statistical Report*, 1955; *Steel*, June 27, 1955.
29. *The Encyclopedia Americana Annual*, 1956, 25; *Aluminum Statistical Review*, 1975, Aluminum Association.

30. *Americana Annual*, 629.
31. *Iron Age*, June 22, 1955; *Bethlehem Review*, 1955. In the midst of heavy steel demand there were fall-offs in two traditional sectors: railroads and shipbuilding. Once the pacesetter of steel sales, total railroad consumption dropped to 2.46 million tons in 1954, half of the 1945 rate. Shipbuilding also had begun to dry up: "Private yards active along the seaboard are not operating at more than 10 to 20 percent of their combined capacities. Were it not for Navy work the plight of these yards would be worse than it is," said *Steel* magazine on June 13, 1955.
32. Auto registration from Bureau of Roads, U.S. Department of Commerce; use of materials in cars from Motor Vehicle Manufacturers' Association; Car advertisements in *Baltimore Sun*, Jan. 15, 1956.
33. *AISI Annual Statistical Report*, 1955.
34. *Iron Age*, June 2, 1955; Sweeney, *Steelworkers*, 149; Herling, *Right to Challenge*, 24–25.
35. Sweeney, 169–170 (steel wages); *New York Times*, July 2 and 6, 1955 (steel prices).

Chapter 21: The Edifice Complex

1. See *Steel and Inflation: Fact vs. Fiction*, (New York: U.S. Steel, 1957).
2. *Administered Prices*, 85.
3. *New York Times*, July 2, 1955, and July 4, 1957.
4. *Study of Monopoly Power*, 930–31.
5. *New York Times* and *Baltimore Sun*, Jan. 27, 1956; *Sparrows Point Management Agenda*, Feb. 1956.
6. *Baltimore Sun*, Jan. 27, 1956.
7. George C. Little, "The Expansion Program at Sparrows Point, Engineering and Mechanical Aspects," *Iron and Steel Engineer*, Feb. 1959.
8. Little.
9. Little.
10. Interview with company engineer, Feb. 1983. Anonymity requested.
11. Little; *Evening Sun*, Dec. 26, 1956; *Baltimore Sun*, Feb. 26, 1958; *Fortune*, April 1962.
12. Little.
13. Baltimore effluent contract from Wolman Papers; Amarillo contract from Edmund B. Besselievre, *The Treatment of Industrial Wastes* (New York: McGraw-Hill, 1969), 3.
14. Inspection report published in *Biennial Report, 1954–55*, Maryland Water Pollution Control Commission, 50–52. Author interviewed Henry Silbermann in Annapolis and reviewed WPC records at the Department of Natural Resources. Other interviews with J. Henry Schilpp, retired engineer for the commission, Steve Lerner, formerly of the Bay Institute, and Dr. L. Eugene Cronin, formerly of Chesapeake Biological Laboratory at Solomons.
15. From correspondence in Wolman Papers and interview with Wolman.

16. From Wolman Papers (1949–53 correspondence); *Baltimore Evening Sun,* Aug. 10, 1952; interviews with Schilpp and Margaret Weinholder of Back River Improvement Association. Mencken from *Happy Days, 1880–1892* (New York: Knopf, 1940), 55.
17. Interview with Charles Renn, March 1980.
18. Besselievre, 3.
19. Wolman Papers and D. W. Pritchard, "Controlled Discharge of Acid Waste Through a Submerged Pipeline" (typescript), July 1, 1957, 1.
20. McKee in *Evening Sun* series on bay pollution.
21. Interview with Silbermann and review of commission records.
22. From *Speaking of Sparrows Point*... (South Bethlehem, Pa: Bethlehem Steel, 1958); *AISI Works Directory,* 1958.
23. From *Works Directory of the World.*
24. *Port of Baltimore Handbook* (Baltimore: Md. Port Authority, 1957).
25. Interviews with Gorman and others.
26. *Baltimore Sun,* Jan. 17, 1956.
27. Interviews with Gorman, McShane, and Roberts.
28. Interviews with workmen and *First Helper's Manual.* Sparrows Point Open Hearth Department, 1959.
29. Interview with workers.
30. "Bethlehem Abroad," Bethlehem Annual Report, 1965.
31. Bethlehem Annual Reports, 1956–57.
32. *New York Times,* June 28, 1957; Annual Reports, 1956–57.
33. "The Big Money-Earners of '56," *Business Week,* May 25, 1957.
34. *Administered Prices,* 107–110.
35. Matthews' famous stories on the discovery of Castro began on Feb. 24, 1957.

Chapter 22: Ruins

1. "How Top Salaries Weathered the Recession," *Business Week,* June 13, 1959; *New York Times,* July 31, August 1, and 6, 1958.
2. From BLS Yearly Price Index, 1945–58, Steel Mill Products and All Commodities.
3. *Business Week,* Nov. 11, 1961, and Sept. 28, 1957.
4. *Business Week,* May 11 and Nov. 2, 1957.
5. *Business Week,* Nov. 11, 1961, and Oct. 13, 1962.
6. Fisher, *Epic of Steel,* 244.
7. "The Private Strategy of Bethlehem Steel," *Fortune,* April 1962; *Business Week,* May 11, 1963.
8. *Aluminum Statistical Reviews* for years 1960–85, Aluminum Association; *AISI Annual Statistical Reports,* 1960–85; *Wall Street Journal,* Oct. 30, 1987; Can Manufacturers Institute and National Can Co.
9. From Charles E. Silberman's assessment of steel's shortcomings, "Steel: It's a Brand-New Industry," *Fortune,* Dec. 1960; also "Private Strategy," *Fortune,* April 1962.

10. Silberman; *AISI Annual Statistical Reports*, 1957–62.
11. *Business Week*, Oct. 16, 1965, and Aug. 3, 1963.
12. Adams, "Steel Industry," in *Structure* (1977 ed.), 118; Bethlehem Annual Report, 1961.
13. *New York Times*, July 26, 1960; Bethlehem Annual Reports, 1961–65.
14. *Fortune*, April 1962.
15. *Fortune*, April 1962.
16. *Business Week*, Nov. 16, 1963.
17. Bethlehem Annual Report, 1966. "The Basic Oxygen Process of Steelmaking."
18. Press release, Sparrows Point, December 1965. The company liked to flatter itself with unfounded boasts like, "Bethlehem's response to technological change has been to put itself in the forefront of steel research and development."
19. During the 1930s, the renowned metallurgist A. J. Townsend had experimented with "rotary casting," a process of rolling liquid steel from the open hearths. When Townsend died in 1935, so did his idea as far as American producers were concerned.
20. Interview with management source, April 1984; Richard Foster discusses the problem of technological stasis in *Innovation: The Attacker's Advantage* (New York: Summit, 1986.)
21. Management source.
22. *New York Times*, July 14 and 15, 1959.
23. George J. McManus, *The Inside Story of Steel Wages and Prices 1959–1967* (Philadelphia, Pa.: Chilton Books, 1967), 13–17.
24. McManus, *Inside*, 13; *New York Times*, July 14, 1959; interviews with Duerbeck and other employees.
25. *New York Times*, Nov. 7, 1959; see Chapter 12 for Perkins's comments.
26. *New York Times*, Dec. 20, 1959.
27. McManus, *Inside*, 19.
28. Import-export figures from *AISI Annual Statistical Reports*.
29. Donald F. Barnett and Louis Schorsch, *Steel: Upheaval in a Basic Industry* (Cambridge, Mass.: Ballinger, 1983), 14–15.
30. *Business Week*, Feb. 3, 1962, and "Special Report—The world battle for steel," June 4, 1966; also see Leonard Lynn, *How Japan Innovates, a Comparison with the U.S. in the Case of Oxygen Steelmaking* (Boulder, Co.: Westview, 1982).
31. *Business Week*, Sept. 29, 1962.
32. *Business Week*, Jan. 27, 1962.
33. *AISI Annual Statistical Reports; Business Week*, Feb. 11 and Dec. 9, 1962; "The Hard Case of Steel," *Fortune*, Oct. 1967.
34. "How Steel Imports Have Hurt Bethlehem" Bethlehem Annual Report, 1968.
35. Paul A. Tiffany, *The Decline of American Steel* (New York: Oxford, 1988), 169.

36. Grace stated in 1946 that Bethlehem would not build an integrated plant in California because it might undermine the value of Sparrows Point.
37. Barnett, *Steel,* 238–42.
38. From BLS Yearly Price Index.
39. For an excellent study of mini-mills, see Donald F. Barnett and Robert W. Crandall, *Up from the Ashes: The Rise of the Steel Minimill in the U.S.,* (Washington, D.C.: Brookings Institution, 1986).
40. From *Project '85,* a report of Bethlehem Steel research department, and statements of Chairman Stewart S. Cort.
41. Address by Lewis W. Foy at L furnace dedication luncheon, Sept. 28, 1978.
42. Bethlehem Annual Report, 1977, and "International Trade Policy and American Industry: A Threatened Consensus," address by Charles Stern of American Institute for Imported Steel, 1977.
43. Interviews with supervisors and employees of rod and wire mill, 1978.
44. Employee interviews, 1978.
45. Mark Reutter, "The Invisible Risk: Why Sparrows Point Workers Should Be Congratulated for Making It Through a Day Without Injury," *Mother Jones,* Aug. 1980; MOSHA safety reports and citations; *Management Conference Agenda,* Sparrows Point Plant, 1978–79.
46. MOSHA citations; Reutter, "Risk."
47. Reutter, "Risk."
48. MOSHA records.
49. *Baltimore Sun,* Sept. 23, 1977, and Dec. 5, 1979; *Environmental Steel Update,* 90–93.
50. *Wall Street Journal,* Oct. 10, 1977.
51. Strohmeyer, *Crisis,* 129, 133. Trautlein joined the company in 1977 as comptroller after spending 13 years as a partner at Price Waterhouse, where he was responsible for auditing the steelmaker's accounts. He was recruited by Foy.
52. *Baltimore Sun,* May 8, 1982; *Wall Street Journal,* May 2, 1982.
53. *Baltimore Sun,* May 14, 1982.
54. From union meeting and *Baltimore News American,* Sept. 30, 1982.
55. Union meeting and interviews, Oct.–Dec. 1982.
56. Soviet steel production first edged ahead of the U.S. in 1971; a decade later, in 1982, it had reached 162 million ingot tons, against 75 million tons in America. Japanese production in 1982 was 110 million tons. In the same year, Communist China passed West Germany to become the world's fourth-largest steel producer with 41 million tons. *AISI Annual Statistical Report, 1982.*
57. *Baltimore Sun,* April 30, 1982.
58. "Eat Your Imports" was another popular slogan. For more on the social impact of unemployment, see David Bensman and Roberta Lynch, *Rusted Dreams: Hard Times in a Steel Community* (New York: McGraw-Hill, 1987) and Jack Metzgar, "Johnstown, Pa.: Ordeal of a Union Town," *Dissent,* Spring 1985.

59. *Forbes*, Feb. 8, 1988; also *New York Times*, June 30, 1986.
60. Interview with management source, and *Wall Street Journal*, May 27, 1986.
61. *Wall Street Journal*, May 27, 1986. Union negotiators accepted limited profit-sharing in return for wage concessions.
62. From John Kenneth Galbraith, "Oil: A Solution" *New York Review of Books*, Sept. 27, 1979; Strohmeyer, article for Alicia Patterson Foundation; *New York Times*, May 14, 1986.
63. *HR Link*, Sparrows Point, March 1988; *Wall Street Journal*, Dec. 30, 1987.
64. For an appraisal of the technological restructuring needed to restore the United States to world-class status, see, "Can Advanced Technology Save the U.S. Steel Industry?" *Scientific American*, July 1987.
65. *HR Link*, Sparrows Point, March 1988; "Topping Out Completed for Slab/Bloom Caster," *Iron and Steel Engineer*, April 1985.
66. Interview with Dave Wilson, March 1988.

INDEX